AN INTRODUCTION TO HEAT PIPES

AN INTRODUCTION TO HEAT PIPES

Modeling, Testing, and Applications

G. P. PETERSON
Department of Mechanical Engineering
Texas A&M University
College Station, Texas

A Wiley-Interscience Publication
JOHN WILEY & SONS, INC.
New York / Chichester / Brisbane / Toronto / Singapore

This text is printed on acid-free paper.

Library of Congress Cataloging in Publication Data:

Peterson, G. P. (George P.)
An introduction to heat pipes : modeling, testing, and applications / G. P. Peterson.
p. cm.
Includes bibliographical references.
ISBM 0-471-30512-X (alk. paper)
1. Heat pipes. I. Title.
TJ264.P48 1994
621.402'5--dc20 93-51014

10 9 8 7 6 5 4

To Valerie—Thank you.

CONTENTS

PREFACE

The concept of a passive two-phase heat transfer device capable of transferring large quantities of heat with a minimal temperature drop was first introduced by Gaugler in 1942. This device received little attention until 1964, when Grover and his colleagues at Los Alamos National Laboratories published the results of an independent investigation and first applied the term *heat pipe*. Since that time, heat pipes have been employed in numerous applications ranging from temperature control of the permafrost layer under the Alaska pipeline to the thermal control of optical surfaces in spacecraft. Heat pipe theory, application, and manufacturing techniques have developed rapidly in recent years. This book presents the fundamental theory that governs the operational limitations and performance characteristics of these devices along with design and manufacturing information.

The book begins with the background and historical development of heat pipes and a discussion of the relative advantages. This introduction is followed by a discussion of the interfacial phenomena that govern the operational characteristics and a discussion of the fundamental operating principles and limits of both heat pipes and thermosyphons. Considerable attention is devoted to modeling of heat pipe performance along with a discussion of several different types or classifications of heat pipes including rotating and revolving heat pipes, micro–heat pipes, cryogenic heat pipes, variable conductance heat pipes, and thermal diodes and switches. Information on the design and manufacturing of these and other types of heat pipes along with a discussion of traditional testing and analytical procedures is also presented. The text closes with a discussion of recent applications of heat pipes to the thermal control of electronic equipment.

The book is intended to serve three purposes: first, as a sourcebook for engineers involved in the design, development, modeling, or testing of heat pipes and/or thermosyphons; second, as a resource for those interested in exploring the possibility of using these devices in new or existing applications; and third, as a textbook for an introductory graduate or undergraduate course dealing with this subject.

Literally thousands of articles and papers on heat pipes and thermosyphons have been published in recent years. An attempt is made here to identify the most current and descriptive articles on both the fundamental phenomena and recent applications. It is the hope of the author that this book will serve the reader well and provide a basis from which a better understanding and more extensive use of heat pipes and thermosyphons can be achieved.

ACKNOWLEDGMENTS

I would like to thank the many people who made the preparation and publication of this book possible. In particular, I acknowledge and thank Ms. Allison Royalty and Ms. Kim Duffer for the many long hours they spent in the preparation of the manuscript. I also thank my students. In their relentless pursuit of knowledge, they have taught me more than I ever thought possible. Finally, I thank my family for their support, their patience, and their understanding.

Chapter 1

INTRODUCTION

1.1 BACKGROUND AND HISTORICAL DEVELOPMENT

The initial formulation of the heat pipe concept can be traced to the patents of A. M. Perkins and J. Perkins in the mid-1800s (King, 1931; Perkins and Buck, 1892). These patents focused on a device referred to as the Perkins tube, shown in Fig. 1.1, which utilized either single- or two-phase processes to transfer heat from a furnace to a boiler. Somewhat later, Gay (1929) obtained a patent on a device similar to the Perkins tube, shown in Fig. 1.2, in which a number of vertical tubes were arranged with the evaporator located below the condenser. These devices, which are actually classified as thermosyphons, laid the groundwork for the later development of what is now commonly referred to as the heat pipe.

A heat pipe typically consists of a sealed container lined with a wicking material. The container is evacuated and backfilled with just enough liquid to fully saturate the wick. Because heat pipes operate on a closed two-phase cycle and only pure liquid and vapor are present within the container, the working fluid will remain at saturation conditions as long as the operating temperature is between the triple point and the critical state. As illustrated in Fig. 1.3, a heat pipe consists of three distinct regions: an evaporator or heat addition region, a condenser or heat rejection region, and an adiabatic or isothermal region. When heat is added to the evaporator region of the container, the working fluid present in the wicking structure is heated until it vaporizes. The high temperature and corresponding high pressure in this region cause the vapor to flow to the cooler condenser region, where the vapor condenses, giving up its latent heat of vaporization. The capillary forces existing in the wicking structure then pump the liquid back to the evaporator.

The concept of utilizing a wicking structure as part of a passive two-phase heat transfer device capable of transferring large quantities of heat with a minimal temperature drop was first introduced by Gaugler

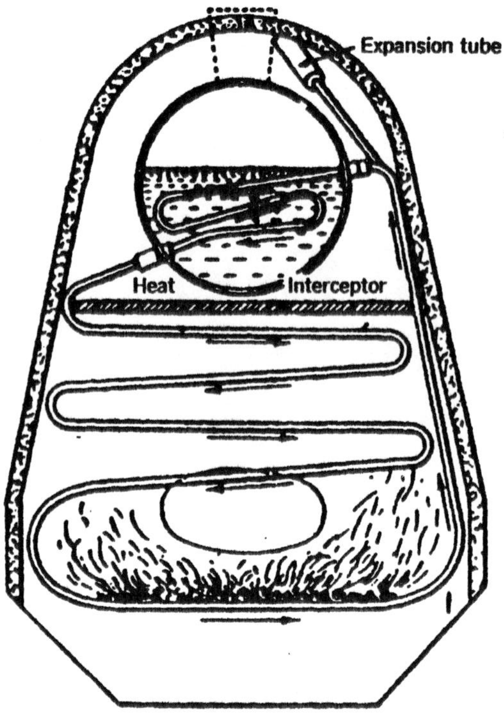

Figure 1.1 First heat pipe type device, the "Perkins tube" (King, 1931; Perkins and Buck, 1892).

(1944). This concept again surfaced in 1962 in connection with the space program (Trefethen, 1962) and also in the form of a patent application submitted by Wyatt in 1963 (Brennan and Kroliczek, 1979). The heat pipe concept received relatively little attention, however, until Grover et al. (1964) published the results of an independent investigation and first applied the term *heat pipe* to describe a "synergistic engineering structure which is equivalent to a material having a thermal conductivity greatly exceeding that of any known metal." Since that time, heat pipes have been employed in numerous applications ranging from temperature control of the permafrost layer under the Alaska pipeline to the thermal control of optical surfaces in spacecraft. Although a more in depth explanation will be undertaken at a later point in the text, the primary difference between the thermosyphons first proposed by Perkins (Perkins and Buck, 1892) and Gay (1929) and the heat pipes proposed by Gaugler (1944) and Grover et al. (1964), is that the heat pipe utilizes some sort of capillary structure to assist or promote flow of the liquid from the condenser to the evaporator, instead of relying solely on the local gravitational acceleration for the return of liquid from the condenser to the evaporator.

Basically, a heat pipe consists of three major components: the container,

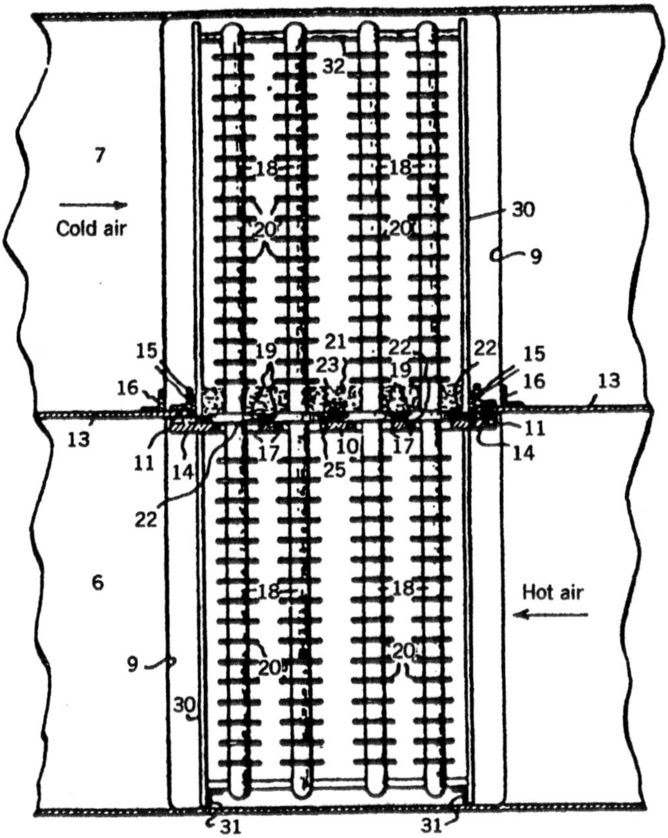

Figure 1.2 Thermosyphon heat exchanger (Gay, 1929).

which can be constructed from glass, ceramics, or metals; a wicking structure, constructed from woven fiberglass, sintered metal powders, screens, wire meshes, or grooves; and a working fluid, which can vary from nitrogen or helium for low temperature (cryogenic) heat pipes to lithium, potassium, or sodium for high temperature (liquid metal) heat pipes. Each of these three components is equally important with careful consideration given to the material type, thermophysical properties, and compatibility. For example, the container material must be compatible with both the working fluid and the wicking structure, strong enough to withstand pressures associated with the saturation temperatures encountered during storage and normal operation, and must have a high thermal conductivity. In addition to these characteristics, which are primarily concerned with the internal effects, the container material must be resistant to corrosion resulting from interaction with the environment and must be malleable enough to be formed into the appropriate size and shape.

The wicking structure has two functions in heat pipe operation: It is the

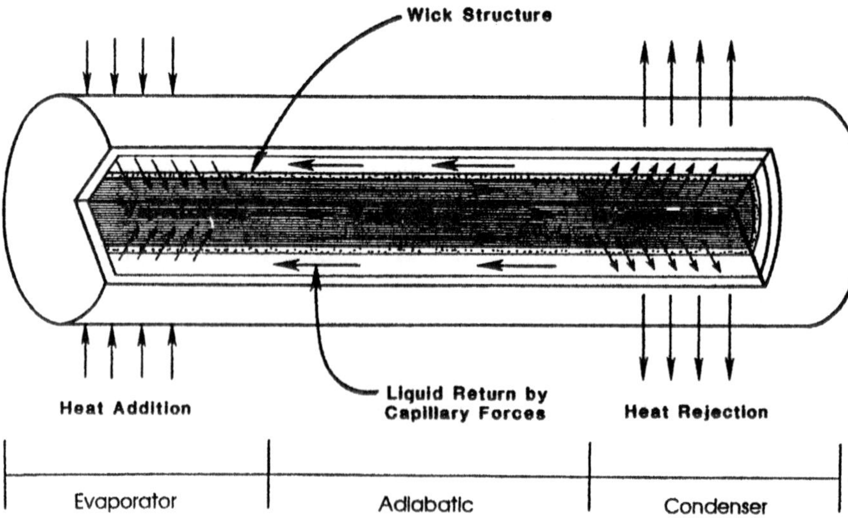

Figure 1.3 Typical heat pipe construction and operation.

vehicle through which, and provides the mechanism by which, the working fluid is returned from the condenser to the evaporator and also ensures that the working fluid is evenly distributed over the evaporator surface. Figure 1.4 illustrates several common wicking structures presently in usé, along with several more advanced concepts under development. In order to provide a flow path with low flow resistance through which the liquid can be returned from the condenser to the evaporator, an open porous structure with a high permeability is desirable. However, to increase the capillary pumping pressure, a small pore size is necessary. Solutions to this apparent dichotomy can be achieved through the use of a nonhomogeneous wick made of several different materials or through composite wicking structures similar to those shown in Fig. 1.4b.

Udell and Jennings (1984) proposed and formulated a model for a heat pipe with a wick consisting of porous media of two different permeabilities oriented parallel to the direction of the axial heat flux. This wick structure provided a large pore size in the center of the wick to reduce the viscous pressure drop associated with the axial liquid flow and a smaller size pore to enhance the capillary pressure occurring at the liquid–vapor interface in the evaporator. Current composite wicking structures utilize the same approach with the capillary pumping and axial fluid transport being handled independently. In addition to fulfilling this dual purpose, several of the advanced wick designs illustrated separate the liquid and vapor flow to reduce the viscous shear force that occurs during the countercurrent, liquid–vapor flow.

Because the basis for operation of a heat pipe is the vaporization and condensation of the working fluid, selection of a suitable fluid is an

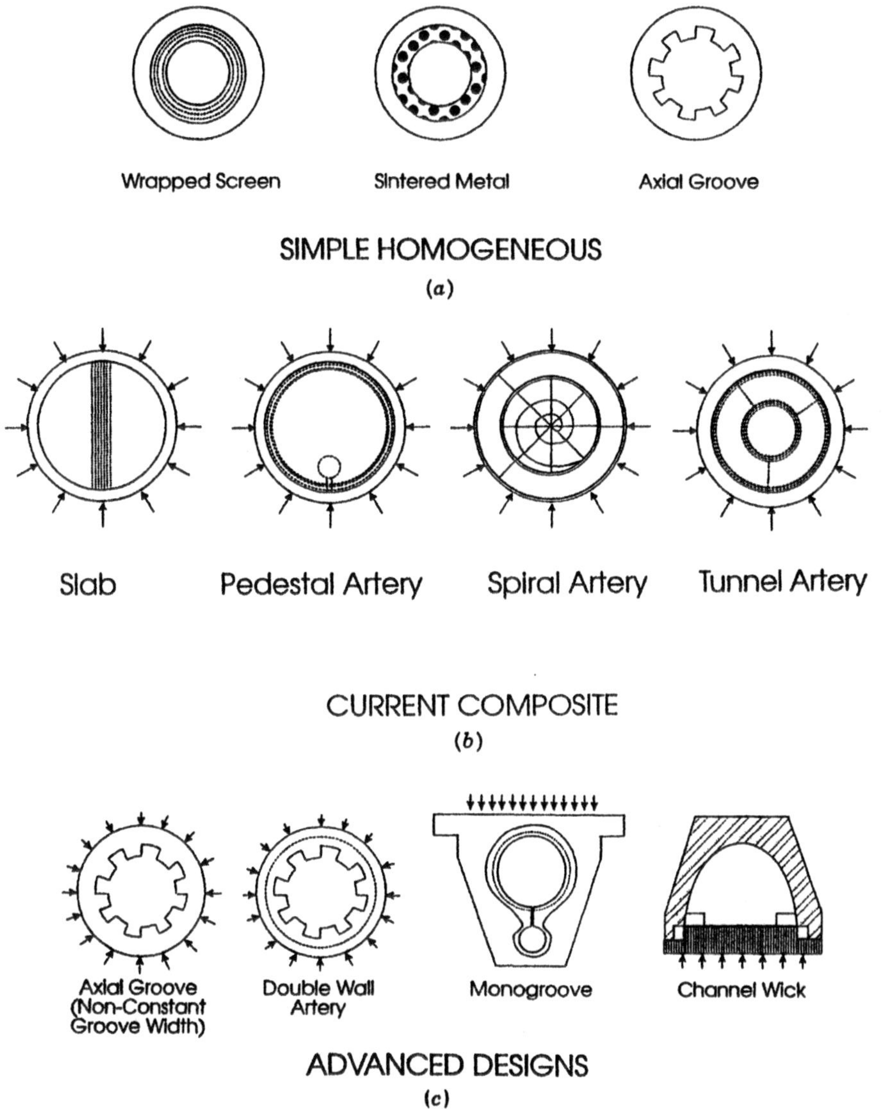

Figure 1.4 Typical heat pipe wicking configurations and structures.

important factor in the design and manufacture of heat pipes. Care must be taken to ensure that the operating temperature range is adequate for the application. While most applications involving the use of heat pipes in the thermal control of electronic devices and systems require the use of a working fluid with boiling temperatures between 250 and 375 K, both cryogenic heat pipes (operating in the 5 to 100 K temperature range) and liquid metal heat pipes (operating in the 750 to 5000 K temperature range) have been developed and used. Figure 1.5 illustrates the possible working

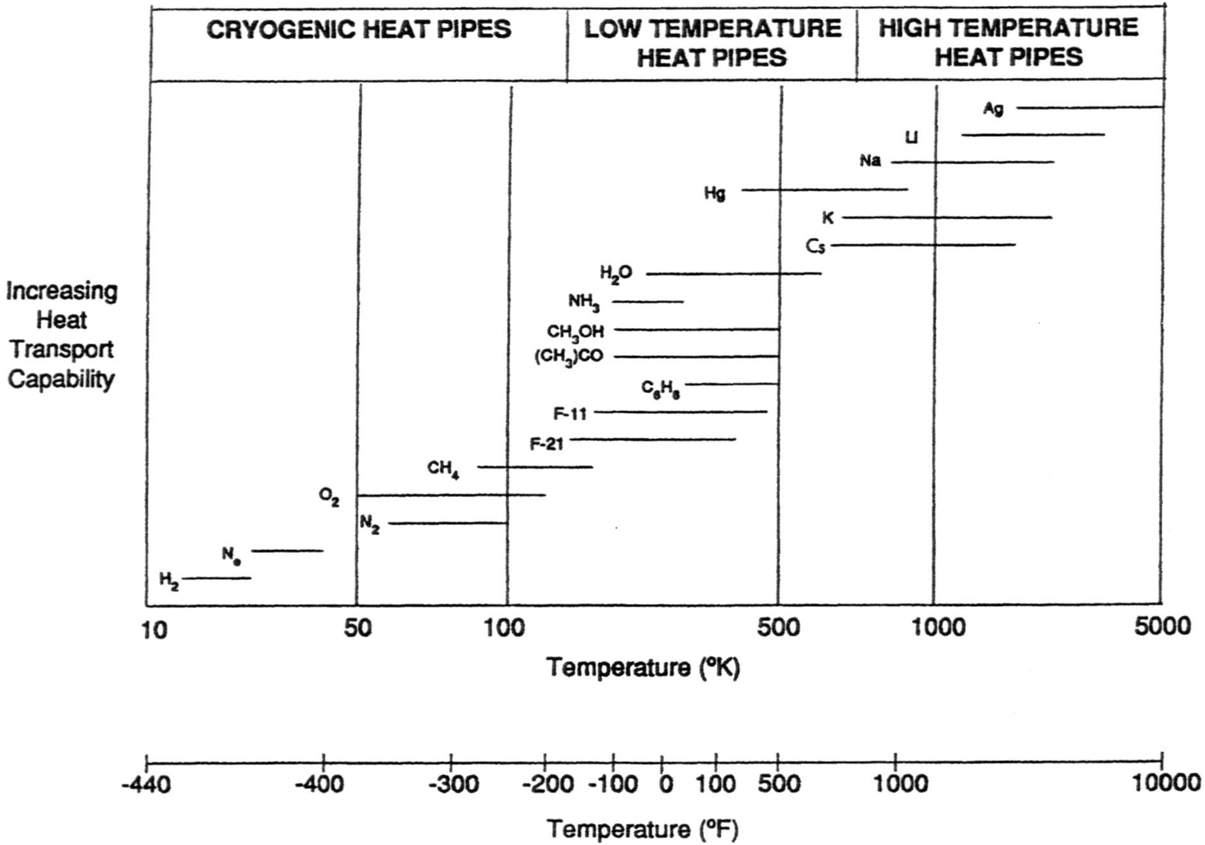

Figure 1.5 Possible working temperature ranges for some of the various heat pipe fluids.

temperature ranges for some of the various heat pipe fluids. In addition to the thermophysical properties of the working fluid, consideration must be given to the ability of the working fluid, to the wettability of the working fluid, and to the wick and wall materials. Further criteria for the selection of the working fluids have been presented by Heine and Groll (1984), in which a number of other factors, including the liquid and vapor pressure and the compatibility of the materials, are considered.

1.2 FUNDAMENTAL OPERATING PRINCIPLES

Both heat pipes and thermosyphons operate on a closed two-phase cycle and utilize the latent heat of vaporization to transfer heat with very small temperature gradients. However, the operation of these two devices is significantly different. As illustrated in Fig. 1.6, heat added to the bottom portion of a thermosyphon vaporizes the working fluid. During this phase change process, the fluid picks up the heat associated with its latent heat of vaporization. Because the vapor in the evaporator region is at a higher temperature and hence at a higher pressure than the vapor in the condenser, the vapor rises and flows to the cooler condenser where it gives up the latent heat of vaporization (buoyancy forces assist this process). Gravitational forces then cause the condensate film to flow back down the inside of the heat pipe wall where it can again be vaporized. Although the inside surface of a thermosyphon may occasionally be lined with grooves or a porous structure to promote return of the condensate to the evaporator or increase the heat transfer coefficient, thermosyphons principally rely upon the local gravitational acceleration for the return of the

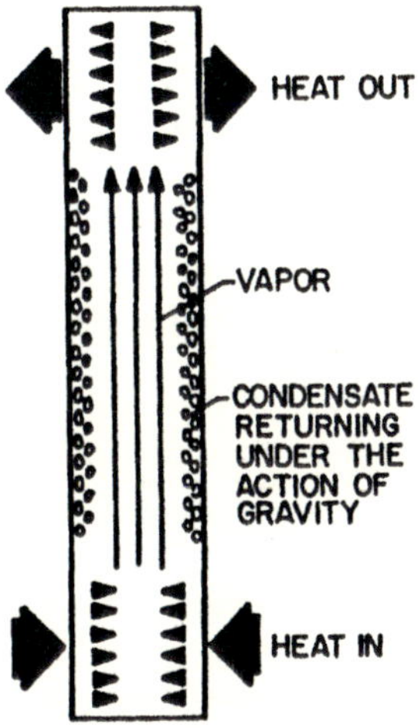

Figure 1.6 Operation of a typical thermosyphon.

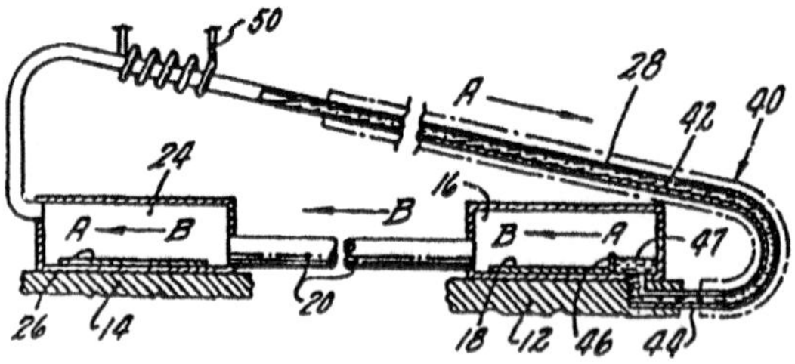

Figure 1.7 Chemical pipe (Kesten and Haught, 1986).

liquid from the evaporator to the condenser. By definition, then, for proper operation the evaporator of a thermosyphon must be located below the condenser or dryout of the evaporator will occur.

Alternatively, heat pipes utilize some sort of capillary wicking structure to promote the flow of liquid from the condenser to the evaporator and as a result can be used in a horizontal orientation, microgravity environments, or even applications where the capillary structure must "pump" the liquid against gravity from the evaporator to the condenser. It is this single characteristic—the dependence of the local gravitational field to promote the flow of the liquid from the condenser to the evaporator—that differentiates thermosyphons from heat pipes.

Recently, several investigations have been conducted on a device referred to as a "chemical pipe" (Kesten and Haught, 1986). While the primary function of the chemical pipe illustrated in Fig. 1.7 is to promote the transfer of heat, it differs from a heat pipe in that chemical pipes utilize the heat of fusion resulting from endothermic reactions occurring in the "evaporator" and exothermic reactions occurring in the "condenser" to transport the energy, as opposed to the latent heat of vaporization. Studies have indicated that several combinations of fluids could be used in such a device, including methane and water, which are converted to carbon monoxide and hydrogen. While several investigations of this concept have been conducted, further discussion is beyond the scope of this book, and the results of these investigations will not be presented.

1.3 LIMITATIONS ON THE TRANSPORT CAPACITY

Although heat pipe performance and operation are strongly dependent on shape, working fluid, and wick structure, the fundamental phenomenon that governs the operation arises from the difference in the capillary pressure across the liquid–vapor interfaces in the evaporator and condenser. When this capillary pressure is not sufficient to promote the flow of liquid from the condenser to the evaporator, the heat pipe is said to have reached its capillary limit and dryout of the evaporator wick occurs.

During steady-state operation, several other important mechanisms can limit the maximum amount of heat that a particular heat pipe can transfer. Among these are the viscous limit, sonic limit, entrainment limit, and boiling limit. The capillary wicking limit and viscous limits deal with the pressure drops occurring in the liquid and vapor phases, respectively. The sonic limit results from the occurrence of choked flow in the vapor passage, while the entrainment limit is due to the high liquid vapor shear forces developed as the vapor passes in a counterflow direction over the liquid-saturated wick. All of these limits are axial heat flux limits, that is, they are a function of the axial heat transport capacity of the heat pipe. The boiling limit, however, is a "radial" heat flux limit and is reached when the heat flux applied in the evaporator portion is so high that nucleate boiling occurs in the evaporator wick. This creates vapor bubbles that partially block the return of fluid and may ultimately lead to premature dryout of the evaporator.

For moderate temperature heat pipes, the most significant of these limits is usually the capillary wicking limit. However, the significance of this limit decreases somewhat for reduced gravity applications. In low temperature applications such as those using cryogenic working fluids, either the viscous limit or capillary limit occurs first, while in high temperature heat pipes, such as those that use liquid metal working fluids, the sonic and entrainment limits are of increased importance. The theory and fundamental phenomena that cause each of these limitations have been the object of a considerable number of investigations and have been presented and discussed by Cotter (1964), Chisholm (1971), Tien (1975), Feldman (1976), Chi (1976), Brennan and Kroliczek (1979), Dunn and Reay (1982), and the proceedings from the eight International Heat Pipe Conferences held during the past 25 years. A detailed discussion of the capillary limit will be presented and examined in detail in Chapter 3 with the remaining limits presented in Chapter 4.

Although transient modeling and startup dynamics have been the subject of numerous recent investigations (Antoniuk, 1987; Chang and Colwell, 1985; Cullimore, 1985; Merrigan et al., 1985), only a brief synopsis of the transient modeling techniques will be presented in Chapter 5.

1.4 ADVANTAGES OF HEAT PIPES

Eastman (1968) identified several characteristics of heat pipes that make them useful in a wide variety of applications. First, because heat pipes operate on a closed two-phase cycle, the heat transfer capacity may be several orders of magnitude greater than even the best solid conductors. This results in a relatively small thermal resistance and allows physical separation of the evaporator and condenser without a high penalty in overall temperature drop. Second, increases in the heat flux in the evaporator may result in an increase in the rate at which the working fluid is vaporized, without significant increases in the operating temperature. Thus the heat pipe can function as a nearly isothermal device, adjusting the evaporation rate to accommodate a wide range of power inputs while maintaining a relatively constant source temperature. Third, the evaporator and condenser portions of a heat pipe function independently, needing only common liquid and vapor streams; for this reason, the area over which heat is introduced can differ in size and shape from the area over which it is rejected, provided that the rate at which the liquid is vaporized does not exceed the rate at which it can be condensed. Hence, high heat fluxes generated over relatively small areas can be dissipated over larger areas with reduced heat fluxes. This is particularly useful in the thermal control of electronic components and systems, because it allows the high heat fluxes generated at the component level to be reduced and allows free or forced convection to be used to dissipate the heat.

One additional advantage not originally identified by Eastman (1986) is the thermal response time. Because heat pipes and thermosyphons utilize a closed two-phase cycle, the thermal response time is considerably less than for other types of heat transfer devices, particularly solid conductors, and is not a function of length.

The high heat transfer characteristics, the ability to maintain constant evaporator temperatures under different heat flux levels, and the diversity

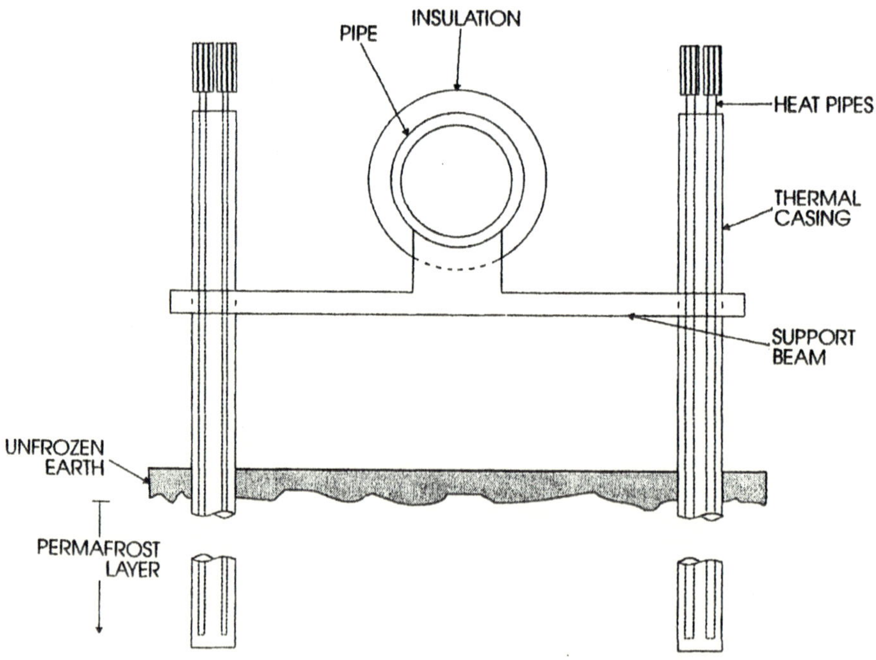

Figure 1.8 Heat pipe / thermosyphon utilized on the Alaska pipeline.

and variability of evaporator and condenser sizes make heat pipes and thermosyphons effective devices for use in many diverse situations. Perhaps the best known application of the heat pipe/thermosyphon concept is on the Alaska pipeline, where heat pipes are used to maintain the permafrost in a stable condition around the pipeline vertical support members. As shown in Fig. 1.8, heat transferred from the oil flowing in the pipe is transferred through the concrete support members to the evaporator section where the working fluid (ammonia) is vaporized. Other applications include the use of heat pipes for air-to-air heat exchangers (Gerak et al., 1987), illustrated in Fig. 1.9; in the thermal control of nuclear reactor cores, illustrated in Fig. 1.10; in electronic devices and components, illustrated in Fig. 1.11 (Murase et al., 1982, 1987); and more recently, as illustrated in Fig. 1.12, to transfer heat from the torso region of the human body to the fingers in order to prevent frostbite (Faghri et al., 1989).

Because the return of liquid from the condenser to the evaporator is dependent upon the surface tension, heat pipes are particularly well suited for use in reduced gravity environments. Early applications included the

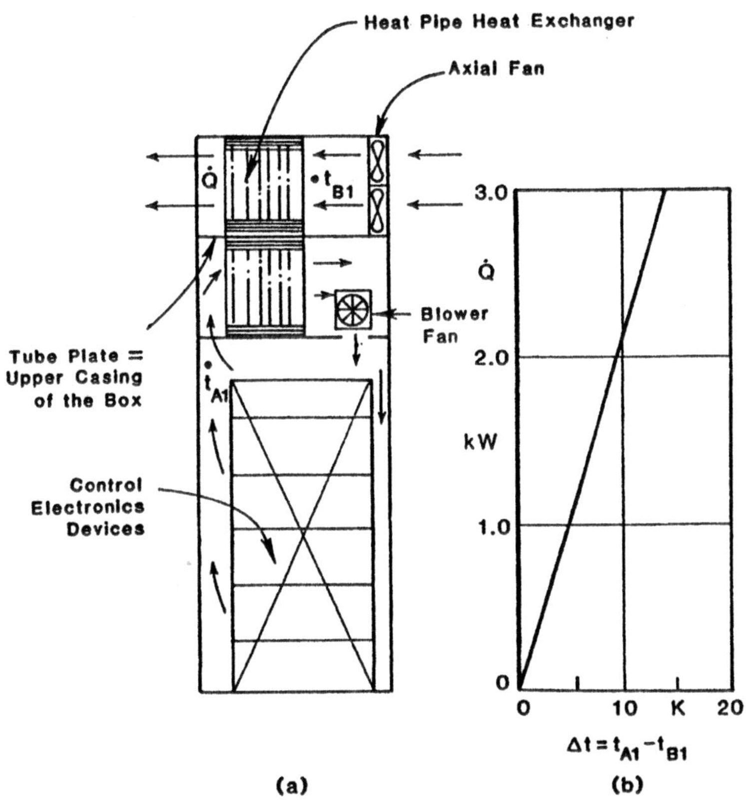

Figure 1.9 Heat pipe heat exchanger to cool electronic equipment in a sealed cabinet (Gerak et al. 1987).

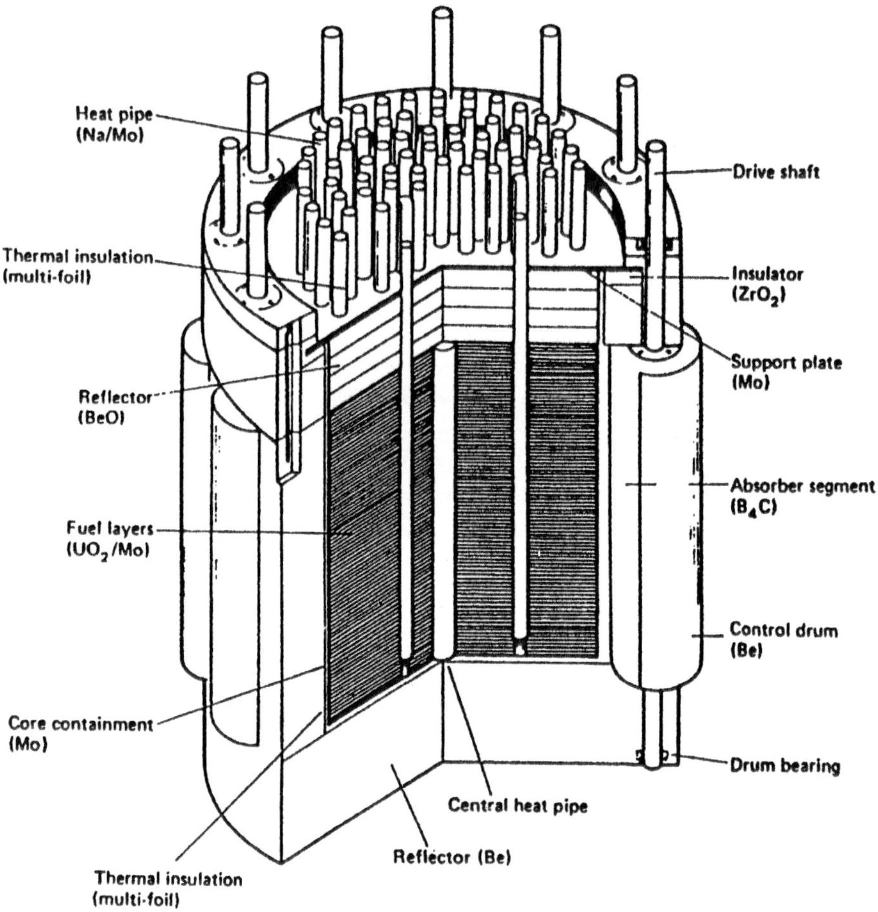

Figure 1.10 Heat pipes designed for use in the thermal control of nuclear reactor cores (Dunn and Reay, 1982).

use of heat pipes in the OAO-C telescope housing to control temperature variations to the monogroove heat pipe proposed for use on the NASA Space Station (Alario et al., 1984). Terpstra and Van Veen (1987) have prepared a summary of all of the patents, both U.S. and international, dealing with heat pipes and thermosyphons. This reference is particularly useful for individuals beginning the study of heat pipe applications and trying to access the current state-of-the-art.

1.4.1 Heat Pipe Sizes and Shapes

The sizes and shapes of heat pipes are almost as varied as the applications. Sizes range from a 15-m-long monogroove heat pipe developed by Alario

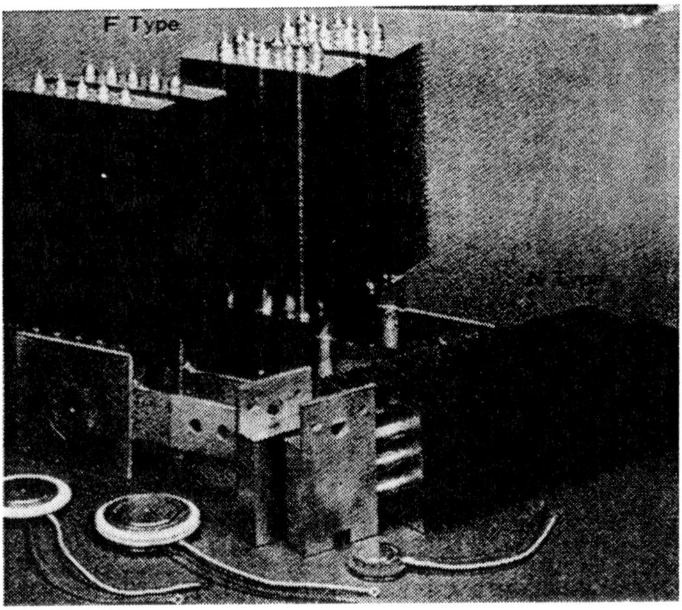

Figure 1.11 Heat pipes used in the thermal control of electronic devices and components (Murase et al., 1982, 1987).

et al. (1984) for spacecraft heat rejection to a 10-mm-long expandable bellows type heat pipe developed by Peterson (1986) for the thermal control of semiconductor devices in multichip modules. Vapor and liquid flow cross-sectional areas also vary significantly, from those encountered in flat-plate heat pipes, which have very large flow areas, to very small micro–heat pipes approximately 35 μm in diameter (Peterson, 1988b, 1992) proposed for use in a wide variety of applications, including the removal of heat from laser diodes (Mrácek, 1988) and other small localized heat-generating devices (Peterson, 1988a; Stulc and Horváth, 1988), the thermal control of photovoltaic cells (Peterson 1987a,b), the removal or dissipation of heat from the leading edge of hypersonic aircraft (Camarda et al., in press), and applications involving the nonsurgical treatment of cancerous tissue through either hyper- or hypothermia (Anonymous, 1989; Fletcher and Peterson, 1993). Heat pipes may be fixed or variable in length and either rigid or flexible, for situations where relative motion or vibration poses a problem.

1.4.2 Effective Thermal Resistance

Because heat pipes utilize the latent heat of vaporization of the working fluid as opposed to sensible heat, the required temperature difference is small, and hence the effective thermal conductivity may be several orders

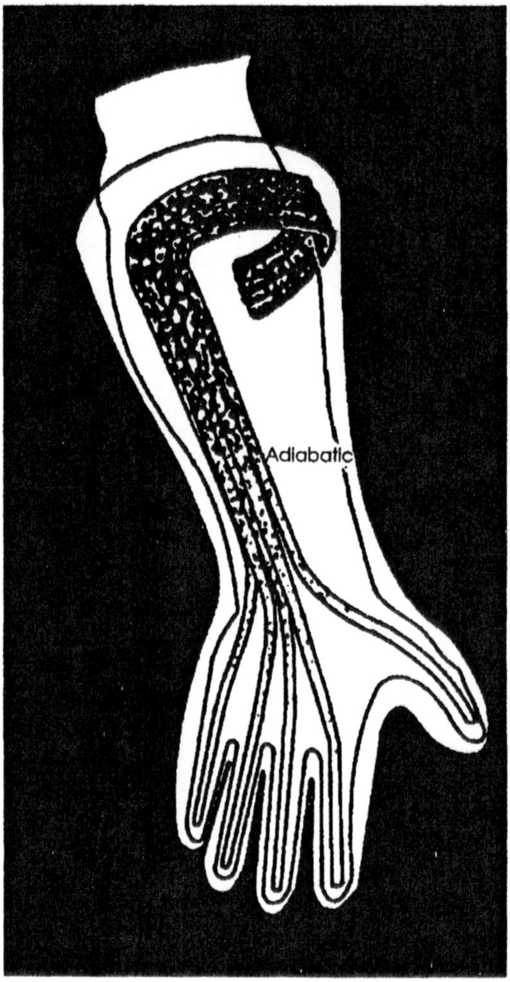

Figure 1.12 Heat pipe gloves for frostbite prevention (Faghri et al., 1989)

of magnitude greater than that of even the best solid conductors. For example, assume that it is necessary to transfer 20 W of thermal energy over a distance of 0.5 m in a device 1.27 cm in diameter. As shown in Fig. 1.13, using Fourier's law, a solid aluminum rod would result in a temperature difference of approximately 460°C. A solid copper rod would result in a temperature difference of 206°C, while use of a simple copper–water heat pipe with a screen wick could result in a temperature difference between the external evaporator and condenser surfaces of only 6°C.

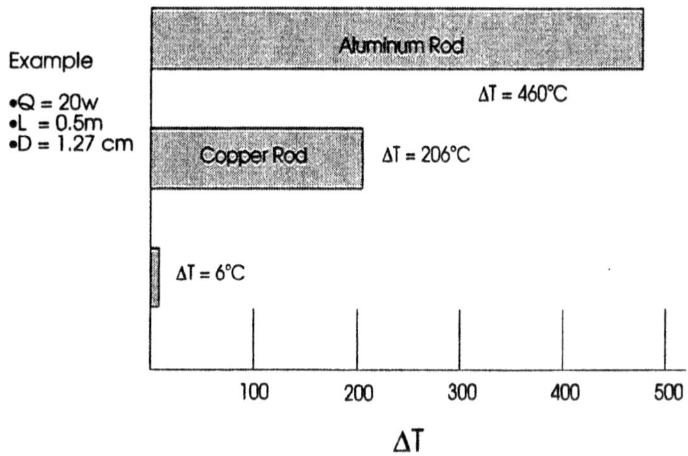

Figure 1.13 Comparison of heat pipes and solid conductors.

1.4.3 Relative Weight

In addition to substantial reductions in the temperature difference, the use of heat pipes normally results in significant decreases in weight when compared to solid conductors. In the previous example, the solid copper and aluminum rods would have weights that would be 13.7 and 4.2 times as heavy as the simple copper–water heat pipe described.

1.5 SUMMARY

As discussed earlier, heat pipe technology has been developing since the mid-1960s. Cylindrical heat pipes are commercially available in a wide range of sizes. They can be utilized in spacecraft applications, in large industrial applications, or as heat sinks to cool electronic components and packages. Heat pipes offer a low weight, long-life solution to many types of thermal problems. Although investigations continue to determine and evaluate new wick materials, working fluids, and geometrical configurations, little additional research is needed on the more simple devices provided that materials with proven compatibility are chosen and that proper cleaning, filling, and sealing procedures are followed.

Heat pipes and thermosyphons can be manufactured for one-of-a-kind applications or in large quantities. As future designs become more standardized and production volumes increase, the cost of these devices will decrease rapidly. The use of heat pipes (or two-phase thermosyphons) to cool semiconductor devices is particularly promising, as increases in power density will require the use of phase change heat transfer for temperature control.

REFERENCES

Alario, J., Brown, R., and Otterstadt, P., 1984, "Space Constructable Radiator Prototype Test Program," Paper No. 84-1793, AIAA. American Institute of Aeronautics and Astronautics, Washington, DC.

Anonymous, 1989, "Application of Micro Heat Pipes in Hyperthermia," Annual Report of the Itoh R and D Laboratory, Osaka, Japan, 1989.

Antoniuk, D., 1987, "Generalized Modeling of Steady-State and Transient Behavior of Variable Conductance Heat Pipes," Paper No. 87-1615, American Institute of Aeronautics and Astronautics, Washington, D.C.

Brennan, P. J., and Kroliczek, E. J., 1979, *Heat Pipe Design Handbook*, B&K Engineering, Towson, MD.

Camarda, C. J., Rummler, D. R., and Peterson, G. P., in press, "Multi Heat Pipe Panels," *NASA Tech Briefs*, LAR-14150.

Chang, W. S., and Colwell, G. T., 1985, "Mathematical Modeling of the Transient Operating Characteristics of a Low-Temperature Heat Pipe," *Numerical Heat Transfer*, Vol. 8, pp. 169–185.

Chi, S. W., 1976, *Heat Pipe Theory and Practice*, McGraw-Hill, New York.

Chisholm, D., 1971, *The Heat Pipe*, Mills and Boon, London, England.

Cotter, T. P., 1964, "Theory of Heat Pipes," Los Alamos National Laboratory Report No. LA-3246-MS, The University of California, Los Alamos, NM.

Cullimore, B., 1985, "Modeling of Transient Heat Pipe Effects Using a Generalized Thermal Analysis Program," Paper No. 85-0938, American Institute of Aeronautics and Astronautics, Washington, D.C.

Dunn, P. D., and Reay, D. A., 1982, *Heat Pipes*, 3rd ed., Pergamon, New York.

Eastman, G. Y. 1968, "The Heat Pipe," *Scientific American*, Vol. 218, No. 12, pp. 38–46.

Faghri, A., Reynolds, D. B., and Faghri, P., 1989, "Heat Pipes for Hands," *Mechanical Engineering*, Vol. III, No. 6, pp. 72–75.

Feldman, K. T., 1976, "The Heat Pipe: Theory, Design and Applications," Technology Application Center, University of New Mexico, Albuquerque, NM.

Fletcher, L. S., and Peterson, G. P., 1993, "A Micro Heat Pipe Catheter for Local Tumor Hyperthermia," U.S. Patent 5,190,539, January 12.

Gaugler, R. S., 1944, "Heat Transfer Devices," U.S. Patent 2,350,348.

Gay, F. W., 1929, "Heat Transfer Means," U.S. Patent 1,725,906.

Gerak, A., Horvath, L., Jelinek, F., Stulc, P., and Zboril, V., 1987, "Examples of Heat Pipe Application in Chemical, Electrical and Other Industries," Proc. 6th Intl. Heat Pipe Conf., Grenoble, France, pp. 522–530.

Grover, G. M., Cotter, T. P., and Erikson, G. F., 1964, "Structures of Very High Thermal Conductivity," *Journal of Applied Physics*, Vol. 35, pp. 1190–1191.

Heine, D., and Groll, M., 1984, "Compatibility of Organic Fluids with Commercial

Structure Materials for Use in Heat Pipes," Proc. 5th Int'l. Heat Pipe Conf., Tsukuba, Japan, pp. 170–174.

Kesten, A. C., and Haught, A. F., 1986, "Chemical Heat Pipe Employing Self-driven Chemical Pump Based on a Molar Increase," U.S. Patent 4,712,610.

King, C. R., 1931, "Perkins Hermetic Tube Boilers," *The Engineer*, Vol. 152, pp. 405–406.

Merrigan, M. A., Keddy, E. S., and Sena, J. T., 1985, "Transient Heat Pipe Investigations for Space Power Systems," Los Alamos Technical Report No. LA-UR-85-3341, Los Alamos National Laboratory, Los Alamos, NM.

Mrácek, P., 1988, "Application of Micro Heat Pipes to Laser Diode Cooling," Annual Report of the SVÚSS, Prague, Czechoslovakia.

Murase, T., Yoshida, K. Fujikake, J., Koizumi, T., and Ishida, N., 1982, "Heat Pipe Heat Sink HEAT KICKER for Cooling of Semi-Conductors," *Furukawa Review*, No. 2, pp. 24–33.

Murase, T., Tanaka, S., and Ishida, S., 1987, "Natural Convection Type Long Heat Pipe Heat Sink "POWERKICKER-N" for the Cooling of GTO Thyristor, Proc. 6th Int'l. Heat Pipe Conf., Grenoble, France, pp. 537–542.

Perkins, L. P., and Buck, W. E., 1892, "Improvements in Devices for the Diffusion or Transference of Heat," UK Patent 22,272, London, England.

Peterson, G. P., 1986, "Analytical Development and Computer Modeling of a Bellows Type Heat Pipe for the Cooling of Electronic Components," Paper No. 86-WA / HT-69, American Society of Mechanical Engineers, New York.

Peterson, G. P., 1987a, "Analysis of a Heat Pipe Thermal Switch," Proc. 6th Int'l Heat Pipe Conf., Vol. 1, Grenoble, France, May 25–28, pp. 177–183.

Peterson, G. P., 1987b, "Heat Removal Key to Shrinking Avionics," *Aerospace America*, October, No. 8, pp. 20–22.

Peterson, G. P., 1988a, "Heat Pipes in the Thermal Control of Electronic Components," invited paper, Proc. 3rd International Heat Pipe Symposium, Tsukuba, Japan, September 12–14, pp. 2–12.

Peterson, G. P., 1988b, "Investigation of Miniature Heat Pipes," Final Report, Contract No. F33615-86-C-2733, Task 9, Wright-Patterson Air Force Base.

Peterson, G.P., 1992, "An Overview of Micro Heat Pipe Research," *Applied Mechanics Review*, Vol. 45, No. 5, May, pp. 317–328.

Stulc, P., Horváth, L., 1988, "Heat Pipes for Cooling of Electronic Elements," Annual Report of the SVÚSS, Prague, Czechoslovakia.

Terpstra, M., and Van Veen, J. G., 1987, *Heat Pipes: Construction and Application*, Elsevier Applied Science, New York.

Tien, C. L., 1975, "Fluid Mechanics of Heat Pipes," *Annual Review of Fluid Mechanics*, Vol. 2, pp. 167–186.

Trefethen, L., 1962, "On the Surface Tension Pumping of Liquids or a Possible Role of the Candlewick in Space Exploration," G. E. Tech. Info., Serial No. 615 D114, Detroit, MI, February 1962.

Udell, K. S., and Jennings, J. D., 1984, "A Composite Porous Heat Pipe," Proc. 5th Int'l. Heat Pipe Conf., Tskuba, Japan, pp. 41–47.

Chapter

2

BACKGROUND INFORMATION

2.1 SURFACE TENSION AND WETTABILITY

In order to analyze and understand the operation of a heat pipe, it is first necessary to understand the phenomena that govern the behavior of the various interfaces, particularly those occurring at the liquid–vapor interface. This behavior, which is primarily governed by the surface tension, wettability, and contact angle, results in a pressure imbalance occurring at or near the interface and provides the capillary pumping action required for proper heat pipe priming and operation. The interfaces existing at the boundaries of the liquid, vapor, and solid phases are typically thought of as a clearly delineated line or, in the case of the interface between the liquid and a vapor phase, as a meniscus. However, as shown in Fig. 2.1, there really exists a very small region surrounding this meniscus where the density of both the liquid and vapor vary gradually. At this length scale, the close range repulsive forces between individual molecules have been shown to vary more rapidly with molecular spacing than the longer range electrostatic, inductive, and dispersive attractive forces (Adamson, 1990).

As a result, the increased molecular spacing caused by this density gradient significantly decreases the repulsive forces but only slightly decreases the longer range attractive forces. Normal to the interface, this force imbalance tends to draw the vapor molecules toward the liquid surface, as illustrated in Fig. 2.2*a*. Parallel to the interface, the forces are balanced, but the relatively large decrease in the repulsive forces and the somewhat smaller decrease in the long range attractive forces cause what appears to be a net tension at the interface that, as illustrated in Fig. 2.2*b*, acts equally in all directions. It is this interfacial tension resulting from the stored energy in the molecules near the interface that is normally referred

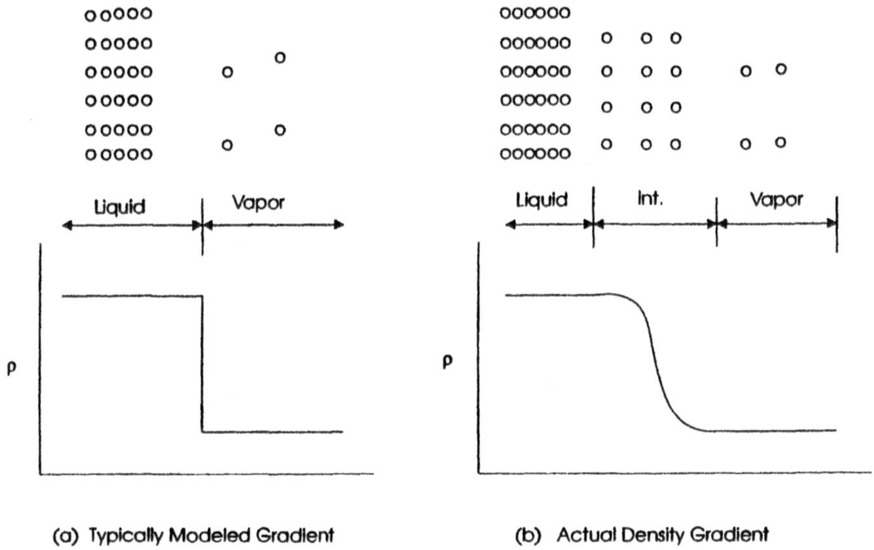

Figure 2.1 Density variation at the liquid–vapor interface.

to as the surface tension σ. This surface tension is a temperature dependent fluid property and has units of force per unit length.

An appreciation for surface tension and the pressure difference caused by the resulting curvature can be gained by performing a fundamental thermodynamic analysis of the interface between the liquid and vapor phases and the region immediately adjacent to this interface (Carey, 1992).

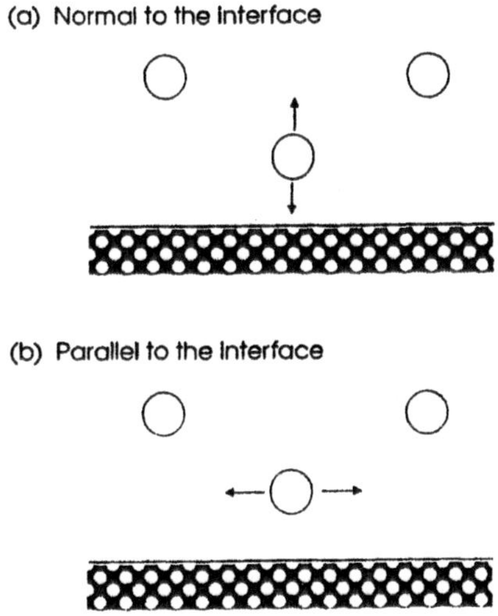

Figure 2.2 Force imbalance occurring at the liquid–vapor interface.

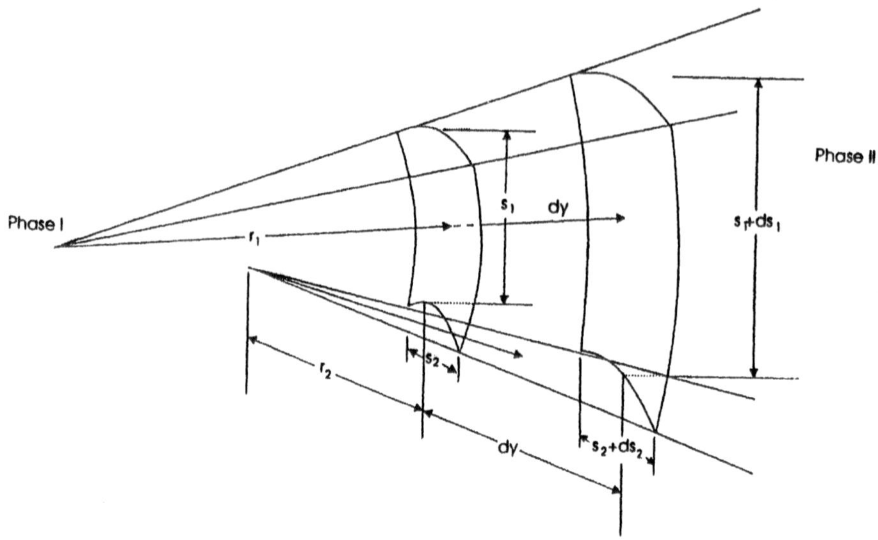

Figure 2.3 Thermodynamic analysis of a liquid–vapor interface (Carey, 1992).

As shown in Fig. 2.3, a liquid–vapor interface can be characterized by two distinct radii of curvature r_1 and r_2 oriented perpendicular to one another. Defining a control volume consisting of the liquid–vapor interface and the region immediately adjacent to this interface, it is clear that any movement of the interface requires that work be done either on or by the system. This work results in a change in the free surface energy, dF, at the interface and may be accompanied by changes in the entropy of the surface, the chemical potential, and / or the boundary work. Applying the first law of thermodynamics to this system yields an expression for the change in the Helmholtz free energy of

$$dF = -S\,dT + \mu\,dN + \Sigma\,\delta W \tag{2.1}$$

Because, as illustrated in Fig. 2.1, there are really three regions within this system (the interfacial region, the region labeled phase I, and the region labeled phase II), the total change in the free surface energy is in fact the sum of the change in the entropy of the surface, the chemical potential, and the boundary work for all three regions (Carey, 1992). The fact that the boundary work term, $P\,dV$, for the interface is equal to the work associated with the stretching of the liquid surface, $\sigma_{lv}\,dA$, yields

$$\begin{aligned} dF = &- S_{\text{Int}}\,dT + \mu\,dN_{\text{Int}} + \sigma\,dA_{\text{Int}} \\ &- S_{\text{I}}\,dT + \mu\,dN_{\text{I}} - (P\,dV)_{\text{I}} \\ &- S_{\text{II}}\,dT + \mu\,dN_{\text{II}} - (P\,dV)_{\text{II}} \end{aligned} \tag{2.2}$$

If the temperature is assumed to be constant (a reasonable assumption during evaporation or condensation) and the total number of moles and total volume do not change, that is,

$$\begin{aligned} dT &= 0 \\ dN_T = dN_{\text{Int}} + dN_{\text{I}} + dN_{\text{II}} &= 0 \\ dV = dV_{\text{I}} + dV_{\text{II}} &= 0 \end{aligned} \tag{2.3}$$

equation 2.2 reduces to

$$P_{\text{I}} - P_{\text{II}} = \sigma \frac{dA_{\text{Int}}}{dV_{\text{I}}} \tag{2.4}$$

It is apparent from Fig. 2.3 that as the interface moves to the right (i.e., away from phase I and toward phase II), the area of the interface, dA_{Int}, and volume of phase I, dV_{I}, both change. These changes can be expressed as

$$dA_{\text{Int}} = S_2 dS_1 + S_1 dS_2 \tag{2.5}$$

and

$$dV_{\text{I}} = S_1 S_2\, dy \tag{2.6}$$

respectively. By similar triangles, the change in length of the two corresponding sides of the interfacial areas can be related to the distance moved by

$$dS_1 = \frac{S_1\, dy}{r_1} \quad \text{and} \quad dS_2 = S_2 \frac{dy}{r_2} \tag{2.7}$$

respectively (Carey, 1992). Combining these two expressions with equations 2.5 and 2.6 and substituting into equation 2.4 allow the pressure difference between the two phases to be expressed as

$$P_{\text{I}} - P_{\text{II}} = \sigma\left(\frac{1}{r_1} + \frac{1}{r_2}\right) \tag{2.8}$$

This expression is typically referred to as the Laplace or Young–Laplace equation and defines the capillary pressure difference occurring across the meniscus separating the liquid and vapor regions as a function of the two principal radii of curvature (Carey, 1992).

An example of this capillary pressure difference can be seen when a cylindrical tube is inserted into a pool of liquid, as illustrated in Fig. 2.4. For the case of a stationary interface, the capillary pressure difference acting over the surface of the interface must be equal to the summation of the forces acting around the circumference of the interface, or

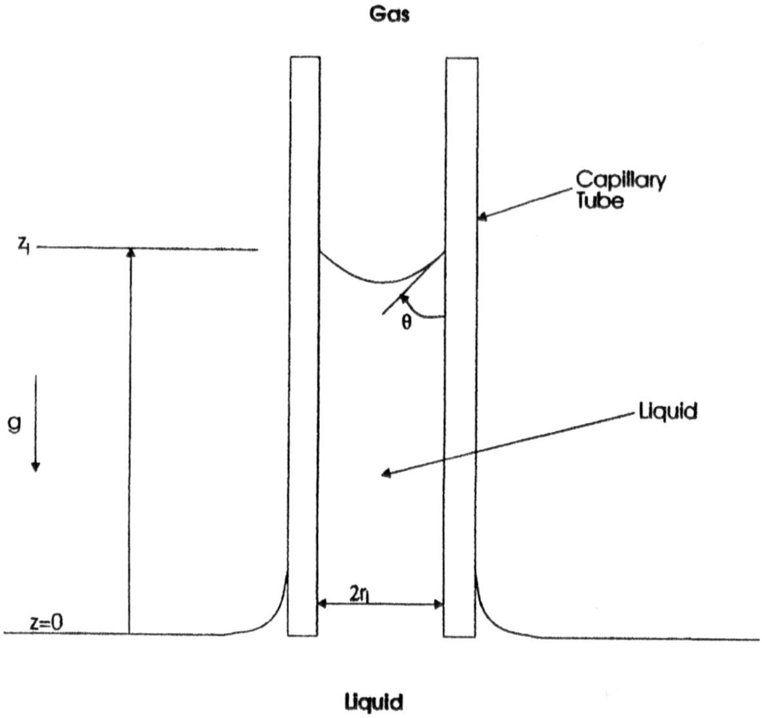

Figure 2.4 Capillary pressure difference occurring in a cylindrical tube inserted in a liquid (Carey, 1992).

$$\Delta P \pi r^2 = (2\pi r)\sigma \cos\theta \tag{2.9}$$

Simplifying this expression yields

$$P_v - P_l = \frac{2\sigma\cos\theta}{r} \tag{2.10}$$

where θ is the angle between the solid and vapor regions. This angle, commonly referred to as the apparent *contact angle* is always measured from the solid to the vapor regions through the liquid and is a somewhat subjective value, which is strongly dependent upon the magnification level.

Example 2.1 A circular dish with a diameter of 1 in. is filled with water at 20°C to a height of 1 in. If three "straws" with diameters of 25.4, 12.7, and 6.35 mm are placed in the dish, what is the equilibrium height of water in each straw? Assume that the contact angle of the water is 45° and that the tubes are of negligible thickness. How will the equilibrium configuration change with respect to temperature over the range 10–70°C?

SOLUTION

$d_1 = 25.4 \text{ mm} = 0.00254 \text{ m}$
$d_2 = 12.7 \text{ mm} = 0.00127 \text{ m}$
$d_3 = 6.35 \text{ mm} = 0.000635 \text{ m}$

Evaluating each straw or tube independently as shown below,

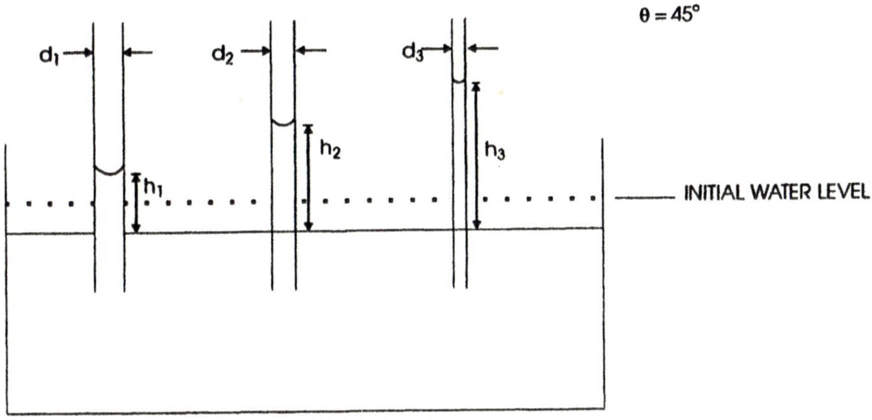

and applying a pressure balance between the hydrostatic and capillary pressures yields

$$\rho_l g h_i = \frac{2\sigma}{r_i}\cos\theta = \frac{4\sigma}{d_i}\cos\theta \text{ assuming } \rho_l \gg \rho_v$$

or

$$h_i = \frac{4\sigma}{\rho_l g d_i}\cos\theta$$

at 20°C

$$\sigma \simeq 72.8 \text{ mN/m} \qquad \rho_l = 998 \text{ kg/m}^3$$

$h_1 = 0.00828 \text{ m} = 0.325 \text{ in.} = 8.28 \text{ mm}$

$h_2 = 0.01656 \text{ m} = 0.652 \text{ in.} = 16.56 \text{ mm}$

$h_3 = 0.03312 \text{ m} = 1.304 \text{ in.} = 33.12 \text{ mm}$

T	σ (mN/m)	ρ_l (kg/m^3)	h_1 (mm)	h_2 (mm)	h_3 (mm)
10°	74.3	1000.0	8.45	16.90	33.80
20°	72.8	998.0	8.28	16.56	33.12
30°	71.7	996.0	8.15	16.31	32.62
40°	70.0	992.1	8.01	16.02	32.04
50°	68.3	988.1	7.85	15.69	31.38
60°	66.6	983.3	7.69	15.38	30.75
70°	64.9	977.5	7.54	15.07	30.15

Note that the increase in temperature results in a decrease in the value of the surface tension and, hence, a decrease in the height of the liquid column.

The behavior of liquids in contact with solids varies significantly from one surface to another and is dependent upon many factors. A common example of this variation, occurs when a water droplet comes in contact with a metallic surface such as the surface of an automobile. If the solid surface has been freshly waxed, the water beads up, that is, the wetting angle is large, forming a discrete droplet. If, however, the surface is unwaxed, the contact angle is small and the liquid spreads out to form a thin film. Generally, this variation in the behavior at the solid–liquid interface can be thought of in terms of the affinity of a liquid for the solid and is referred to as the *wettability*. As seen in the previous example, the wettability of a liquid on a particular surface is a function not only of the thermophysical properties of the liquid but also of the physical characteristics of the solid. These characteristics include things such as the surface roughness, surface temperature, surface cleanliness and also other more subtle factors such as the presence of oxide layers.

The wettability of a particular liquid or liquid–solid combination is typically quantified by measuring the contact angle discussed previously, which may be significantly different depending upon whether a liquid is advancing or receding across a solid surface. This difference results in a phenomenon referred to as capillary or *contact angle hysteresis*. This hysteresis may be caused by contamination of the liquid or solid surface, surface roughness, or surface immobility on an intermolecular scale (Adamson, 1990). While an in-depth discussion of this phenomenon is beyond the scope of this text and will not be addressed, it results from the molecular forces occurring at the common line and may be significant in determining the actual contact angle for a specific liquid–vapor–solid interface.

2.2 CAPILLARITY AND FREE SURFACE ENERGY

To completely understand the phenomena that govern the capillarity in a heat pipe, it is necessary to understand not only the interaction between the liquid and vapor regions but also the interactions occurring at the boundary between the solid and liquid regions and the solid and vapor regions. Figure 2.5 illustrates a single droplet at rest on a solid surface surrounded by a pure vapor. The point at which the three phases meet (which is actually a circle at the base of the drop) is called the *common* or *contact line*. If the droplet volume is fixed, that is, there is no vaporization or condensation, then increasing the contact angle will decrease the solid–liquid interface area. Likewise, decreasing the contact angle will allow the liquid to spread, forming a thin film and thereby increasing the solid–liquid interfacial area. As the limit of $\theta \rightarrow 0$, the liquid spreads over the entire available surface, forming a thin film.

Previously, when discussing the surface tension, only the surface tension existing between the liquid and vapor phases of a substance, σ_{lv}, was considered. In reality there are three surface tensions that exist: one between the liquid and vapor, σ_{lv}; one between the solid and liquid, σ_{sl}; and one between the solid and vapor, σ_{sv}. Returning to the previous force balance, it is clear that the summation of forces in the horizontal direction must be equal to zero for a droplet in static equilibrium, hence,

$$\sigma_{sv} = \sigma_{sl} + \sigma_{lv} \cos \theta \tag{2.11}$$

which is referred to as *Neumann's formula* or *Young's equation* (Adamson, 1990).

A similar force balance in the vertical direction indicates the solid surface must be exerting a force equal and opposite to the vertical force exerted by the droplet, $\sigma_{lv} \sin \theta$. Because this force is small for fluids typically of interest in heat pipes and the modulus of elasticity of the solid is typically large, no measurable deformation of the solid surface is observed.

If, as before, the interfaces are thought of as regions instead of clearly delineated lines and the volume, number of moles, and temperature of the

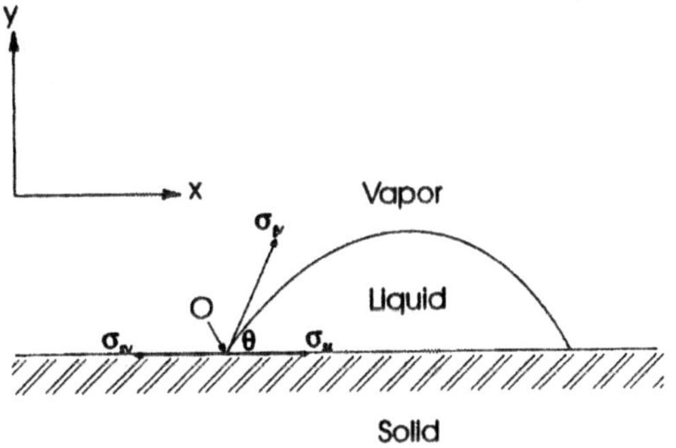

Figure 2.5 A single droplet at rest on a solid surface surrounded by a pure vapor (Carey, 1992).

system are held constant, (i.e., equation 2.3), an energy balance can be written for the system. To be completely accurate, however, this energy balance must include all energy terms from all three interfacial regions and, when simplified, takes the form

$$dF = \sigma_{lv}\, dA_{lv} + \sigma_{sl}\, dA_{sl} + \sigma_{sv}\, dA_{sv} \tag{2.12}$$

where dF is referred to as the change in the *Helmholtz free energy* (Carey, 1992).

For the droplet pictured in Fig. 2.5, any change in the position of the common line will result in a change in the area of the solid–liquid interface equal and opposite to the change in the area of the solid–vapor interface area. This implies that

$$dA_{sv} = -dA_{sl} \tag{2.13}$$

If the droplet is assumed to be hemispherical in shape (i.e., gravity is negligible), it can also be shown that

$$dA_{lv} = dA_{sl} \cos\theta \tag{2.14}$$

Combining equations 2.12–2.14 and rearranging yield

$$\frac{\partial F}{\partial A_{sl}} = \sigma_{lv} \cos\theta + \sigma_{sl} - \sigma_{sv} \tag{2.15}$$

which describes the rate of change of the total free energy with respect to the change in the interfacial area (Myshkis et al., 1976).

The preceding development demonstrates that Young's equation can be derived from a thermodynamic analysis of the interface regions and bulk phases in which the minimization of the free energy of a system at equilibrium is imposed. Often in heat pipe modeling and design, it is important to know or determine the equilibrium shape of the liquid–vapor interface. This is particularly true for the case of very small heat pipes where the effect of the local gravitational acceleration on the liquid–vapor interfacial configuration is small or negligible or when the heat pipes are to be utilized in microgravity environments. Determination at the shape of the liquid–vapor interface can be accomplished by recognizing that the most stable position of an interface will occur when the free surface energy is minimized. Therefore, by taking the first derivative of the free surface energy, F, with respect to the solid–liquid interfacial area, A_{sl}, and setting it equal to zero, that is,

$$\frac{\partial F}{\partial A_{sl}} = 0 \tag{2.16}$$

the most stable equilibrium condition of the interface can be obtained. This procedure has been utilized previously for microgravity applications involving the priming of heat pipe radiators (Peterson, 1984) and more recently for determining the stable liquid configuration of micro-heat pipes in 1 g environments. It should be noted that this procedure assumes constant volume, constant temperature conditions and therefore may only be applied to determine the stable interfacial configuration prior to the addition or rejection of heat.

Example 2.2 The heat pipe illustrated below was designed and patented by Grumman Aerospace Corporation (Alario et al., 1984) for use on the NASA space station radiator panels. In order to operate properly, the liquid artery must be filled with working fluid; however, during launch loads or orbital maneuvers the liquid may become displaced, as presented by Peterson (1984). To determine if this heat pipe will operate properly, it is first necessary to find the stable equilibrium configuration as a function of the liquid contact angle, the radius of the liquid artery, and the radius of the vapor passage when placed in a zero-gravity environment.

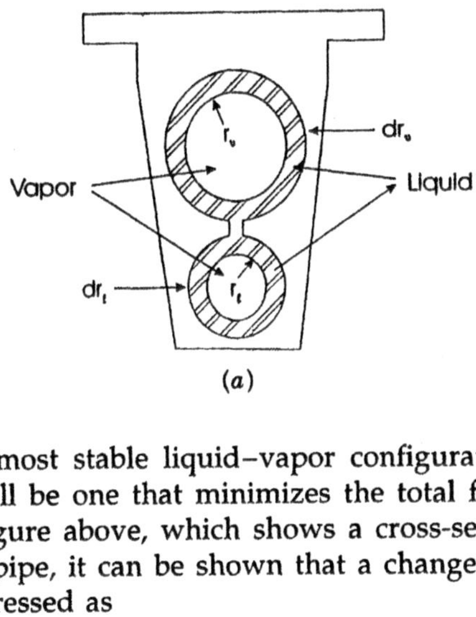

(a)

SOLUTION The most stable liquid–vapor configuration in this or any other heat pipe will be one that minimizes the total free surface energy. Referring to the figure above, which shows a cross-sectional view of the monogroove heat pipe, it can be shown that a change in the free surface energy can be expressed as

$$dF = (P_v - P_l)A_l\, dx$$

or

$$dF_l = \left(2\sigma_{lv} \cos\theta \frac{1}{r_l + dr_l}\right)[2\pi L(r_l + dr_l)]\, dr_l$$

where r_l is the radius of the vapor space in the liquid channel and dr_l represents the change in the radius of the vapor space that exists in the liquid channel.

Similarly for the vapor space located in the vapor channel

$$dF_v = \left(2\sigma_{lv} \cos\theta \frac{1}{r_v + dr_v}\right)[2\pi L(r_v + dr_v)]\, dr_v$$

and the total change in the free surface energy in the heat pipe can be expressed as

$$dF = dF_l + dF_v.$$

Changes in the volume of the vapor space in the liquid channel can be defined as the final volume minus the initial volume, or

$$dV_l = \pi L(r_l + dr_l)^2 - \pi L r_l^2$$

and similarly for the vapor channel

$$dV_v = \pi L(r_v + dr_v)^2 - \pi L r_v^2$$

Under steady-state conditions the total volume of vapor in the heat pipe will remain constant, and hence a decrease in the volume of the vapor space in the liquid channel will result in a corresponding increase in the vapor space located in the vapor channel. This can be expressed as

$$-dV_l = dV_v$$

because the volume of the liquid will remain constant. Substituting and simplifying yield

$$-(r_l + dr_l)^2 - (r_l)^2 = (r_v + dr_v)^2 - (r_v)^2$$

which can be further simplified to

$$-(2r_l\, dr_l) - (dr_l)^2 = (2r_v\, dr_v) + (dr_v)^2$$

Neglecting the second-degree terms and simplifying yield

$$dr_v = -\frac{r_l\, dr_l}{r_v}$$

and finally substituting expressions for dF_l and dF_v yields

$$dF = 4\pi L\sigma_{lv} \cos\theta(dr_l + dr_v)$$

which describes the total change in the free surface energy as a function of changes in the radius of the vapor space in the liquid and vapor channels. Combining this with the expression for dr_v yields

$$dF = 4\pi L\sigma_{lv} \cos\theta\, dr_l\left(1 - \frac{r_l}{r_v}\right).$$

The premise that the liquid–vapor system will seek a configuration of minimum total free surface energy requires that the total change in the free surface energy, dF, be less than or equal to zero because an increase would require that work be done on the system by some external force. Because L, σ_{lv} and $\cos\theta$ are all positive, either dr_l or $1 - r_l/r_v$ must be negative.

When dr_l is negative and $1 - r_l/r_v$ is positive, the radius of the vapor space in the liquid channel is decreasing and the heat pipe will prime properly. When dr_l is positive, $1 - r_l/r_v$ must be negative, which implies that r_l, the radius of the vapor space in the liquid channel must be larger than the radius of the vapor space in the vapor channel. Thus, the liquid will move into the liquid channel when $r_l < r_v$ and move out of the liquid channel when $r_l > r_v$. When $r_l = r_v$, the fluid will be at equilibrium but will be highly unstable.

In other words, as long as the radius of the vapor space in the vapor channel is never smaller than the liquid channel radius as shown above in (a), the liquid will flow toward the smaller characteristic radius of the liquid channel. This tendency, which results in the correct configuration for proper operation, will be governed by the amount of fluid or the initial charge in the heat pipe and the ratio of the diameters of the two channels.

This brings up a fairly subtle factor in the determination of the initial diameters of the liquid and vapor channels. If due to some sudden acceleration the liquid was momentarily contained in the vapor channel, it would assume the shape of a hollow cylinder with vapor space in the center. If the initial charge was such that this vapor space radius was smaller than the radius of the liquid channel as shown in (b), the liquid configuration would be stable but the heat pipe would be improperly primed. Therefore, it is essential that the initial charge be such that the diameter of the vapor space in the vapor channel never be smaller than the diameter of the liquid channel.

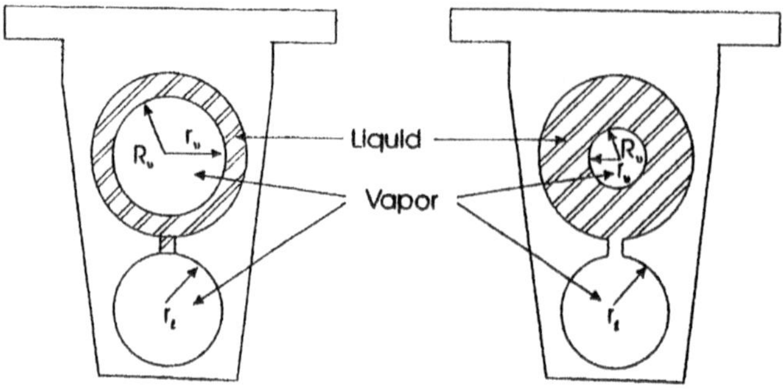

(b)

(c)

One of the assumptions made in the development of equations 2.12–2.15 is that the solid surface is smooth. For most heat pipe applications this may not be valid as axial grooves or permeable layers are typically used as wicking structures. It is therefore useful to determine the effect of variations in the surface roughness on wettability, contact angle, and interfacial behavior at the common line. As presented by Carey (1992), a uniformly rough surface can be described as the ratio of the actual solid surface area to the area obtained when the surface is projected onto a horizontal plane. This ratio can be expressed in terms of the solid–liquid interfacial area as

$$\gamma = \frac{dA_{sl}}{(dA_{sl})_{\text{Horz}}} \tag{2.17}$$

Using this relationship and replacing the wetting angle of a smooth surface, θ, with the wetting angle of the rough surface, θ_R, equation 2.17 can be rewritten as

$$dA_{lv} = dA_{sl} \frac{\cos \theta_R}{\gamma} \tag{2.18}$$

Combining this expression with equation 2.12 yields

$$\frac{\partial F}{\partial A_{sl}} = \left(\frac{\sigma_{lv}}{\gamma}\right) \cos \theta_R + \sigma_{sl} - \sigma_{sv} \tag{2.19}$$

As mentioned previously, at equilibrium the change in the free surface energy with respect to area $\partial F / \partial A_{sl} = 0$, therefore

$$\cos \theta_R = \frac{\sigma_{sv} - \sigma_{sl}}{\sigma_{lv}} \gamma \tag{2.20}$$

Combining this expression with Young's equation yields

$$\cos \theta_R = \gamma \cos \theta \tag{2.21}$$

where θ is the contact angle of the smooth surface. Because $\gamma > 1$ and $\cos \theta_R > \cos \theta$, therefore $\theta_R < \theta$; the contact angle for rough surfaces is typically smaller than for smooth surfaces. Rearranging and substituting into equation 2.17 yields

$$\frac{\partial F}{\partial A_{sl}} = \left(\frac{\sigma_{lv}}{\gamma}\right)(\cos \theta_R - \gamma \cos \theta) \tag{2.22}$$

This implies that if $\gamma > \sec \theta$ the right-hand side of equation 2.21 is negative and the liquid will spread (Carey, 1992). This spontaneous spreading of the liquid is referred to as *wicking* and is the underlying principle on which the operation of heat pipes depends.

It is helpful at this point to introduce the concept of cohesive forces, which are the forces occurring between two liquids, and adhesive forces, which are the forces between a liquid and a solid. It is clear from Fig. 2.6 that if the body forces acting on the system (i.e., gravity) are small, the droplet shape approaches that of a sphere with contact at only a single point, resulting in a contact angle that approaches 180°. Alternatively, as the liquid spreads out, the contact angle decreases and approaches zero. As illustrated in Fig. 2.6, liquids with wetting angles between 0° and 90° are referred to as wetting liquids and liquids with wetting angles between 90° and 180° are called nonwetting liquids. For wetting liquids, the cohesive forces (i.e., the liquid–liquid forces) are greater than the adhesive forces (i.e., the solid–liquid forces). In a nonwettting liquid, the opposite is true.

Example 2.3 A meniscus is supported between two plates forming a wedge of angle ϕ. If the fluid is water and has a contact angle, θ, of 55°C, determine the angle ϕ at which the shape changes from concave to convex.

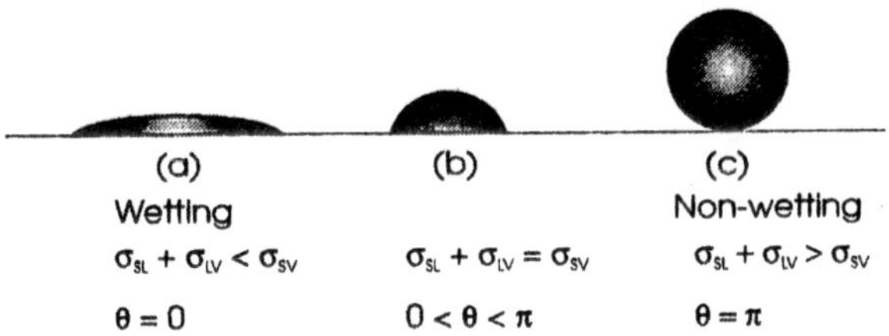

Figure 2.6 Wetting and non-wetting fluids (Dunn and Reay, 1982).

SOLUTION Given a groove as shown below,

(a)

ϕ for concave, convex

$\theta = 55°$

At the equilibrium point (i.e., where the meniscus is "flat"):

(b)

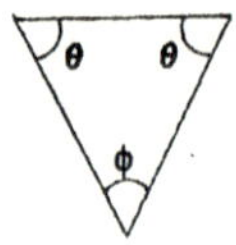

$\phi + 2\theta = 180°$

$\Rightarrow \phi = 70°$

At an angle $\phi = 70°$, the meniscus makes a transition from concave to convex. If $\phi > 70°$, the meniscus will be convex:

(c)

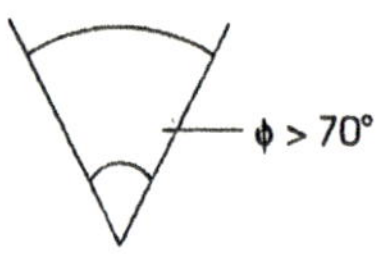

If $\phi < 70°$, the meniscus will be concave:

(d)

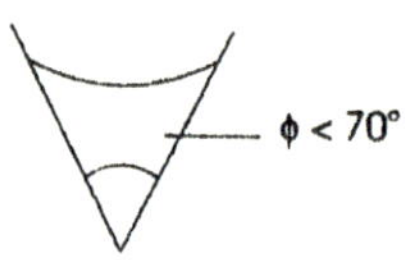

Several important aspects of the physical system have been omitted in the previous discussions and development. Of these, perhaps the most important is that the effects of any adsorbed liquid films on the solid surface have been neglected. The presence of these adsorbed films will affect the force balance at the common line and, hence, the shape of the liquid–vapor interface (Adamson, 1990). As a result, this form of Young's equation is typically valid only when there is no adsorbed liquid film on the solid surface. For the case of most operating heat pipes, a properly charged or primed wicking structure is normally assumed, and hence neglecting the effects of the adsorbed film introduces only small errors. This is not true, however, for the case of very small micro-heat pipes where the adsorbed film region may make up a significant portion of the liquid–vapor interfacial area. For this case, Young's equation may not be a true physical representation of the physics of the system.

A second simplifying assumption that is not entirely accurate is the assumption that the interfacial tensions are constant along each of the interfaces. Clearly, because of the variation in the density gradients, the interfacial tension near the contact line may be different from that occurring some distance from the contact line. These differences, however, are usually small, and the measured contact angles for systems with no adsorbed film on the solid surface are typically in good agreement with the values predicted by Young's equation (Slattery, 1990).

Finally, because of the magnitude and the difficulty encountered in measuring the solid–liquid and solid–vapor interfacial tensions σ_{sl} and σ_{sv}, these terms are often neglected in the development of heat pipe models. While in many cases this leads to only minimal errors in the predicted values, it can and in the case of micro-heat pipe modeling does have a significant effect. More rigorous derivations of Young's equation and its limitations are available in Buff and Saltzburg (1957), Johnson (1959), Adamson (1990), and Carey (1992).

2.3 VAPORIZATION AND CONDENSATION AT A LIQUID–VAPOR INTERFACE

The previous discussions have demonstrated that the capillarity, wettability, surface tension, and contact angle all play an important role in determining the shape of the liquid vapor interface in operating heat pipes. While up to this point, only equilibrium conditions have been considered, it is now necessary to discuss the effects of nonequilibrium conditions on the interfacial behavior. It is principally these nonequilibrium conditions that exist between the evaporator and condenser liquid–vapor interfaces that are responsible for the operation of heat pipes. As shown in Fig. 2.7, vaporization from a liquid–vapor interface causes the liquid meniscus to recede into the wicking structure. Alternatively, condensation results in a liquid meniscus whose radius of curvature is significantly greater than the curvature of the interface in equilibrium. It is essentially this difference in the radius of curvature, the decreased radius in the evaporator portion caused by vaporization of the liquid from the interface and the increased radius of curvature in the condenser portion caused by condensation at the

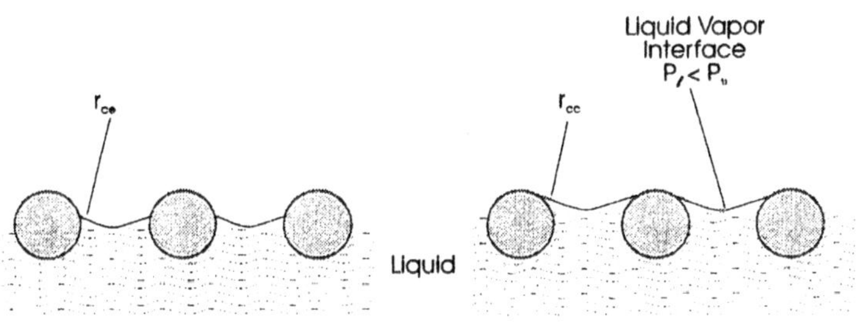

Figure 2.7 Nonequilibrium conditions occurring in the evaporator and condenser of a heat pipe.

liquid interface, that provides the capillary pumping pressure necessary for heat pipe operation.

It is helpful at this point to discuss how this vaporization and condensation affects the equilibrium conditions at the liquid–vapor interface. During the vaporization process molecules escaping from the liquid surface must overcome a force of attraction exerted by the other molecules in the liquid. If the liquid under discussion wets the wick or solid surface (which is normally the case for heat pipe working fluids), then the interfacial shape is concave. This concave interface results in a molecular attraction that is greater than that occurring at a plane surface in equilibrium. As a result, the energy level required for molecules to break away from a concave liquid surface is greater than that required for a plane surface, and hence, the pressure and density over the concave curved liquid surface are less than for a flat liquid surface. Alternatively, a convex liquid surface requires an energy level that is slightly less than that required for a plane surface. This results in a vapor pressure that is slightly greater for a convex surface than for a plane surface.

Using a simple hydrostatic pressure balance, the difference in the saturation pressure over a concave meniscus and a flat or planar meniscus can be shown to be equal to

$$\Delta P_{sat} = \frac{2\sigma}{r} \cos\theta \left(\frac{\rho_{v,sat}}{\rho_l} \right) \tag{2.23}$$

Likewise the magnitude of the difference in the saturation pressure between convex and flat interfaces would be equal to the same amount. While seemingly insignificant, this pressure difference can be appreciable when the radius of curvature at the liquid–vapor interface is on the order of 1 μm or less. For this case, it may account for as much as 20–30% of the vapor pressure over a plane interface. For radii of curvature on the order of 10 μm or more, this difference may amount to less than several percent. And finally, for a capillary pore radius of more than 10 μm this effect may be negligible (Ivanovskii et al., 1982).

2.4 EFFECT OF TEMPERATURE AND CONTAMINANTS ON INTERFACIAL BEHAVIOR

There are many factors that affect the surface tension and wettability of various heat pipe working fluids. Among the most significant are the effect of temperature and dissolved contaminants. As discussed previously, the surface tension is to some extent determined by the density gradients that exist at the three-phase common line. Because increases in the temperature of a liquid in contact with a heated solid will result in increases in the density difference, the surface tension is a strong function of the liquid and vapor temperatures. Typically, increases in temperature will result in decreases in surface tension until the critical temperature is reached, at which point the saturated liquid and saturated vapor densities are equal and the surface tension approaches zero.

Many equations are available that describe this phenomenon. For example, the relationship between surface tension and temperature for liquid water in contact with pure water vapor can be expressed as

$$\sigma = 235.8\left(1 - \frac{T_{\text{sat}}}{T_{\text{cr}}}\right)^{1.256}\left[1 - 0.625\left(1 - \frac{T_{\text{sat}}}{T_{\text{cr}}}\right)\right] \tag{2.24}$$

where the temperature T_{cr} is the critical temperature in kelvins and the surface tension σ is expressed in mN/m. This expression matches the available data for water to within $\pm 0.5\%$ for $0.01 \leqslant T \leqslant 300°\text{C}$ or $\pm 13\%$ for $300 \leqslant T \leqslant 374.16°\text{C}$. Additional expressions are available for many other types of working fluids typically used in heat pipe applications and are listed in Appendix A.

The presence of dissolved contaminants in a liquid, whether deliberate or accidental, can have a significant effect on the surface tension characteristics. Miller and Neogi (1985) have shown that for a binary mixture consisting of two species, A and B, the surface excess mass of species A can be expressed as

$$\Gamma_{\text{A}} = -\left(\frac{x_{\text{A}}}{RT}\right)\left(\frac{\partial \sigma}{\partial x_{\text{A}}}\right)_T \tag{2.25}$$

where A is the solute, B is the solvent, and x_{A} is the concentration of solute A. Examination of this equation indicates that the surface tension decreases as the concentration of the solute increases. As a result, $\partial \sigma / \partial x_{\text{A}}$ is negative and the surface excess mass Γ_A is positive. For these conditions, the solute A has a higher concentration at the liquid surface and, hence, is referred to as a surfactant. Surfactants that are commonly used to alter the surface tension of various fluids can either increase or decrease the surface tension. The most common example is a combination of soap and water, for which the soap (solute A) accumulates at the surface, causing the surface tension to decrease. Because water is a polar molecule and therefore accepts other polar molecules, the hydrocarbon chains align at the liquid–vapor interface to form a monolayer. These hydrocarbon chains, which are nonpolar in nature, prevent the liquid water from dissolving substances such as oil and grease.

Another factor that may play a significant role in the behavior of surface tension and, hence, the performance of heat pipes, particularly in zero-gravity conditions, is a phenomenon referred to as the Marangoni effect. Because the surface tension is a function of both temperature and concentration, variations in either of these values occurring at a liquid–vapor

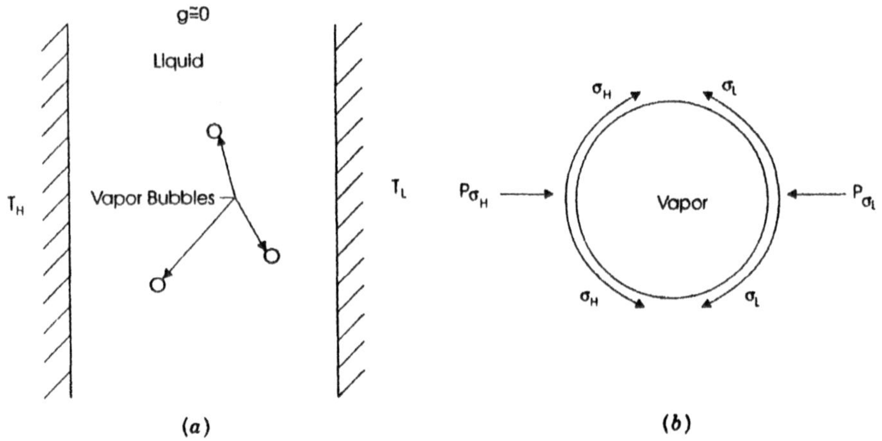

Figure 2.8 (*a*) Vapor bubble behavior in zero gravity conditions. (*b*) Force balance of a single vapor bubble.

interface will cause the liquid near the interface to move toward regions of higher surface tension. This can be explained by observing vapor bubbles in a liquid pool in zero gravity in which one wall is heated and the other wall is cooled, as shown in Fig. 2.8*a*. An examination of a single vapor bubble in this situation, shown in Fig. 2.8*b*, reveals that the surface tension on the right side is greater than that occurring on the left side. As a result, the pressure difference across the right surface is greater than that across the left surface. Assuming that the pressure in the liquid and the pressure in the vapor are uniform, a force imbalance exists that forces the bubble toward the hot surface.

This effect is significant in that it results in a net force in the opposite direction of the density gradient. In zero or reduced gravity environments, this Marangoni effect may result in the formation of a thin vapor film on a heated surface significantly decreasing the heat transfer rate from the solid surface.

Example 2.4 A gas / vapor bubble is suspended in a fluid contained by a rectangular container as shown below. The upper surface of the container is held at a temperature greater than that of the lower surface, such that a linear temperature gradient is imposed across the fluid. The sides may be assumed to be insulated. If the system is in a microgravity environment, will the bubble move? If so, which direction and why?

(a)

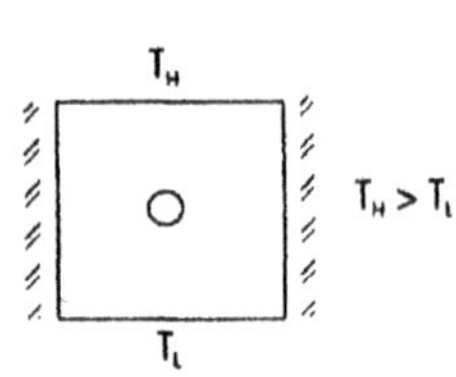

SOLUTION

(b)
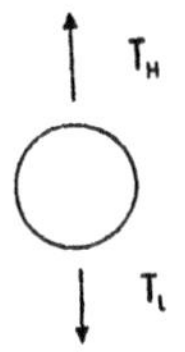

Note the differential element of fluid on the bubble surface, $d\theta$:

(c)
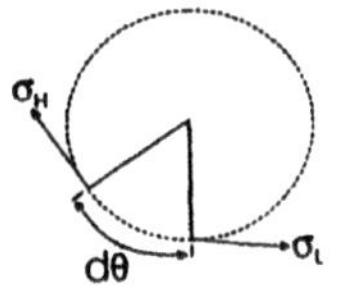

Knowing that $\sigma_L(T_L) > \sigma_H(T_H)$ implies that the surface tension on the lower portion of the bubble is greater than on the upper surface, or $\sigma_L > \sigma_H$. Thus, a force imbalance exists and "drags" the fluid along the surface from the hot region to the cold region. This results in a net force toward the *hot* surface.

Also, looking on a macroscopic level, the capillary pressure is given by

$$P_v - P_l = \frac{2\sigma}{r}$$

(d)
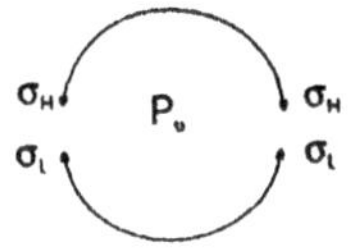

and $\sigma_L > \sigma_H$ so that the pressure difference across the lower surface is greater than that across the upper surface. With P_l and P_v assumed uniform, a force imbalance exists that forces the bubble toward the *hot* surface.

One final factor which should be noted is what is referred to as *contact angle hysteresis*. When a solid surface is immersed in a liquid pool, the liquid meniscus will form a characteristic contact angle. While typically assumed constant, this contact angle may vary depending upon whether the surface is pushed into or withdrawn from the liquid pool. This variation in the contact angle for advancing interfaces (the case where the solid is pushed into the liquid pool) and the receding interface (the case where the solid is withdrawn from the liquid pool) is typically referred to as the contact angle hysteresis and can be expressed as the difference between the advancing and receding contact angles, $\theta_A - \theta_R$. There are numerous causes for this phenomenon, the most common of which are surface contaminants, surface roughness, or surface inhomogeneities.

Contact angle hysteresis provides an explanation for how the small capillary tube shown in Fig. 2.9, can support a liquid slug. The capillary pressure between points 1 and 2 can be expressed as

$$P_1 - P_2 = \frac{2\sigma}{r_{c,top}} \tag{2.26}$$

while the capillary pressure between points 3 and 4 can be expressed as

$$P_4 - P_3 = \frac{2\sigma}{r_{c,bottom}} \tag{2.27}$$

The pressure difference between points 4 and 1 and 3 and 2 can be expressed simply as the hydrostatic pressures of the gas and liquid, respectively. Combining these relationships yields the expression

$$2\sigma\left(\frac{1}{r_{c,top}} - \frac{1}{r_{c,bottom}}\right) = (P_L - P_g)gL \tag{2.28}$$

which illustrates the mechanism by which the liquid slug can be supported against the force of gravity. Differences in the radii of curvature resulting from differences in the contact angle at the upper and lower interfaces cause a pressure imbalance and, hence, a force imbalance, which supports the hydrostatic head.

Example 2.5 A 3.0-mm-diameter capillary tube is used to support a liquid (water) slug against gravity, as shown in Fig. 2.9. If the upper meniscus

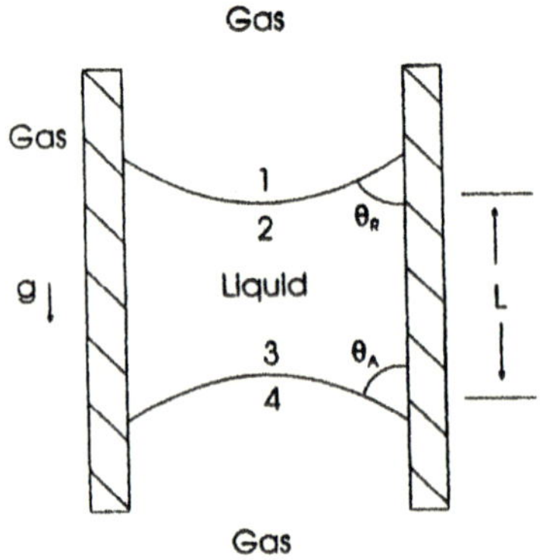

Figure 2.9 A liquid slug supported in a capillary tube (Carey, 1992).

contact angle is 20° and the lower meniscus contact angle is 65°, determine the maximum length of the slug that may be supported by this tube. How does the length of the slug change with respect to temperatures between 10 and 85°C.

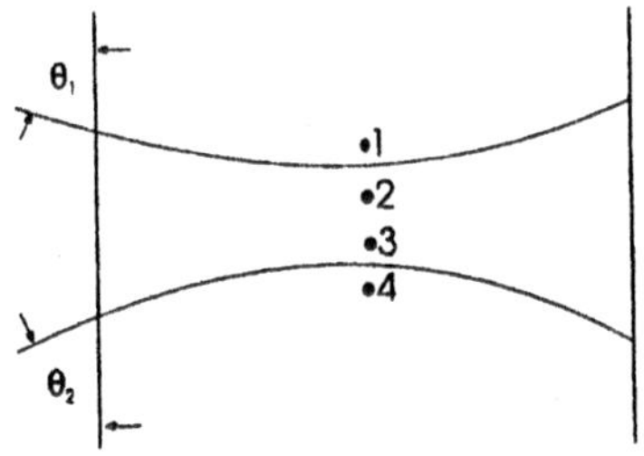

(*a*)

SOLUTION

$$P_1 - P_2 = \frac{2\sigma \cos \theta_1}{r}$$

$$P_4 - P_3 = \frac{2\sigma \cos \theta_2}{r}$$

Neglecting the hydrostatic pressure difference between P_1 and P_4 due to the large density differences (i.e., $\rho_{air} \ll \rho_l$) implies

$$P_1 = P_4$$

and

$$P_3 - P_2 = \rho_l g L$$

Substituting and solving for L yield

$$L = \frac{2\sigma}{\rho_l r g}(\cos \theta_1 - \cos \theta_2)$$

and

$$\sigma = f(\mathrm{T}) \qquad \rho_l = f(\mathrm{T})$$

$T(k)$	σ (mN / m)	ρ_l (kg / m^3)	L (mm)
285	74.3	1000.0	5.23
295	72.7	998.0	5.12
300	71.7	997.0	5.06
310	70.0	993.1	4.96
320	68.3	989.1	4.86
340	64.9	979.4	4.66
360	61.4	967.1	4.47

which can be represented graphically as

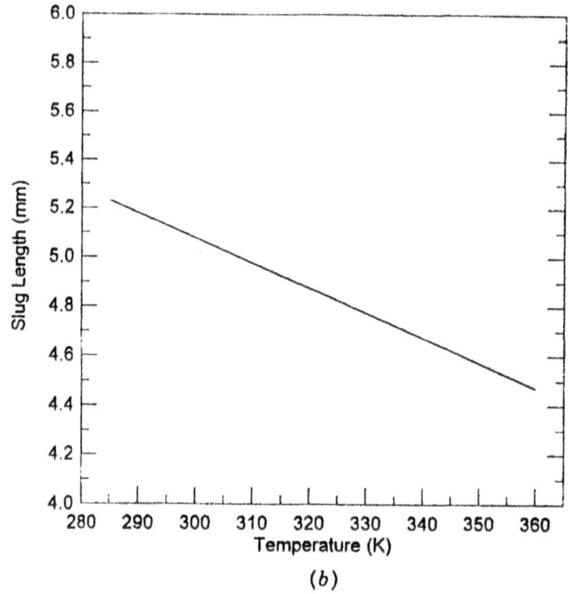

(*b*)

A somewhat similar phenomenon results from the presence of surface roughnesses and can be seen by examining a liquid drop on a vertical surface. Because of the local gravitational acceleration and variations in the advancing and receding wetting angles, the liquid droplet shown in Fig. 2.10 is not symmetric about a horizontal axis but assumes a teardrop shape.

Finally, it is important to note that the actual liquid–vapor interface is conventionally divided into three regions, as shown in Fig. 2.11. The intrinsic meniscus, which is the region governed by the Young–Laplace equation and where the disjoining pressure effects are negligible; the thin-film region, or that region above the intrinsic meniscus where the disjoining pressures are important; and the extended meniscus, which typically describes the entire meniscus. While the behavior in the thin-film

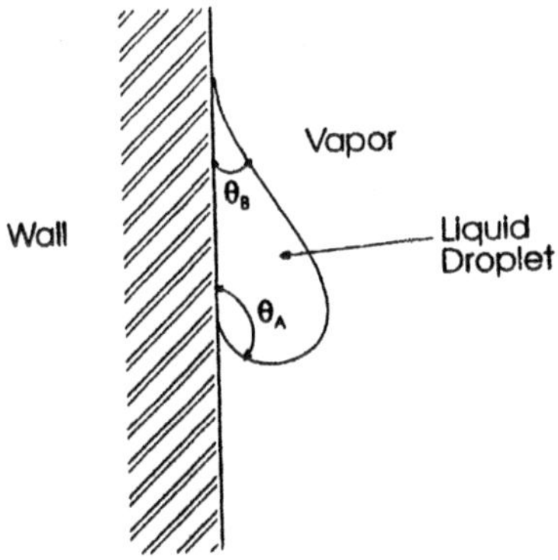

Figure 2.10 Effect of surface roughness on droplet shape on a vertical surface.

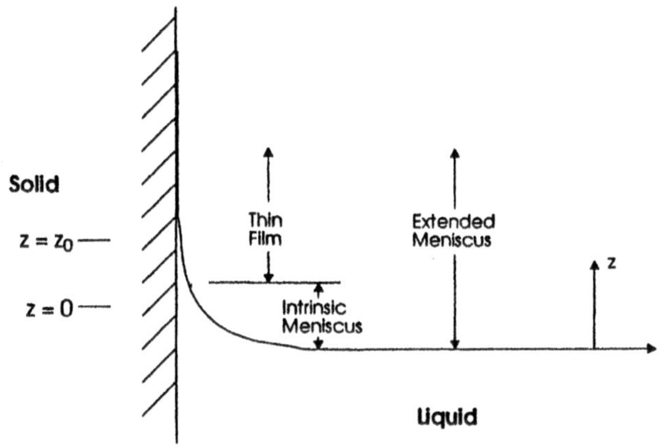

Figure 2.11 Interfacial regions.

region may be significant in determining the performance characteristics and operational limits of micro-heat pipes, in-depth discussion of such effects as disjoining pressure, van der Waals forces, and so on, are beyond the scope of this book. For additional information, the readers are referred to Adamson (1990) or Carey (1992).

NOMENCLATURE

A	Area (m^2)
d	Diameter (m)
g	Gravitational acceleration (m/s^2)
Γ	Surface excess mass (kg)
h	Height (m)
F	Free surface energy (J)
L	Length (m)

N	Number of moles
P	Pressure (Pa)
r	Radius (m)
S	Entropy (J / K)
s	Length (m)
T	Temperature (K)
V	Volume (m^3)
W	Work (J)
x	Distance, position coordinate (m)
y	Distance, position coordinate (m)

Greek Symbols

γ	Area ratio
μ	Chemical potential (J / mol or J / kg)
ρ	Density (kg / m^3)
σ	Surface tension (N / m)
θ	Wetting angle (deg)
ϕ	Groove angle (deg)

Subscripts

A	Advancing
c	Capillary
cr	Critical
H	High temperature
Horz	Horizontal surface
Int	Interface
l	Liquid
L	Low temperature
R	Rough, receding
S	Solid
sat	Saturation
v	Vapor
I	Phase I
II	Phase II

REFERENCES

Adamson, A. W., 1990, *Physical Chemistry of Surfaces*, 5th ed., Wiley, New York.

Buff, F. P., and Saltzburg, H., 1957, "Curved Fluid Interfaces. II. The Generalized Neumann Formula," *Journal of Chemical Physics*, Vol. 26, pp. 23–31.

Carey, V. P., 1992, *Liquid–Vapor Phase Change Phenomena*, Hemisphere, Washington, D.C.

Dunn, P. D., and Reay, D. A., 1982, *Heat Pipes*, 3rd ed., Pergamon, New York.

Ivanovskii, M. N., Sorokin, V. P., and Yagodkin, I. V., 1982, *The Physical Principles of Heat Pipes*, Clarendon, Oxford, England.

Johnson, R. E., Jr., 1959, "Conflicts between Gibbsian Thermodynamics and Recent Treatments of Interfacial Energies in Solid–Liquid–Vapor Systems," *Journal of Chemical Physics*, Vol. 63, pp. 1655–1658.

Miller, C. A., and Neogi, P., 1985, Interfacial Phenomena, Marcel Dekker, New York.

Myshkis, A. D., Babskii, V. G., Kopachevskii, N. D., Slobozhanin, L. A., and Tyuptsov, A. D., 1976, *Low-Gravity Fluid Mechanics*, Springer-Verlag, New York.

Peterson, G. P., 1984, "Capillary Priming Characteristics of High Capacity Dual Passage Heat Pipes," *Chemical Engineering Communications*, Vol. 27, No. 1, pp. 119–126.

Slattery, J. C., 1990, *Interfacial Transport Phenomena*, Springer-Verlag, New York.

Chapter

3

CAPILLARY LIMITATION IN HEAT PIPES AND THERMOSYPHONS

3.1 CAPILLARY PRESSURE BALANCE

Although heat pipe performance and operation are strongly dependent on the shape, working fluid, and wick structure, the fundamental phenomenon that governs the operation of these devices arises from the difference in the capillary pressure across the liquid–vapor interfaces in the evaporator and condenser regions. As explained in Chapter 2, vaporization from the evaporator section of a heat pipe causes the meniscus to recede into the wick and condensation in the condenser causes flooding. The combined effect of this vaporization and condensation is illustrated in Fig. 3.1*a* and results in a meniscus radius of curvature that varies along the axial length of the heat pipe. The point at which the meniscus has a minimum radius of curvature is referred to as the "dry" point and usually occurs in the evaporator at the point furthest from the condenser region. The "wet" point occurs at that point where the vapor pressure and liquid pressure are approximately equal or where the radius of curvature is at a maximum. As will be discussed, the location of this point can occur at any point in the condenser or adiabatic sections but typically occurs at a point located near the end of the condenser farthest from the evaporator.

Figure 3.1*b* illustrates the relationship between the static liquid and static vapor pressures in an operating heat pipe. As shown, the capillary pressure gradient across a liquid–vapor interface is equal to the pressure difference between the liquid and vapor phases at any given axial position.

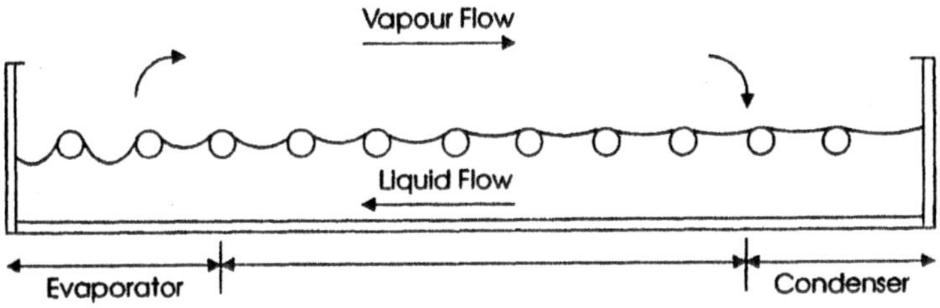

(a)

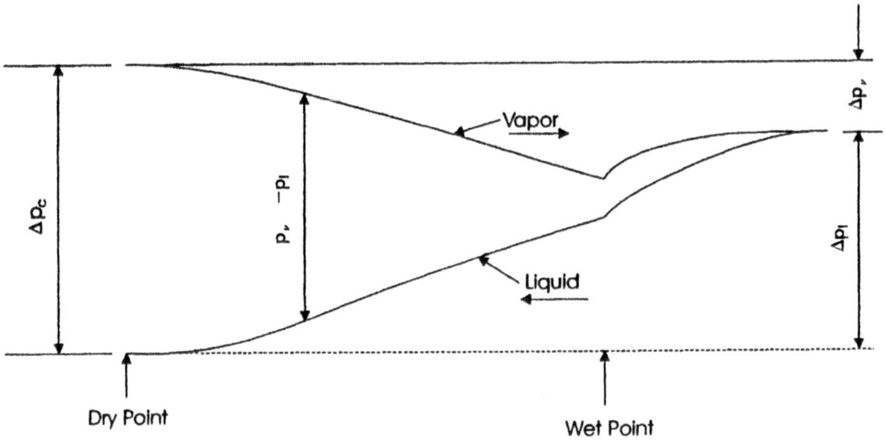

(b)

Figure 3.1 (a) Variation of meniscus curvature as a function of axial position. (b) Typical liquid and vapor pressure distributions in a heat pipe.

For a heat pipe to function properly, the net capillary pressure difference between the wet and dry points, identified in Fig. 3.1b, must be greater than the summation of all the pressure losses occurring throughout the liquid and vapor flow paths. This relationship, referred to as the capillary limitation, can be expressed mathematically as

$$(\Delta P_c)_m \geq \int_{L_{\text{eff}}} \frac{\partial P_v}{\partial x} dx + \int_{L_{\text{eff}}} \frac{\partial P_l}{\partial x} dx + \Delta P_{\text{ph},e} + \Delta P_{\text{ph},c} + \Delta P_{+} + \Delta P_{\|} \tag{3.1}$$

where $(\Delta P_c)_m$ = maximum capillary pressure difference generated within capillary wicking structure between wet and dry points

$\frac{\partial P_v}{\partial x}$ = sum of inertial and viscous pressure drops occurring in vapor phase

$\frac{\partial P_l}{\partial x}$ = sum of inertial and viscous pressure drops occurring in liquid phase

$\Delta P_{\text{ph},e}$ = pressure gradient across phase transition in evaporator

$\Delta P_{\text{ph},c}$ = pressure gradient across phase transition in condenser
ΔP_{+} = normal hydrostatic pressure drop
$\Delta T_{\parallel}$ = axial hydrostatic pressure drop

The first two terms on the right-hand side of this equation, $\partial P_v / \partial x$ and $\partial P_l / \partial x$, represent the summation of viscous and inertial losses in the vapor and liquid flow paths, respectively. The next two, $\Delta P_{\text{ph},e}$ and $\Delta P_{\text{ph},c}$, represent the pressure gradients occurring across the phase transition in the evaporator and condenser; and the last two, ΔP_{+} and $\Delta P_{\parallel}$, represent the normal and axial hydrostatic pressure drops. When the maximum capillary pressure is equal to or greater than the summation of these pressure drops, the capillary structure is capable of returning an adequate amount of working fluid to prevent dryout of the evaporator wicking structure. When the summation of all six pressure drops exceeds the maximum capillary pumping pressure, the working fluid is not supplied rapidly enough to the evaporator to compensate for the liquid loss through vaporization and the wicking structure becomes starved of liquid and dries out. This condition, referred to as capillary limitation, varies according to the wicking structure, working fluid, evaporator heat flux, and operating temperature.

Figure 3.2 gives a more accurate representation of the actual pressure distribution occurring in an operating heat pipe. Three separate curves are

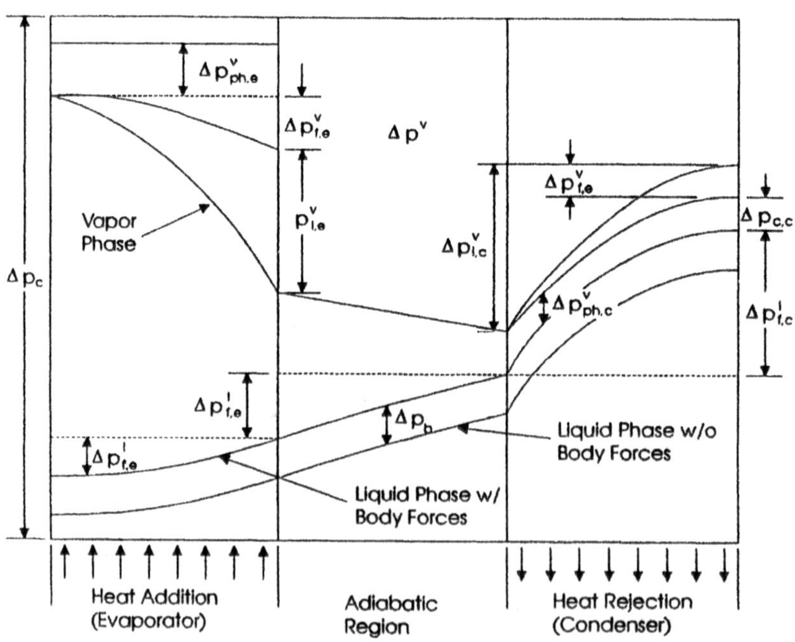

Figure 3.2 Graphical representation of pressure drops in an operating heat pipe.

presented; the first of these represents the vapor pressure distribution. Factors contributing to the pressure variation in the evaporator are the pressure drop across the liquid–vapor phase transition $\Delta P_{\text{ph},e}$, the frictional pressure drop $\Delta P_{f,e}$, and the inertial pressure gradient $\Delta P_{i,e}$. In the adiabatic region the only factor contributing to variations in the pressure distribution is the frictional pressure drop $\Delta P_{f,a}$ resulting in a linear variation of pressure with axial position. The factors affecting the pressure variation in the condenser are similar to those occurring in the evaporator, namely, the pressure drop across the liquid–vapor phase transition $\Delta P_{\text{ph},c}$, the frictional pressure drop $\Delta P_{i,c}$, which in the condenser provides some pressure recovery.

Two separate curves are shown in Fig. 3.2 for the pressure distribution in the liquid phase. The lower one is for a heat pipe in which the body forces are negligible, and the upper one includes the effect to a uniform and constant body force. While gravitational acceleration is the most obvious body force, additional forces occurring in rotating or revolving heat pipes or in heat pipes undergoing acceleration forces, such as those encountered in high performance aircraft, may also be significant. As illustrated in Fig. 3.2, the factors affecting the pressure distribution in the liquid phase in the condenser region consist of the pressure drop occurring across the liquid–vapor phase transition $\Delta P_{\text{ph},c}$ and the frictional pressure drop $\Delta P_{f,c}$. In the adiabatic region the only factor is the frictional pressure drop $\Delta P_{f,a}$, and in the evaporator, the primary factors are the liquid–vapor phase transition pressure drop $\Delta P_{\text{ph},c}$ and the frictional pressure drop $\Delta P_{f,c}$. As mentioned previously, the inertial pressure gradient in the liquid phase is typically negligible and therefore has been omitted.

As discussed previously and illustrated in Fig. 3.2, the total capillary pressure across the liquid–vapor interface must be greater than the summation of all of the pressure drops occurring throughout the liquid–vapor flow paths. If this is not true, the working fluid will not be returned to the evaporator, causing the liquid level in the evaporator wicking structure to be depleted, leading to dryout. Therefore, in order to effectively understand the behavior of the vapor and liquid flow in an operating heat pipe, each of the factors contributing to the overall pressure gradient must be clearly understood. Following is a detailed explanation of each.

3.2 CAPILLARY PRESSURE AT THE LIQUID–VAPOR INTERFACE

As discussed in Section 2.1, the capillary pressure difference at a single liquid–vapor interface, defined as $P_v - P_l$, or ΔP_c, can be found from the Laplace–Young equation,

$$\Delta P_c = \sigma\left(\frac{1}{r_1} + \frac{1}{r_2}\right) \tag{3.2}$$

where r_1 and r_2 are the principal radii of curvature and σ is the surface tension. For many heat pipe wicking structures, the maximum capillary pressure may be written in terms of a single radius of curvature, r_c. In this manner, the maximum capillary pressure between the wet and dry points can be expressed as the difference between the capillary pressure across the meniscus at the wet point and the capillary pressure at the dry point, or

$$\Delta P_{c,m} = \frac{2\sigma}{r_{c,e}} - \frac{2\sigma}{r_{c,c}} \tag{3.3}$$

As discussed previously and illustrated in Fig. 3.1*a*, the vaporization occurring in the evaporator causes the liquid meniscus to recede into the wick, reducing the local capillary radius $r_{c,e}$, and in the condenser section, condensation causes flooding of the wick, which in turn increases the local capillary radius $r_{c,c}$. The resulting difference in the two radii of curvature causes a pressure difference and, hence, pumping of the liquid from the condenser to the evaporator. During steady-state operation, it is generally assumed that the capillary radius in the condenser or at the wet point, $r_{c,c}$, approaches infinity, so that the maximum capillary pressure for a heat pipe operating at steady state in many cases can be expressed as a function of only the effective capillary radius of the evaporator wick,

$$\Delta P_{c,m} = \frac{2\sigma}{r_{c,e}} \tag{3.4}$$

Table 3.1 gives values for the effective capillary radius r_c for some of the more common wicking structures (Chi, 1976). For other simple geometries, values for the effective capillary radius can be found theoretically using the methods proposed by Chi (1976) or experimentally for pores or structures of more complex geometries using the methods described by Ferrell and

TABLE 3.1 Effective Capillary Radius for Several Wick Structures

Structure	r_c	Data
Circular cylinder (artery or tunnel wick)	r	r = radius of liquid flow passage
Rectangular groove	ω	ω = groove width
Triangular groove	$\omega / \cos\beta$	ω = groove width β = half-included angle
Parallel wires	ω	ω = wire spacing
Wire screens	$(\omega + d_\omega)/2 = 1/2N$	d = wire diameter N = screen mesh number ω = wire spacing
Packed spheres	$0.41 r_s$	r_s = sphere radius

Source: Chi, 1976.

Alleavitch (1969), Freggens (1969), or Eninger (1975). In addition, limited information on the transient behavior of capillary structures is also available (Colwell and Chang, 1984).

3.3 VAPOR PRESSURE DROP

The vapor pressure distribution as a function of axial position is illustrated in Fig. 3.2. This variation in vapor pressure is principally the result of the viscous pressure drop occurring along the vapor flow path. However, in the evaporator region, the viscous pressure drop may be coupled with the momentum changes required to accelerate the vapor molecules escaping from the liquid meniscus, which further decreases the pressure. For very high vapor flow rates, these momentum or inertial effects may be the dominant factor in determining the pressure gradient and can amount to as much as 80% of the total vapor pressure drop (Ivanovskii et al., 1982). In the adiabatic section of the heat pipe, the vapor pressure gradient is typically linear and results only from the viscous friction occurring between the flowing vapor and the pipe walls and wicking structure. In the condenser region, the inertial effects caused by the deceleration of the vapor molecules can be recovered (although not completely) because of the decreasing mass flow rate caused by condensation of some of the vapor. This condensation results in some pressure recovery and effectively decreases the rate at which the pressure decreases.

While calculation of the liquid pressure drop occurring in the wicking structure of a heat pipe is relatively straightforward, calculation of the vapor pressure gradient is significantly more difficult. As mentioned previously, the pressure gradient in the vapor passage not only is a function of the viscous pressure drop but also may be significantly influenced by the inertial effects resulting from the addition and rejection of mass in the evaporator and condenser, respectively. The vapor flow in heat pipes is further complicated by specific situations such as two-dimensional flow, the low vapor pressures occurring in cryogenic heat pipes, and the sonic or near sonic velocities that may be encountered in liquid metal heat pipes.

3.3.1 Two-Dimensional Vapor Flow Models

The vapor pressure within an operating heat pipe can be described by the Navier–Stokes equations, given the appropriate boundary conditions. In cylindrical coordinates, these equations can be written as

$$\frac{\partial P}{\partial x} = -\rho\left(u\frac{\partial v}{\partial r} + v\frac{\partial v}{\partial x}\right) - \frac{\mu}{r}\frac{\partial}{\partial r}r\left(\frac{\partial u}{\partial x} - \frac{\partial v}{\partial r}\right) \tag{3.5}$$

and

$$\frac{\partial P}{\partial r} = -\rho\left(u\frac{\partial u}{\partial r} + v\frac{\partial u}{\partial x}\right) + \mu\frac{\partial}{\partial x}\left(\frac{\partial u}{\partial x} - \frac{\partial v}{\partial r}\right) \tag{3.6}$$

where r and x represent the radial and axial coordinates of the heat pipe, respectively. Conservation of mass requires that the radial and axial pressure components be related by the continuity equation, or

$$\frac{\partial(ru)}{\partial r} + \frac{\partial(rv)}{\partial x} = 0 \tag{3.7}$$

This flow is dynamically similar to laminar pipe flow with injection and suction through a porous wall (Cotter, 1965). Solutions of the Navier–Stokes equations for this particular problem with boundary conditions similar to those occurring in heat pipes have been presented by several authors. Notable among these are the presentations by Wageman and Guevara (1960) for cylindrical pipes and Knight and McInteer (1965) for two parallel porous plates.

$$\left.\begin{aligned} u &= 0 \\ \partial v / \partial r &= 0 \end{aligned}\right\} \text{ for } r = 0 \qquad \left.\begin{aligned} v &= 0 \\ u &= u_0 \end{aligned}\right\} \text{ for } r = r_0 \tag{3.8}$$

In both of these analyses, laminar incompressible flow with uniform injection and suction was assumed. To distinguish between the various flow regimes, a radial Reynolds number, Re_r was defined in terms of the vapor density ρ_v and vapor viscosity μ_v as

$$\mathrm{Re}_r = -\frac{\rho_v u_v r_0}{\mu_v} = \frac{1}{2\pi\mu_v}\frac{d\dot{m}_v}{dx} \tag{3.9}$$

It is important to note that for evaporation, the radial Reynolds number is positive and for condensation it is negative. Regardless of the algebraic sign on the radial Reynolds number, the axial velocity profile is symmetric about the channel axis and differs only by a scaling factor based upon the origin of the flow. For radial Reynolds numbers much smaller than unity the viscous effects dominate and the axial velocity profile assumes a parabolic shape typical of Poiseuille flow, with the pressure decreasing in the direction of the flow. For vaporization, the pressure gradient assumes a value greater than that normally encountered for Poiseuille flow; for condensation, the pressure gradient is somewhat less.

Other investigators (Knight and McInteer, 1965; Wageman and Guevara, 1960) have found that the axial pressure distribution in injection increases as a parabolic function of the axial coordinate for radial Reynolds numbers much less than unity ($\mathrm{Re}_r \ll 1$). For large values of radial Reynolds number ($\mathrm{Re}_r \gg 1$), the axial pressure drop has been shown to be proportional to $\cos[(\pi/2)(r/r_0)^2]$ and to decrease in the direction of flow. In the extraction region or condenser portion of a heat pipe, the axial velocity was found to be constant across most of the channel with the pressure decreasing to zero rapidly near the pipe wall. For the liquid flow, the pressure increases in the direction of flow due to partial dynamic recovery (Cotter, 1965).

Utilizing the boundary conditions given in equation 3.8 and the definition of the radial Reynolds number given in equation 3.9, an expression for the axial pressure gradient was found and expressed in terms of a series expansion (Yuan and Finkelstein, 1956).

$$\frac{dP_v}{dx} = -\frac{8\mu_v \dot{m}}{\pi \rho_v r_0^4}\left[1 + \tfrac{3}{4}\mathrm{Re}_r - \tfrac{11}{270}\mathrm{Re}_r^2 + \cdots\right]\frac{x^2}{L_e^2} \tag{3.10}$$

Somewhat later, Busse (1967) solved the Navier–Stokes and continuity equations for laminar incompressible flow in a long cylindrical heat pipe with vaporization and condensation regions that had constant heat addition and rejection rates. The axial velocity was expressed as a fourth-order polynomial

$$v(r, x) = 2\bar{v}(x)\left(1 - \frac{r^2}{r_0^2}\right)\left[1 + a(x)\left(\frac{r^2}{r_0^2} - \frac{1}{3}\right)\right] \tag{3.11}$$

where $\bar{v}$ is the mean axial flow velocity of the vapor, r_0 is the radius of the vapor channel, and $a(x)$ is the correction factor to compensate for deviation from Poiseuille flow.

In the evaporator section of a heat pipe, the Poiseuille flow approximation is reasonable for Reynolds numbers approaching zero; however, as the Reynolds number approaches infinity, the velocity profile can be more closely approximated by a cosine function. The corresponding solutions to the Navier–Stokes equations with the velocity profile approximation given in equation 3.11 results in an expression for the pressure profile of

$$P(x) \simeq P(x=0) - 4\mu \frac{\bar{v} L_e}{r_0^2}\left[1 + 0.61\,\mathrm{Re}_r + \frac{0.61\,\mathrm{Re}_r}{3.6 + \mathrm{Re}_r}\right]\frac{x^2}{L_e^2} \tag{3.12}$$

where $\bar{v}$ is the mean vapor velocity at the end of the evaporator section (Busse, 1967). Note that the first term in the brackets in this equation is dominate for radial Reynolds numbers $\mathrm{Re}_r \ll 1$, while the second term becomes dominate for $\mathrm{Re}_r \gg 1$. In the intermediate range, where $\mathrm{Re}_r \approx 1$, the third term contributes approximately 10% of the total value. This expression has been shown to provide estimations of the vapor flow velocity in the evaporator portion of a heat pipe to within ±1%.

Busse found that the velocity profile in the adiabatic section of heat pipes can be closely approximated by Poiseuille flow as

$$P(x) \simeq P(x=0) - \frac{8\mu\bar{v}x}{r_0^2}\left[1 + \frac{0.106\,\mathrm{Re}_r}{18 + 5\,\mathrm{Re}_r}\,\frac{1 - e^{(-30x/\mathrm{Re}_a r_0)}}{x/\mathrm{Re}_a r_0}\right] \tag{3.13}$$

where x is the coordinate measured from the beginning of the adiabatic region and Re_a is the axial Reynolds number. This expression has been shown to yield a vapor flow velocity error of less than ±1%.

In the condenser region of an operating heat pipe, use of a fourth-order polynomial to approximate the axial flow velocity is valid only for radial Reynolds numbers $\mathrm{Re}_r \lesssim 1$. For these conditions solution of the axial pressure distribution in the vapor region of the heat pipe can be approximated as

$$P(x) = P(x=L_c) + \frac{4\mu\bar{v}L_c}{r_0^2}\left[1 - \mathrm{Re}_{r,c}\left(\frac{7}{9} - \frac{8a}{27} + \frac{23a^2}{405}\right)\right]\left(1 - \frac{x}{L_c}\right)^2 \tag{3.14}$$

where a is a velocity profile correction factor that compensates for the deviation from Poiseuille flow.

In the evaporator section, the velocity profile correction factor a takes the form

$$a = 0.68\left[\left(5 + \frac{18}{\mathrm{Re}_r}\right) - \left[\left(5 + \frac{18}{\mathrm{Re}_r}\right)^2 - 8.8\right]^{1/2}\right] \tag{3.15}$$

and varies from $a = 0$ for the Poiseuille velocity profile where the radial Reynolds number approaches zero, to $a = 0.665$ for radial Reynolds numbers approaching infinity. In the adiabatic region this correction factor is dependent upon the velocity profile at the entrance to the adiabatic region. If a profile can be approximated by a cosine function, the correction factor varies from 0.665 and decreases exponentially by a factor of approximately 10 over a distance of one diameter. In the condenser region, the correction factor a is again dependent upon the value at the entrance to the condenser region. Figure 3.3 illustrates the correction as a function of the radial Reynolds number and the distance measured from the entrance to the condenser region for the two cases $a_0 = 0$ (i.e., Poiseuille velocity at the entrance to the condenser) and $a_0 = 0.665$ (a cosine velocity profile at the entrance of the condenser region).

The maximum axial pressure gradient in a heat pipe occurs between the end of the evaporator farthest from the condenser and the point at which the liquid and vapor pressures are equal, or the so-called wet point. As mentioned previously, this point will occur at the entrance to the condenser for heat pipes with sufficiently high cooling rates. For these cases, it is only necessary to know the combined pressure drop in the evaporator and the adiabatic sections, which can be found from

$$\Delta P_e + \Delta P_a = 8\mu\bar{v}r_0^{-2}[0.5L_e(1 + F\,\mathrm{Re}_r) + L_a] \tag{3.16}$$

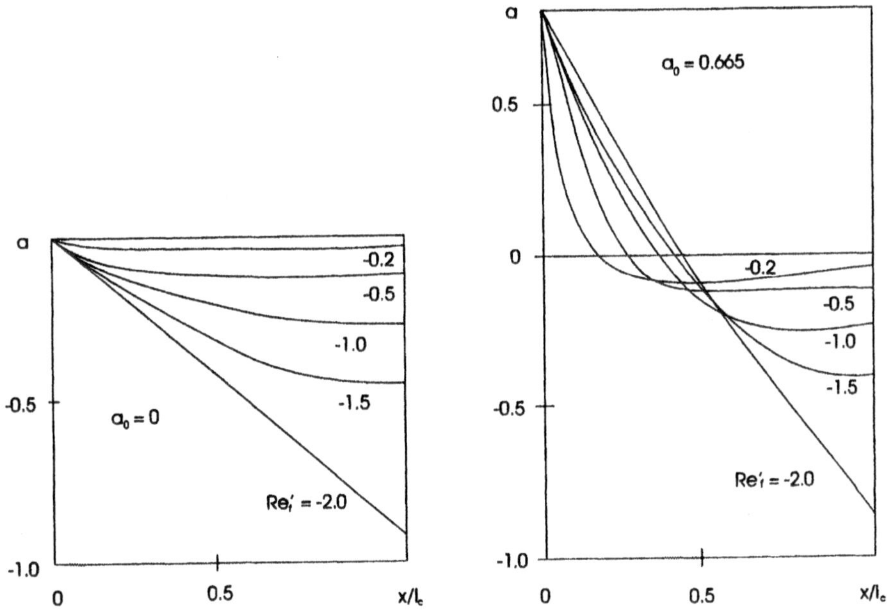

Figure 3.3 Correction to the velocity profile along the length of the condenser (Busse, 1967).

where F is a correction factor that can be approximated as

$$F = \frac{7}{9} - \frac{1.7\,\mathrm{Re}_r}{36 + 10\,\mathrm{Re}_r}\,e^{-\Lambda} \tag{3.17}$$

where

$$\Lambda = \frac{7.5L_a}{\mathrm{Re}_r L_e}$$

and represents the inertial pressure drop plus a term to compensate for the frictional pressure drop resulting from the deviation from Poiseuille flow.

When the cooling rate is low, the wet point may occur at the end of the condenser farthest from the evaporator, in which case it is necessary to combine the individual pressure drops from all three sections, or

$$\Delta P_e + \Delta P_a + \Delta P_c = 8\mu\bar{v}r_0^{-2}[0.5L_e + L_a + 0.5L_c] \tag{3.18}$$

Examination of this equation indicates that the inertial pressure gradients have all canceled out, resulting in an approximation that is identical to that obtained for Poiseuille flow and is valid for any radial Reynolds number (Busse, 1967).

3.3.2 One-Dimensional Vapor Flow Approximations

Perhaps the most widely accepted one-dimensional vapor pressure drop expression is the one presented by Chisholm (1971) and Chi (1976). This expression, which uses the control volume shown in Fig. 3.4, was derived in both references from conservation of momentum as follows:

$$F_S + F_B = \frac{\partial}{\partial t}\int_{c_v} u_{xyz}\rho\, dV + \int_{cs} u_{xyz}\rho \bar{V}_{xyz}\, d\bar{A} \tag{3.19}$$

$$A_v P_v - \frac{d(A_v P_v)}{dx}dx + A_v P_v - \tau_v 2\pi r_v\, dx$$
$$= \int_A \rho_v v_v^2(r)\, dA + \frac{d}{dx}\int_A \rho_v v_v^2(r)\, dA\, dx - \int_A \rho v_v^2(r)\, dA \tag{3.20}$$

or

$$\left(P_v \frac{dA_v}{dx} + A_v \frac{dP_v}{dx}\right)dx - \tau_v 2\pi r_v\, dx = \frac{d}{dx}\int_A \rho_v v_v^2(r)\, dA\, dx \tag{3.21}$$

$$\frac{dP_v}{dx} = -\frac{\tau_v 2\pi r_v}{A_v} - \frac{1}{A_v}\frac{d}{dx}\int_A \rho v_v^2(r)\, dA \tag{3.22}$$

Looking at the second term on the right side of equation 3.22,

$$\text{Rate of change in momentum} = \frac{1}{A_v}\frac{d}{dx}\int_A \rho v_v^2(r)\, dA \tag{3.23}$$

and recognizing that the actual momentum across the area A_v is not equal

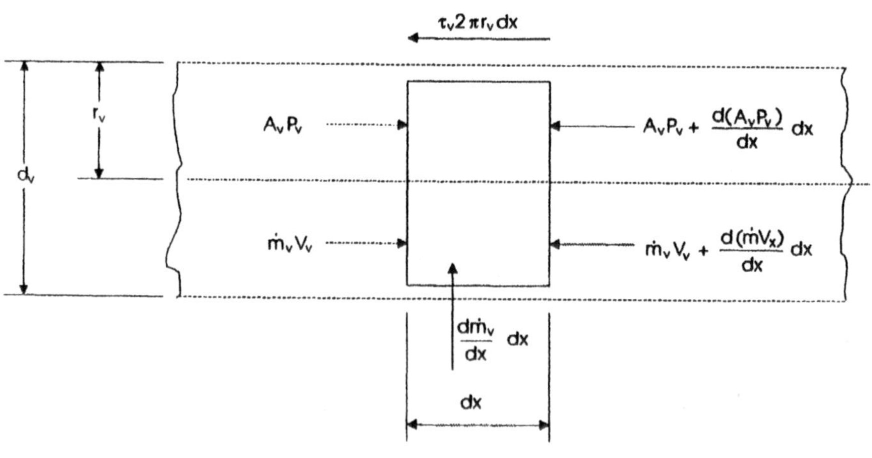

Figure 3.4 Control volume utilized by Chi (1976).

to the momentum calculated by the product of the average values (of $\dot{m}_v$ and v_v), or $\dot{m}_v v_v$, a *momentum correction factor* β can be introduced, where

$$\beta = \frac{\int_A \rho v_v^2(r)\, dA}{\rho v_v^2 A_v} = \frac{\text{actual momentum}}{\text{momentum calculated by average } v_v} \tag{3.24}$$

If the vapor velocity is constant and does not vary over the cross section, then $\beta = 1$. Substituting yields

$$\frac{1}{A_v}\frac{d}{dx}(\beta \rho v_v^2 A_v) \tag{3.25}$$

where the average velocity is equal to

$$\rho v_v^2 \frac{d}{dx}(\beta) + \beta \rho v \frac{d}{dx}(v_v^2) \tag{3.26}$$

Assuming that the velocity profile $v_v(r)$ is a weak function of the position x implies that $d\beta/dx \simeq 0$ and the second term on the right-hand side of equation 3.22 reduces to

$$\beta \rho v \frac{d}{dx}(v_v^2) = \beta \rho v \frac{d}{dx}\left(\frac{\dot{m}_v^2}{\rho_v^2 A_v^2}\right) = \frac{2\beta \dot{m}_v}{\rho_v A_v^2}\frac{d\dot{m}_v}{dx} \tag{3.27}$$

Looking at the first term on the right-hand side of equation 3.22,

$$\frac{\tau_v 2\pi r_v}{A_v} \tag{3.28}$$

and recognizing that

$$\tau_v = \tfrac{1}{2} f_v \rho_v v_v^2 \qquad v_v = \frac{\dot{m}}{\rho_v A_v} \qquad \mathrm{Re}_v = \frac{2r_v \rho_v \bar{v}_v}{\mu_v} \qquad A_v = \pi r_v^2 \tag{3.29}$$

and substituting yield

$$\frac{\tau 2\pi r_v}{A_v} = \frac{(f_v\,\mathrm{Re}_v)\mu_v \dot{m}_v}{2A_v r_v^2 \rho_v} \tag{3.30}$$

and finally,

$$\frac{dP_v}{dx} = -\frac{(f_v\,\mathrm{Re}_v)\mu_v \dot{m}_v}{2A_v r_v^2 \rho_v} - \beta \frac{2\dot{m}_v}{\rho_v A_v^2}\frac{d\dot{m}_v}{dx} \tag{3.31}$$

where

$$\beta = \frac{1}{\rho \bar{v}_v^2 A_v} \int_A \rho v_v^2(r)\, dA = \frac{\rho_v^2 A_v}{\dot{m}_v^2} \int_A v_v^2(r)\, dA \tag{3.32}$$

compensates for variations in the vapor velocity across the cross section (Chi, 1976).

Because determination of the vapor pressure drop in heat pipes is complicated by the mass addition and removal in the evaporator and condenser, respectively, and by the compressibility of the vapor phase, an alternative procedure for evaluating the vapor pressure drop has been developed. A mass balance on a section of the adiabatic region of the heat pipe ensures that for continued operation, the liquid mass flow rate and vapor mass flow rate must be equal. Because of the large difference in the density of these two phases, the vapor velocity must be significantly higher than the velocity of the liquid phase. For this reason, in addition to the pressure gradient resulting from frictional drag, the pressure gradient due to variations in the dynamic pressure must also be considered even in one-dimensional flow. Both Chi (1976) and Dunn and Reay (1982) have addressed this problem. Chi (1976) found that upon integration of the vapor pressure gradient, the dynamic pressure effects cancel. The result is the expression

$$\Delta P_v = \left(\frac{C(f_v \, \mathrm{Re}_v)\mu_v}{2(r_{h,v})^2 A_v \rho_v \lambda} \right) L_{\mathrm{eff}}\, q \tag{3.33}$$

where $r_{h,v}$ is the hydraulic radius of the vapor space and C is a constant that depends on the Mach number.

As mentioned previously, during steady-state operation, the liquid mass flow rate at any axial position, $\dot{m}_l$, must equal the vapor mass flow rate $\dot{m}_v$, and while the liquid flow regime is always laminar, the vapor flow may be either laminar or turbulent. It is therefore necessary to determine the vapor flow regime as a function of the heat flux. This can be accomplished by evaluating the local axial Reynolds number in the vapor, defined as

$$\mathrm{Re}_v = \frac{2(r_{h,v})q}{A_v \mu_v \lambda} \tag{3.34}$$

In addition, it is necessary to determine if the flow should be treated as compressible or incompressible by evaluating the local Mach number, defined as

$$\mathrm{Ma}_v = \frac{q}{A_v \rho_v \lambda (R_v T_v \gamma_v)^{1/2}} \tag{3.35}$$

where R_v is the gas constant, T_v is the vapor temperature, and γ_v is the ratio of specific heats, which is equal to 1.67, 1.4 and 1.33 for monatomic, diatomic, and polyatomic vapors, respectively (Chi, 1976).

Previous investigations summarized by Kraus and Bar-Cohen (1983) have demonstrated that the following combinations of these conditions can be used with reasonable accuracy:

$$\mathrm{Re}_v < 2300 \qquad \mathrm{Ma}_v < 0.2$$

$$(f_v\,\mathrm{Re}_v) = 16 \qquad C = 1.00 \tag{3.36}$$

$$\mathrm{Re}_v < 2300 \qquad \mathrm{Ma}_v > 0.2$$

$$(f_v\,\mathrm{Re}_v) = 16 \qquad C = \left[1 + \left(\frac{\gamma_v - 1}{2}\right)\mathrm{Ma}_v^2\right]^{-1/2} \tag{3.37}$$

$$\mathrm{Re}_v > 2300 \qquad \mathrm{Ma}_v < 0.2$$

$$(f_v\,\mathrm{Re}_v) = 0.038\left(\frac{2(r_{h,v})q}{A_v \mu_\lambda \lambda}\right)^{3/4} \qquad C = 1.00 \tag{3.38}$$

Because the equations used to evaluate both the Reynolds number and the Mach number are functions of the heat transport capacity, it is first necessary to assume the conditions of the vapor flow. Using these assumptions, an iterative procedure must be used to determine the maximum heat capacity, $q_{c,m}$. Once the value of $q_{c,m}$ is known, it can be substituted into the expressions for the vapor Reynolds number and Mach number to determine the accuracy of the original assumption. Using this iterative approach, which is covered in more detail in Chapter 4, accurate values for the capillary limitation as a function of the operating temperature can be determined in units of Watt-meters and watts for $(qL)_{c,m}$ and $q_{c,m}$, respectively.

3.3.3 Inertial Pressure Gradient

The inertial pressure gradient in the vapor phase of heat pipes may make a significant contribution to the flow characteristics and the pressure gradient and therefore has been included in equation 3.19. However, in most heat pipe applications the liquid velocity is so small that the inertial effects caused by vaporization and condensation can typically be ignored. This is not true, however, in micro–heat pipes, where due to the very small quantities of mass contained in the liquid passages, the inertial effects may be significant. Although relatively few investigations have focused solely only on the inertial effects, Busse (1967) considered laminar flow of an incompressible vapor stream in a cylindrical heat pipe with uniform heat addition and rejection. The inertial contribution in the evaporator was found to be

$$\Delta P_i = \frac{\rho_v \bar{V}_v^2}{2g}\left[\frac{28}{9} - \frac{0.68\,\mathrm{Re}}{(29L_e/d_v) + \mathrm{Re}}\exp\left(-\frac{60L_a}{\mathrm{Re}\,d_v}\right)\right] \tag{3.40}$$

and for many cases this expression provides a reasonably accurate estimation of the importance of the inertial effects.

3.4 LIQUID PRESSURE DROP ΔP_l

The liquid pressure distribution as a function of axial position is illustrated in Fig. 3.2. As was the case for the vapor pressure gradient, the liquid pressure gradient is also the result of the combined effect of both viscous and inertial forces. The increase in the pressure gradient in the evaporator section is nonlinear, due to the mass injection of the working fluid into the vapor stream with the pressure in the liquid increasing along the length of the heat pipe moving from the evaporator toward the condenser. Determination of the frictional pressure drop occurring in axial grooves or open arteries of heat pipes is a fairly straightforward procedure. However, determination of the frictional flow characteristics in the porous wicking structures that comprise heat pipe wicks is considerably more complicated. For both cases, the viscous forces in the liquid results in a pressure drop, dP_l/dx, which resists the capillary flow through the porous wick. Because the liquid pressure gradient may vary along the longitudinal axis of the heat pipe, the total liquid pressure drop can be determined by integrating the pressure gradient over the length of the flow passage, or

$$\Delta P_l(x) = -\int_0^x \frac{dP_l}{dx}\,dx \tag{3.41}$$

where the limits of integration are from the evaporator end ($x = 0$) to the condenser end ($x = L$) and dP_l/dx is the gradient of the liquid pressure resulting from frictional drag (Chi, 1976). This frictional drag, due to the shear stress, can be written as

$$\frac{dP_l}{dx} = -\frac{2\tau_l}{(r_{h,l})} \tag{3.42}$$

where τ_l is the frictional shear stress at the liquid–solid interface and $(r_{h,l})$ is the hydraulic radius, defined as twice the cross-sectional area divided by the wetted perimeter.

At this point, the Reynolds number Re_l and drag coefficient, f_l can be introduced as

$$\mathrm{Re}_l = \frac{2(r_{h,l})\rho_l V_l}{\mu_l} \tag{3.43}$$

and

$$f_l = \frac{2\tau_l}{\rho_l V_l^2} \tag{3.44}$$

where V_l is the local liquid velocity, which is related to the local heat flow by

$$V_l = \frac{q}{\epsilon A_w \rho_l \lambda} \tag{3.45}$$

In this expression, A_ω is the wick cross-sectional area, ϵ is the wick porosity, and λ is the latent heat of vaporization.

Combining equations (3.31)–(3.33) with (3.30) yields

$$\frac{dP_l}{dx} = \left(\frac{(f_l \, \mathrm{Re}_l)\mu_l}{2\epsilon A_w (r_{h,l})^2 \lambda \rho_l}\right) q \tag{3.46}$$

This equation can be written as

$$\frac{dP_l}{dx} = \left(\frac{\mu_l}{K A_w \lambda \rho_l}\right) q \tag{3.47}$$

(Chi, 1976), where K is the permeability, expressed as

$$K = \frac{2\epsilon (r_{h,l})^2}{f_l \, \mathrm{Re}_l} \tag{3.48}$$

For laminar flow, $f_l \, \mathrm{Re}_l$ is constant and depends only on the passage shape. Thus, the permeability K is independent of the flow path x. While the permeability is independent of the flow path, it is strongly dependent upon the shape of the flow passage. For circular passages (i.e., arterial or tunnel wicks), the hydraulic radius is equal to the radius of the liquid flow passage r, the porosity is equal to unity, and the friction factor Reynolds number product $f_l \, \mathrm{Re}_l$, is equal to 16. Simplifying equation 3.15 results in a permeability for a circular passage of

$$K = \tfrac{1}{8} r^2 \tag{3.49}$$

Kays (1966) has presented a method for calculating the friction factor Reynolds number product for rectangular and annular flow passages. These methods have been presented in graphical form and are shown in Figs. 3.5*a*,*b*, respectively. Using the aspect ratio, which is the groove width divided by the groove height, the friction factor Reynolds number product can be determined from Fig. 3.5*a* for a broad range of groove sizes. Likewise, Fig. 3.5*b* illustrates the effect of the ratio of annulus radii to the

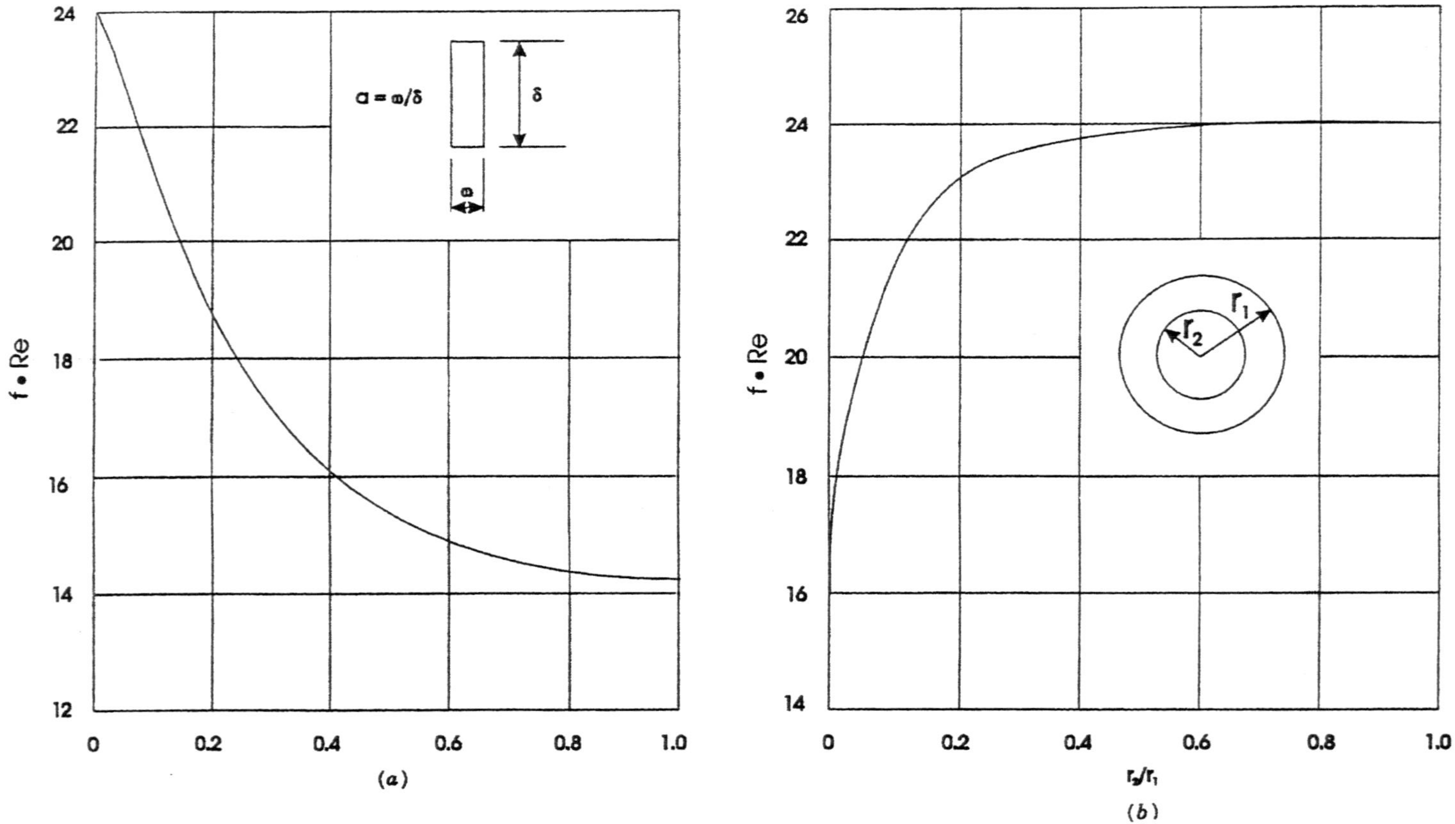

Figure 3.5 Frictional coefficients for laminar flow in (*a*) rectangular channels (Kays, 1966) and (*b*) circular annulus (Kays, 1966).

friction factor Reynolds number product and provides a method by which this value can be computed.

3.4.1 Liquid Flow in Porous Media

The method developed by Kays (1966) for estimating the friction factor Reynolds number product can also be used when calculating the porosity of wrapped, screened wicks given the screen layer spacing. Alternatively, Marcus (1972) has described a method for calculating the permeability of wrapped, screened wicks. This expression, which is a modified form of the Blake–Kozeny equation, can be expressed as

$$K = \frac{d^2 \epsilon^3}{122(1-\epsilon)^2} \tag{3.50}$$

In this expression, d is the wire diameter and ϵ is the wick porosity, which can be determined as

$$\epsilon = 1 - \tfrac{1}{4}\pi SNd \tag{3.51}$$

where N is the mesh number and S is the crimping factor (approximately 1.05) (Chi, 1976). For sintered metal powders, this equation takes the form

$$K = \frac{r_s^2 \epsilon^3}{37.5(1-\epsilon)^2} \tag{3.52}$$

where r_s is the average sphere radius (Chi, 1976). Alternatively, the porosity typically referred to as the *volume porosity* can be defined as the ratio of the pore volume V_{por} to the total volume V_{tot}, or

$$\epsilon = \frac{V_{por}}{V_{tot}} \tag{3.53}$$

Because the pores in a wicking structure may not be contiguous, it is helpful at this point to introduce two new terms, the *permeability*, defined as the property of a porous material that describes the ability of the material to transmit the liquid under some applied pressure gradient, and the *tortuosity*, the ratio of the mean length of the liquid path in a given wicking structure to the actual distance traveled along the wick. For some of the more complicated composite wicking structures discussed in Chapter 6, it may be necessary to determine the wick permeability and / or tortuosity, either experimentally or analytically, using one of the techniques outlined by Kaviany (1991). Values of K for several common wick structures have been summarized by Chi (1976) and are provided in Table 3.2. For more complex wicks, the permeability will have to be determined experimentally.

TABLE 3.2 Wick Permeability for Several Wick Structures

Structure	K	Data
Circular cylinder (artery or tunnel wick)	$r^2/8$	r = radius of liquid flow passage
Open rectangular grooves	$2\epsilon(r_{h,l})^2/(f_l\,\mathrm{Re}_l)$	ϵ = wick porosity ω = groove width s = groove pitch δ = groove depth $(r_{h,l}) = 2\omega\delta/(\omega+2\delta)$
Circular annular wick	$2(r_{h,l})^2/(f_l\,\mathrm{Re}_l)$	$(r_{h,l}) = r_1 - r_2$
Wrapped screen wick	$\dfrac{d^2\epsilon^3}{122(1-\epsilon)^2}$	d_w = wire diameter $\epsilon = 1 - (1.05\pi N d_w/4)$ N = mesh number
Packed sphere	$\dfrac{r_s^2\epsilon^3}{37.5(1-\epsilon)^2}$	r_s = sphere radius ϵ = porosity dependent on packing mode)

Source: Chi, 1976.

For constant heat addition and removal, equation (3.35) can be substituted into equation (3.29) and integrated over the length of the heat pipe to yield

$$\Delta P_l = \left(\frac{\mu_l}{KA_w\lambda\rho_l}\right)L_{\mathrm{eff}}q \tag{3.54}$$

where L_{eff} is the effective heat pipe length defined as

$$L_{\mathrm{eff}} = 0.5L_e + L_a + 0.5L_c \tag{3.55}$$

Equation (3.26) assumes that the rates of vaporization in the evaporator and condensation in the condenser are uniform. This yields a linearly increasing value in the evaporator and a linearly decreasing value in the condenser, as shown in Fig. 3.6.

One final term that may be of significance when determining the liquid pressure drop in capillary wicking structures is the degree of saturation of the porous material. This degree of saturation, S_m, is typically defined as

$$S_m = \frac{\text{volume of liquid filled voids}}{\text{total void volume}}\,. \tag{3.56}$$

For a fully saturated wick in which the liquid wets the wicking structure, the degree of saturation would be equal to unity. However, near the capillary limit, this may not be the case since the wicking structure may be partially dried out. There are many other factors that can influence this

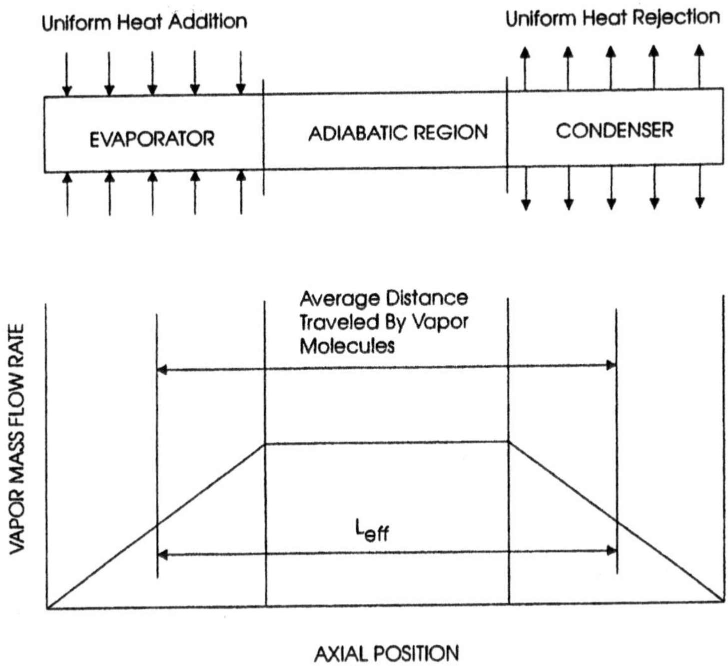

Figure 3.6 Illustration of effective length for constant heat addition and heat rejection.

parameter, including impurities within the porous media, such as oil or oxide films which prevent the liquid from fully wetting the porous surface; noncondensible gas bubbles, which become trapped or are generated because of impurities; and vapor bubbles that result from boiling at or near the heat pipe container wall. The latter two, entrapped noncondensible gas and / or vapor bubbles can result in values for degree of saturation that are significantly less than 1.

Example 3.1 A simple horizontal copper–water heat pipe is constructed from a 1.5-cm-internal-diameter, 0.75-m-long tube. The evaporator and condenser lengths of the heat pipe are 0.25 m each. The wicking structure consists of two wraps of 100-mesh copper screen. Given that the maximum heat transport capacity of the heat pipe at 30°C is 24.5 W, plot the vapor and liquid pressures as a function of axial position for the heat pipe. For the vapor pressure, use both the one-dimensional approximation and the two-dimensional results (equations 3.12–3.16) and compare the results.

SOLUTION

I. One-dimensional Vapor. The pressure gradient is given by equation 3.19:

$$\frac{dP_v}{dx} = \frac{-(f_v \operatorname{Re}_v)\mu_v \dot{m}_v}{2A_v r_{h,v}^2 \rho_v} - \beta \frac{2\dot{m}_v}{A_v^2 \rho_v} \frac{d\dot{m}_v}{dx}$$

Assume that $\beta \simeq 1$ and

$$\frac{d\dot{m}}{dx} = 4.04 \times 10^{-5}\, kg/s \quad \text{(evaporation)}$$

$$\frac{d\dot{m}}{dx} = 0 \quad \text{(adiabatic)}$$

$$\frac{d\dot{m}}{dx} = -4.04 \times 10^{-5}\, \text{kg/s} \quad (\text{condenser}, L_c = L_e)$$

$$f_v \operatorname{Re}_v = 16 \quad \text{(laminar, incompressible)}$$

and with the assumption of uniform mass addition and rejection, the mass flow rate is

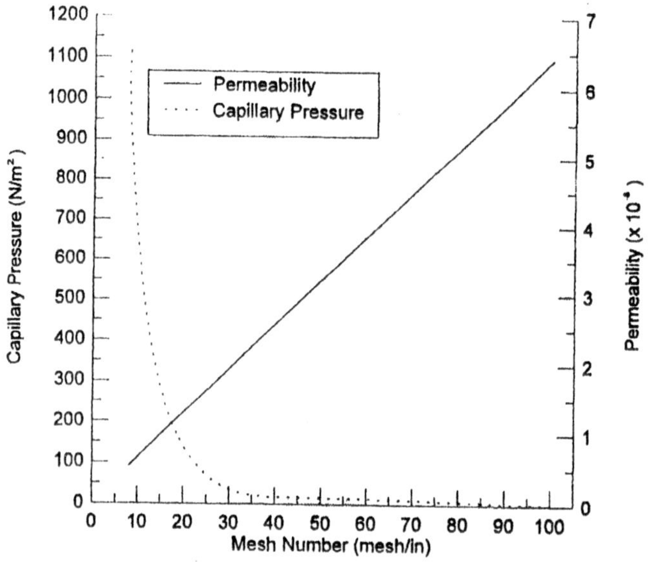

$$P_v(x) = \int_0^x \frac{dP_v}{dx}\, dx$$

Evaporator:

$$P_v(x) = \int_0^x \left(-F_{v,1}\dot{m}_v - F_{v,2}\dot{m}_v \frac{d\dot{m}}{dx} \right) dx$$

where

$$F_{v,1}=\frac{f_v\,\mathrm{Re}_v\mu_v}{2A_v r_v^2\rho_v}=\frac{16(9.29\times10^{-6})}{2(1.56\times10^{-4})(0.00705)^2(0.035)}$$

$$=2.739\times10^5$$

$$F_{v,2}=\frac{2}{A_v^2\rho_v}=\frac{2}{(1.56\times10^{-4})^7(0.035)}=2.35\times10^9$$

$$P_v(x)=\int_0^x-2.739\times10^5(1.62\times10^{-4}x)-2.35$$

$$\times10^9(4.04\times10^{-5})(1.62\times10^{-4})$$

$$=\int_0^x-59.75x\,dx=-29.88x^2\int_0^x$$

$$=P(x=0)-29.88x^2$$

Adiabatic: With $\dot{m}_v=\text{constant}$ and $d\dot{m}_v/dx=0$

$$P_v(x)=P(x=L_e)-F_{v,1}\dot{m}_v(x-L_e)$$

$$=P(x=L_e)-11.06(x-L_e)$$

Condenser:

$$P_v(x)=P_v(x=L_e+L_a)+29.88(x^2-(L_e+L_a)^2)-44.98(x-L_e-L_a)$$

$$=\int_{x=L_e+L_a}^x-(2.739\times10^5+2.35\times10^9(4.04\times10^{-5}))$$

$$\times(-1.62\times10^{-4}x+1.22\times10^{-4})\,dx$$

$$=\int_{x=L_e+L_a}^x 59.73x-44.98\,dx$$

II. Two-dimensional Vapor
Evaporator (equation 3.12):

$$P(x)\simeq P(x=0)-4\mu\frac{\bar{v}L_e}{r_0^2}\left[1+0.61\,\mathrm{Re}_r+\frac{0.61\,\mathrm{Re}_r}{3.6+\mathrm{Re}_r}\right]\frac{x^2}{L_e^2}$$

where

$$\bar{v}=\text{average velocity}$$

$$\mathrm{Re}_r=-\frac{\rho_v u_v r_0}{\mu_v}=\frac{1}{2\pi\mu_v}\frac{d\dot{m}}{dx}$$

From before

$$\bar{v} = \frac{\dot{m}}{\rho_v A_v} = \frac{Q}{\lambda \rho_v A_v} = \frac{24.5}{(2425 \times 10^3)(0.035)(1.56 \times 10^{-4})} = 1.85 \text{ m/s}$$

and Re_r is calculated from the overall mass flow rate divided by the length of the evaporator wall area (assuming that mass addition is uniform throughout the evaporator):

$$\frac{d\dot{m}}{dx} = \frac{\dot{m}}{L_e} = \frac{Q}{\lambda L_e} = \frac{24.5w}{(2425 \times 10^3)(0.25)} = 4.04 \times 10^{-5} \text{ kg/s} \cdot \text{m}$$

and

$$\mathrm{Re}_r = \frac{1}{2\pi\left(9.29 \times 10^{-6} \frac{NS}{r^2}\right)} \times 4.04 \times 10^{-5} \text{ kg/s} \cdot \text{m} = 0.692$$

Thus,

$$P(x) = P(x=0) - 4(9.29 \times 10^{-6}) \frac{1.85(0.25)}{(0.00705)^2}$$

$$\times \left[1 + 0.61(0.692) + \frac{0.61(0.692)}{3.6 + 0.692}\right] \frac{x^2}{(0.25)^2}$$

or

$$P(x) = P(x=0) - 8.4x^2(\text{Pa}) \quad \text{and} \quad \frac{dP}{dx} = -16.8x$$

Adiabatic section (equation 3.13):

$$P(x) \simeq P(x=0) - \frac{8\mu\bar{v}x}{r_0^2}\left[1 + \frac{0.106\,\mathrm{Re}_r}{18 + 5\,\mathrm{Re}_r} \frac{1 - e^{(-30x/\mathrm{Re}_a r_0)}}{x/\mathrm{Re}_a r_0}\right]$$

where

$$\mathrm{Re}_a = \frac{4\dot{m}}{\pi q_v \mu_v} = 97.9$$

and $x = 0$ is the entrance of the adiabatic section:

$$P(x) \simeq P(x=0) - \frac{8(9.29 \times 10^{-6})(1.85)x}{(0.00705)^2}$$

$$\times \left[1 + \frac{0.106(0.692)}{18 + 5(0.692)} \frac{1 - e^{[-30x/(97.9)(0.00705)]}}{x/(97.9)(0.00705)}\right]$$

or

$$P(x) = P(x=0) - 2.76x\left[1 + 0.00236\,\frac{1 - e^{(-43.4x)}}{x}\right]$$

or

$$P(x) = P(x=0) - 2.76x - 0.0065[1 - e^{(-43.4x)}]$$

and

$$\frac{dP}{dx} = -2.76 - 0.283\,e^{(-43.4x)}$$

Condenser section (equation 3.14):

$$P(x) = P(x=L_c) + \frac{4\mu\bar{v}L_c}{r_0^2}\left[1 - \mathrm{Re}_{r,c}\left(\frac{7}{9} - \frac{8a}{27} + \frac{23a^2}{405}\right)\right] \times \left(1 - \frac{x}{L_c}\right)^2 \tag{3.14}$$

The value of a is determined from Fig. 3.3. Since $a = 0.142$ (relatively low value) in the evaporator and $L_a = 0.25$ m, the vapor flow may be assumed to return to a Poiseuille velocity profile such that $a_o = 0$. Thus, with $a_o = 0$, $\mathrm{Re}_{r,c} = -0.692$:

$$0 > a > -0.15 \quad \text{(Fig. 3.3)}$$

$$P(x) = P(x=L_c) + \frac{4(9.29 \times 10^{-6})(1.85)(0.25)}{(0.00705)^2}$$

$$\times\left[1 - (-0.692)\left(\frac{7}{9} - \frac{89}{27} + \frac{23a^2}{405}\right)\right]\left(1 - \frac{x}{L_c}\right)^2$$

$$= P(x=L_c) + 0.346\left[1 + 0.692\left(\frac{7}{9} - \frac{8a}{27}\right)\right]\left(1 - \frac{x}{L_c}\right)^2$$

Differentiation of this term requires the knowledge of the function $a(x)$.

III. Liquid Channel. The pressure gradient is given by equation 3.35 with the substitution $1 = \dot{m}\lambda$:

$$\frac{dP_l}{dx} = \frac{\mu_l \dot{m}_l}{KA_w\rho_l} = \frac{(769 \times 10^{-6})\dot{m}_l}{1.94 \times 10^{-10}(2.057 \times 10^{-5})(995.3)}$$

$$= 1.94 \times 10^8 \dot{m}_l$$

Thus, by substitution and integration,

$$P_l(x) = P_l(x = 0) + 1.571 \times 10^4 x^2 \quad \text{(evaporator)}$$

$$P_l(x) = P_l(x = L_e) + 7.84 \times 10^3 (x - L_e) \quad \text{(adiabatic)}$$

$$P_l(x) = P_l(x = L_a) - 1.571 \times 10^4 (x^2 - (L_e + L_a)^2) + 2.37 \times 10^4 (x - L_e - L_a) \quad \text{(condenser)}$$

3.5 PRESSURE DROP AT THE PHASE TRANSITION

Typically, the interfacial region of a liquid at saturation conditions is thought of as being in equilibrium and therefore static. On a microscopic scale, however, this interfacial region is actually very active, with vapor molecules and liquid molecules continually exchanging places through vaporization and condensation. However, as long as the rates of vaporization and condensation are equal, the system is considered to be in thermal equilibrium. Because the molecules leaving the surface as a vapor and those encroaching on the surface as a liquid possess a certain momentum, a pressure gradient occurs across the liquid–vapor phase transition. As illustrated in Fig. 3.2, this phase transition pressure gradient results in a pressure drop in both the evaporator and condenser.

This pressure gradient is different from the one previously discussed and described in equation 2.8 by the Young–Laplace equation and may in some cases significantly affect the free molecular energy of a substance. To understand this effect, it is first necessary to examine the vaporization and condensation occurring at a plane liquid–vapor interface. From kinetic theory, it has been shown that the mass flow rate of molecules across an interface can be expressed as

$$j = \left(\frac{M}{2\pi R}\right)^{1/2} \frac{P}{T^{1/2}} \tag{3.57}$$

However, because the conditions close to the surface are not truly those of static equilibrium, the thermostatic pressure and temperatures cannot be used (Collier, 1972). For small differences in pressure and temperature, such as those existing at the liquid–vapor interface in a heat pipe, this expression may be differentiated to obtain

$$j = \left(\frac{M}{2\pi RT}\right)^{1/2} P\left(\frac{dP}{P} - \frac{dT}{2T}\right) \tag{3.58}$$

Because the net value of the molecular flux j is equal to zero for an interface in thermal equilibrium, the pressure P in this expression is typically the saturation pressure corresponding to the liquid temperature T. This expression can be further simplified to

$$j = \left(\frac{M}{2\pi RT}\right)^{1/2} (P_g - P_v) \tag{3.59}$$

by assuming that there is no temperature jump across the interface.

The second factor that affects the pressure gradient occurring at the phase transition is due to the curvature of the interface and has been presented in detail by Adamson (1990). Based upon a thermodynamic analysis, the change in mechanical pressure at constant temperature on the molecular free energy of a substance can be expressed as

$$\Delta F = \int \bar{V}\, dP \tag{3.60}$$

If the molar volume $\bar{V}$ is constant and the Young–Laplace expression (equation 2.8) is used to evaluate the pressure gradient dP,

$$\Delta F = \sigma \bar{V}\left(\frac{1}{r_1} + \frac{1}{r_2}\right) \tag{3.61}$$

Assuming the vapor behaves as an ideal gas (i.e., $F = F_o + RT \ln P$) and that the free energy can be related to the vapor pressure yields

$$RT \ln \frac{P}{P^o} = \sigma \bar{V}\left(\frac{1}{r_1} + \frac{1}{r_2}\right) = \frac{\sigma \bar{V}}{r_m} \tag{3.62}$$

where P^o is the normal vapor pressure of the liquid phase and P is the pressure over the curved surface. For the special case of a spherical surface of radius r, this expression simplifies to

$$RT \ln \frac{P}{P^o} = \frac{2\sigma \bar{V}}{r} \tag{3.63}$$

Because the radius of the bubble is positive, $\ln P/P^o$ is also positive, which implies that the vapor pressure inside of the vapor bubble must be greater than that of the surrounding liquid. For water the ratio P/P^o is approximately 1.001 for a bubble radius of 10^{-4} cm and 1.011 for a bubble radius of 10^{-5} cm (Adamson, 1990). This expression, frequently called the *Kelvin equation*, is one of the fundamental relationships of surface chemistry.

While these terms may be important in the modeling and analysis of very small heat pipes, the combined effect of the mass flux at the interface and interfacial surface curvature produces a phase transition pressure gradient that typically can be neglected. Notable exceptions are cases where the vaporization or condensation is extremely intense, such as those occurring in porous structures at very low temperatures (Ivanovskii et al., 1982).

3.6 NORMAL HYDROSTATIC PRESSURE

The hydrostatic pressure gradient occurring in heat pipes is due to gravitational or body forces within the vapor and liquid phases. For the case of thermosyphons that use gravitational forces alone or for heat pipes with the evaporator located below the condenser, the gravitational body force would result in a linearly increasing contribution to the pressure in the liquid phase and a linearly decreasing contribution to the pressure in the vapor phase. It is also important to note the effect of gravitational body forces on the location of the dry point within the cross-sectional configuration of the heat pipe. Because some hydrostatic pressure gradient exists in a heat pipe, even one operating in a horizontal orientation, the first point to experience dryout in the wicking structure would typically occur at the end of the evaporator furthest from the condenser and at the highest elevation.

The normal hydrostatic pressure drop ΔP_+ occurs only in heat pipes in which circumferential communication of the liquid in the wick is possible. It is a result of the component of the body force acting perpendicular to the longitudinal axis of the heat pipe. The normal hydrostatic pressure drop can be expressed as

$$\Delta P_+ = \rho_l g d_v \cos \psi \tag{3.64}$$

where ρ_l is the density of the liquid, g is the gravitational acceleration, d_v is the diameter of the vapor portion of the pipe, and ψ is the angle the heat pipe makes with respect to the horizontal.

3.7 AXIAL HYDROSTATIC PRESSURE

The axial hydrostatic pressure drop $\Delta P_{\|}$ results from the component of the body force acting along the longitudinal axis. It can be expressed as

$$\Delta P_{\|} = \rho_l g L \sin \psi \tag{3.65}$$

where L is the overall length of the heat pipe.

In a gravitational environment, the axial hydrostatic pressure term may either assist or hinder the capillary pumping process depending upon whether the tilt of the heat pipe promotes or hinders the flow of liquid back to the evaporator (i.e., the evaporator lies either below or above the condenser). In a zero-gravity environment, both this term and the normal hydrostatic pressure drop term can be neglected because of the absence of body forces.

3.8 PRESSURE DISTRIBUTION IN HEAT PIPES

Ivanoskii et al. (1982) have discussed several combinations of these pressure gradients and how these combinations affect the pressure distribution occurring along the liquid and vapor flow paths in an operating heat pipe. As shown in Fig. 3.7, the location of the wet point, that is, that point where the pressure in the vapor stream is equal to the pressure in the liquid stream, can shift throughout the condenser region. This shift is primarily caused by the inertial effects of the vapor stream but is also significantly affected by local gravitational accelerations. In some extreme cases the wet point may be shifted into the adiabatic region or even into the evaporator.

The first pressure distribution illustrated in Fig. 3.7, labeled (*a*), is characteristic of heat pipes with negligible inertial effects. This situation typically occurs in heat pipes with long condenser regions or heat pipes operating at sufficiently high vapor pressures. As shown, the wet point occurs at the end of the condenser region furthest from the evaporator. For

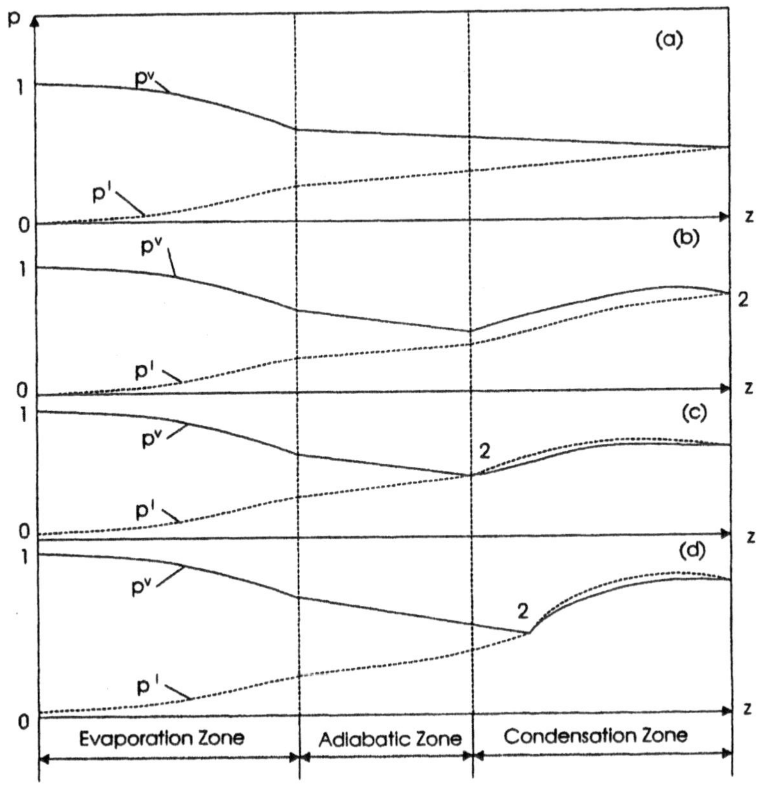

Figure 3.7 Location of wet point (Ivanovskii et al., 1982).

this situation, the inertial effects can typically be ignored and only the viscous pressure drops in the liquid and vapor flow streams need be considered.

Figure 3.7*b* illustrates the case that occurs at low vapor pressures. This situation may occur for heat pipes operating at the low end of their temperature range or during startup conditions. For these situations, the inertial effects begin to become more apparent and the pressure distribution in both the liquid and vapor flow stream of the condenser start to become nonlinear.

As shown in Fig. 3.7*b*, when these inertial effects are small, the wet point occurs at the far end of the condenser; however, as the inertial effects increase (Fig. 3.7*c*), the wet point begins to shift toward the entrance of the condenser region, and the inertial effects become more noticeable. This pressure profile occurs in heat pipes operating at low vapor pressures or arterial pipes in which the liquid viscous pressure drop is small in comparison to the inertial pressure drops (Ivanoskii et al., 1982). Because equation 3.1 only calculates the maximum capillary pressure gradient between the dry point and the wet point, for situations such as those illustrated in Fig. 3.7*c*, it is only necessary to include the pressure gradients in the evaporator and adiabatic regions.

Figure 3.7*d* illustrates the intermediate case where the wet point is located somewhere between the end of the condenser furthest from the evaporator and the condenser entrance. For this situation it is necessary to determine the relative magnitudes of the viscous pressure drops and the inertial pressure drops occurring in the liquid and vapor flow streams.

Example 3.2 A method often used to increase the transport capacity of a given heat pipe design is to substitute a higher mesh number screen (i.e., one with a smaller capillary radius) to increase the maximum capillary pressure. However, this also increases the liquid flow resistance. Using the dimensions for copper screen obtained from handbooks, plot the capillary pressure and wick permeability versus the mesh number. How does varying the mesh number affect the maximum transport capacity?

SOLUTION The maximum capillary pressure is given by (assuming that the contact angle is zero)

$$(\Delta P_c)_{\max} = \frac{2\sigma}{r_c} \quad \text{where} \quad r_c = \frac{1}{2N} \quad \text{for wire screens}$$

The permeability and porosity are obtained from equations 3.38 and 3.39, respectively, as

$$K = \frac{d^2\epsilon^3}{122(1-\epsilon)^2} \quad \text{and} \quad \epsilon = 1 - \frac{\pi SNd}{4}$$

Assuming the fluid is water at 33°C such that $\sigma = 70.9 \times 10^{-3}$ N/m and the mesh dimensions are as shown, the individual values for capillary pressure, porosity, and permeability can be tabulated as follows:

Mesh / in.	Mesh / m	dw (in.)	dw (m)	P_c (Pa)	ϵ	K
8.0000	314.96	0.0280	0.00071	89.32	0.8153	6.58×10^{-8}
12.0000	472.44	0.0230	0.00058	133.98	0.7724	2.49×10^{-8}
16.0000	629.92	0.0180	0.00046	178.65	0.7625	1.35×10^{-8}
18.0000	708.66	0.0170	0.00043	200.98	0.7477	1.00×10^{-8}
24.0000	944.88	0.0140	0.00036	267.97	0.7229	5.10×10^{-9}
30.0000	1181.10	0.0135	0.00034	334.96	0.6660	2.55×10^{-9}
40.0000	1574.80	0.0100	0.00025	446.61	0.6701	1.46×10^{-9}
50.0000	1968.50	0.0090	0.00023	558.27	0.6289	7.74×10^{-10}
60.0000	2362.20	0.0075	0.00019	669.92	0.6289	5.37×10^{-10}
80.0000	3149.60	0.0055	0.00014	893.23	0.6371	3.14×10^{-10}
100.0000	3937.00	0.0045	0.00011	1116.53	0.6289	1.93×10^{-10}

These values can be represented graphically as shown below.

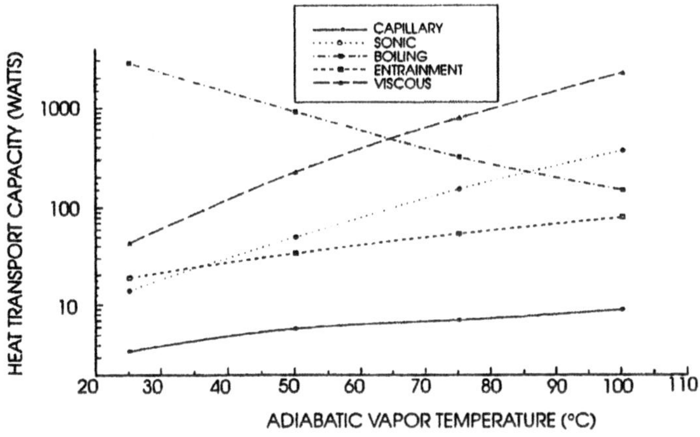

Looking at the graph of capillary pressure and permeability versus mesh number, the two values are seen to be inversely proportional. Thus, by increasing the mesh number, the capillary pressure is correspondingly increased. However, the permeability decreases, and thus the liquid frictional pressure drop increases.

To determine exactly whether changing the mesh number will increase the maximum heat transport capability, the length of the heat pipe must be considered. For a "short" heat pipe the gain in capillary pressure will be much greater than the increased liquid pressure drop. The opposite will be true for a "long" heat pipe.

NOMENCLATURE

A	Area
a	Correction factor to compensate for deviation from Poiseuille flow
C	Constant
d	Diameter (m)
f	Friction factor
F	Molecular free energy, correction factor in equation 3.16 (J)
G	Free energy of system (J)
g	Gravitational acceleration (m/s^2)
j	Molecular flux ($kg/m^2 \cdot s$)
k	Thermal conductivity ($W/m \cdot K$)
K	Wick permeability (m^2)
L	Length (m)
M	Molecular weight (kg/kmol)
Ma	Mach number
$\dot{m}$	Mass flow rate (kg/s)
N	Screen mesh number (m^{-1})
N_l	Liquid figure of merit (W/m^2)
P	Pressure (N/m^2)
P^o	Normal vapor pressure (N/m^2)
q	Heat transferred (W)
R	Thermal resistance, or universal gas constant $\left(\frac{m^2 \cdot K}{W} \text{ or } \frac{J}{mol \cdot K}\right)$
Re	Reynolds number
r	Radius, radial coordinate (m)
S	Crimping factor, degree of saturation
s	Groove pitch (m)
T	Temperature (K)
t	Time (s)
u	Radial component of velocity (m/s)
V	Volume (m^3/mol)
$\bar{v}$	Mean axial velocity (m/s)
v	Axial component of velocity (m/s)
w	Groove width, wire spacing (m)
We	Weber number
x	Axial position, coordinate (m)
y	Position coordinate (m)
z	Position coordinate (m)

Greek Symbols

β	Half of included angle, momentum correction factor (deg)
γ	Ratio of specific heats
δ	Groove depth (m)
ϵ	Wick porosity
λ	Latent heat of vaporization (J / kg)
μ	Absolute viscosity ($N \cdot s / m^2$)
ρ	Density (kg / m^3)
σ	Surface tension (N / m)
τ	Shear stress (N / m^2)
Λ	Dummy variable
Ψ	Angle with respect to horizontal (deg)
ν	Vapor
ω	Groove width (m)

Subscripts

a	Adiabatic section, air
B	Body
b	Boiling
c	Capillary, capillary limitation, condenser
cs	Control surface
e	Entrainment, evaporator section
eff	Effective
f	Fin, frictional
h	Hydraulic
i	Inner, inertial
l	Liquid
m	Maximum
n	Nucleation
o	Outer
p	Pipe
por	Porous or pore
ph	Phase transition
r	Radial
s	Sonic or sphere
tot	Total
ν	Vapor
ω	Wire spacing, wick
$\parallel$	Axial hydrostatic pressure
$+$	Normal hydrostatic pressure

REFERENCES

Adamson, A. W., 1990, *Physical Chemistry of Surfaces*, 5th ed., Wiley, New York.

Busse, C. A., 1967, "Pressure Drop in the Vapor Phase of Long Heat Pipes," *Proceedings of the IEEE International Thermionic Conversion Specialist Conference*, IEEE, New York.

Chi, S. W., 1976, *Heat Pipe Theory and Practice*, McGraw-Hill, New York.

Chisholm, D., 1971, *The Heat Pipe*, Mills and Boon, London, England.

Collier, J. G., 1972, *Convective Boiling and Condensation*, 2nd ed., McGraw-Hill, New York.

Colwell, G. T., and Chang, W. S., 1984, "Measurements of the Transient Behavior of a Capillary Structure under Heavy Thermal Loading," *International Journal of Heat and Mass Transfer*, Vol. 27, No. 4, pp. 541–551.

Cotter, T. P., 1965, "Theory of Heat Pipes," Report No. LA-3246-MS, Los Alamos National Laboratory, Los Alamos, NM.

Dunn, P. D., and Reay, D. A., 1982, *Heat Pipes*, 3rd ed. Pergamon, New York.

Eninger, J. E., 1975, "Capillary Flow through Heat Pipe Wicks," American Institute of Aeronautics and Astronautics, Washington, D.C. Paper No. 75-661, AIAA.

Ferrell, J. K., and Alleavitch, J., 1969, "Vaporization Heat Transfer in Capillary Wick Structures," Preprint No. 6, ASME-AIChE Heat Transfer Conf., Minneapolis, MN.

Freggens, R. A., 1969, "Experimental Determination of Wick Properties for Heat Pipe Applications," Proc. 4th Intersoc. Energy Conversion Eng. Conf., Washington, DC, pp. 888–897.

Ivanovskii, M. N., Sorokin, V. P., and Yagodkin, I. V., 1982, *The Physical Principles of Heat Pipes*, Clarendon, Oxford.

Kaviany, M., 1991, *Principles of Heat Transfer in Porous Media*, Springer-Verlag, New York.

Kays, W. M., 1966, *Convective Heat and Mass Transfer*, McGraw-Hill, New York.

Knight, B. K., and McInteer, B. B., 1965, "Laminar Incompressible Flow in Channels with Porous Walls," Report No. LA, DC, 5309, USAEC–Los Alamos National Laboratory, Los Alamos, NM.

Kraus, A. D., and Bar-Cohen, A., 1983, *Thermal Analysis and Control of Electronic Equipment*, McGraw-Hill, New York.

Marcus, B. D., 1972, "Theory and Design of Variable Conductance Heat Pipes," Report No. NASA CR, 2018, National Aeronautics and Space Administration. Washington, D.C. April.

Wageman, W. E., and Guevara, F. A., 1960, "Fluid Flow through A Porous Channel," *Physics of Fluids*, Vol. 3, p. 878.

Yuan, S. W., and Finkelstein, A. B., 1956, "Laminar Flow with Injection and Suction at a Porous Wall," *ASME Transactions Journal*, Vol. 78, p. 719.

Chapter

4

OTHER OPERATING LIMITS

4.1 INTRODUCTION

In addition to the capillary limitation discussed in Chapter 3, several other physical situations arise that may limit or control the axial heat transport capacity of heat pipes and thermosyphons. These physical limitations are typically referred to as the viscous, sonic, entrainment, and boiling limits. By identifying and characterizing each of these limits independently, the heat transport capacity as a function of the mean adiabatic vapor temperature, or *operating temperature,* can be determined. In this manner, it is possible to ensure that the heat pipe can transport the required thermal load, improve the design and material selection process, and provide a heat pipe that can function within a specified operating temperature range both effectively and reliably.

4.2 VISCOUS LIMITATION

At very low operating temperatures, the vapor pressure difference between the evaporator and the condenser regions of a heat pipe or thermosyphon may be extremely small. In some cases, the viscous forces within the vapor region may actually be larger than the pressure gradients caused by the imposed temperature field. When this occurs, the pressure gradients within the vapor region may not be sufficient to generate flow and the vapor may stagnate. This no-flow or low flow condition in the vapor portion of a heat pipe is referred to as the viscous limitation. Because the vapor pressures typically must be very low for this to occur, the viscous limit is most often observed in cryogenic heat pipes, heat pipes with extremely long condenser regions, or heat pipes undergoing startup from a frozen state.

Early investigations of the viscous limit were performed by assuming an isothermal vapor that behaved as an ideal gas. With these assumptions, a two-dimensional model, of the vapor flow was developed. The results of this investigation indicated that the radial velocity component was a significant factor in determining the maximum axial heat transfer a given heat pipe could transport prior to reaching the viscous limit. Mathematically, this expression can be written as

$$q_v = \frac{A_v r_0^2 \lambda \rho_v P_v}{16 \mu_v L_e} \tag{4.1}$$

where P_v and ρ_v are the vapor pressure and density at the evaporator end of the heat pipe. The values predicted by this expression have been compared with the results of experimental investigations and have been shown to agree reasonably well (Busse, 1973).

Additional experimental work presented in the Engineering Science Data Unit (ESDU) data item on capillary-driven heat pipes (Anonymous, 1979) has indicated that the vapor flow can in fact be limited by the pressure gradient occurring between the exit to the evaporator and the minimum pressure existing in the condenser. As a result of this investigation, a relationship between the overall pressure drop occurring in the vapor phase and the absolute pressure in the vapor phase was used to establish a criteria by which the viscous limit could be avoided:

$$\frac{\Delta P_v}{P_v} < 0.1 \tag{4.2}$$

While this criteria can be used to determine the viscous limit during normal operating conditions, during startup conditions from a cold or frozen state or during rapid transients at or near the triple state, the viscous limit may still act as a limiting condition.

4.3 SONIC LIMITATION

As discussed by Cotter (1967), compressible flow in a duct of constant cross section with mass addition and removal (i.e., vapor flow in a heat pipe) and constant mass flow in a duct of variable cross section (compressible flow in a converging–diverging nozzle) share a number of common properties. Figure 4.1 illustrates the mass flow rate as a function of axial position throughout the evaporator and condenser along with the corre-

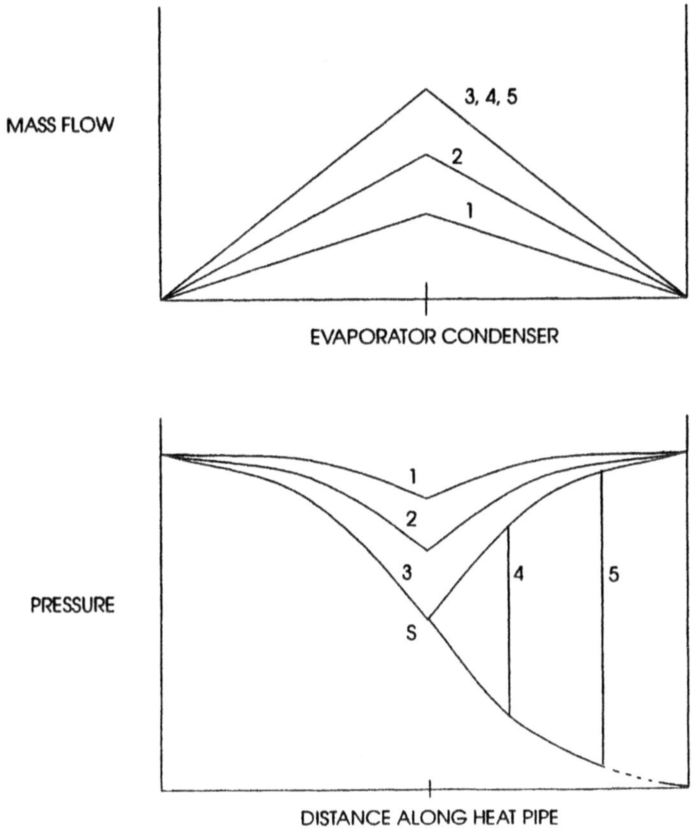

Figure 4.1 Axial pressure distribution and mass flow (Cotter, 1967).

sponding pressure distribution also as a function of axial distance along the pipe for five thermal loads. In curves 1, 2, and 3 the thermal load is increased in a stepwise manner. As a result, the vapor mass flow rate increases with corresponding reductions in the condenser pressure. Eventually, however, the vapor velocity reaches the velocity of sound at the end of the evaporator, shown as point S on curve 3. Because the flow has reached sonic velocity, further reductions of the condenser pressure will not result in increases in the mass flow rate. As explained by Cotter (1967), during the startup phase of a heat pipe, the vapor undergoes a transition from continuum flow to molecular flow that gradually works its way through the condenser region. Once the continuum flow regime reaches the condenser end, a shock wave is formed, as illustrated in curve 5. This shock wave then slowly progresses upstream and grows weaker until it eventually disappears at the exit to the evaporator. At this point the entire flow regime has been restored to a subsonic condition.

Figure 4.2 illustrates the relationship between the vapor temperature

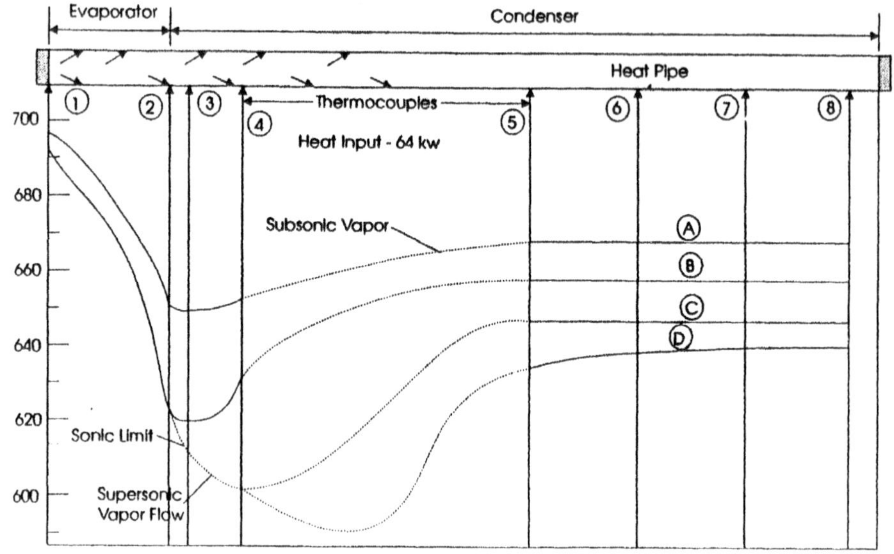

Figure 4.2 Temperature as a function of axial position (Dunn and Reay, 1982).

and the axial position in a heat pipe. Curve *A* represents the temperature distribution typical of a heat pipe with subsonic flow conditions and partial pressure recovery. As was the case for pressure in Fig. 4.1, increasing the heat rejection rate and lowering the condenser temperature will decrease the evaporator temperature, as shown in curve *B*. Continued increases in the heat flux and reductions in the condenser temperature will result in a decrease in the overall average vapor temperature. However, at some point the vapor velocity at the evaporator exit approaches the sonic velocity and a critical or choked flow condition exists. For this situation, continued reductions in the condenser temperature only serve to decrease the temperature in the condenser region and have no effect on the vapor generation rate or vapor temperature distribution in the evaporator.

The various factors that affect the vapor pressure drop have been discussed in detail in Section 3.3, where it was found that the vapor pressure drop is the result of two components: the inertial and viscous terms. In a reversible process, the inertial terms in the evaporator and condenser would have opposite signs and would effectively cancel one another out, as shown in Fig. 4.3. In an actual situation, it is reasonable to assume that only partial pressure recovery will occur. This assumption has been verified experimentally by Grover et al. (1968) using a sodium heat pipe. In this investigation, pressure recovery levels of 60% were achieved for radial Reynolds numbers greater than 10.

Unlike the other previously discussed heat transport limits, the sonic limitation actually serves as an upper bound to the axial heat transport

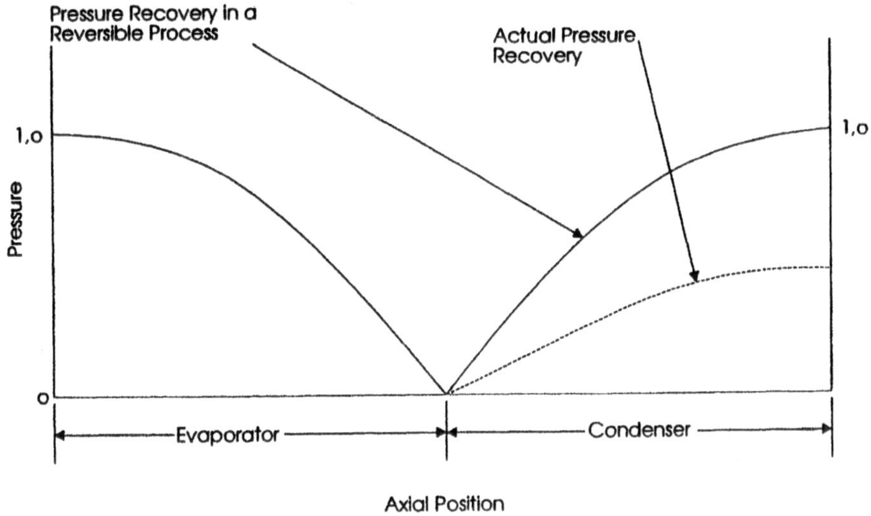

Figure 4.3 Complete pressure recovery (Cotter, 1967).

capacity and does not necessarily result in dryout of the evaporator wick or total heat pipe failure. Attempts to exceed the sonic limit do, however, result in increases in the axial temperature gradient along the heat pipe, eliminating or reducing the isothermal characteristics typically found in the vapor flow region.

4.3.1 Two-dimensional Sonic Models

Sonic limitations in heat pipes are analogous to the sonic limitations in converging–diverging nozzles (Chi, 1976). In a converging–diverging nozzle, the mass flow rate is constant and the vapor velocity varies because of the changing cross-sectional area. In heat pipes, the reverse occurs; the area is constant and the vapor velocity varies because of the evaporation and condensation along the heat pipe. As in nozzle flow, decreased outlet pressure, or in the case of heat pipes, condenser temperature, results in a decrease in the evaporator temperature until the sonic limit is reached. Any further increases in the heat rejection rate do not reduce the evaporator temperature or the maximum heat transfer capability but only reduce the condenser temperature, due to the existence of choked flow.

Several attempts to describe the sonic limit from first principles have been developed. Bankston and Smith (1971) and Rohani and Tien (1974) have all solved the Navier–Stokes equations using numerical methods. The former study indicated that axial flow reversal occurred for $Re_r < -2.3$ at the end of the condenser during conditions of high evaporation and condensation. Comparison with the predicted results of a one-dimensional model developed by Busse indicated good agreement up to $Re_r > 10$ in the condenser region in spite of this flow reversal (Rohani and Tien, 1974).

4.3.2 One-dimensional Sonic Models

Levy (1968) developed a closed form expression for the sonic limit derived from one-dimensional vapor flow theory assuming that the frictional effects may be neglected, inertial effects dominate, and the vapor behaves as an ideal gas. Combining these assumptions with the conservation of energy and momentum equations, results in expressions for the temperature and pressure ratio of

$$\frac{T_o}{T_v} = 1 + \frac{v_v^2}{2c_p T_v} \tag{4.3}$$

and

$$\frac{P_o}{P_v} = 1 + \frac{\rho_v v_v^2}{2P_v} \tag{4.4}$$

where the subscripts o and v indicate the stagnation and static states of the vapor, respectively. Substituting the local Mach number and relating the axial heat flux to the density and velocity

$$v_s = (\gamma_v R_v T_v)^{1/2} \tag{4.5}$$

these expressions can be rewritten as

$$\frac{T_o}{T_v} = 1 + \frac{\gamma_v - 1}{2} \mathrm{Ma}_v^2 \tag{4.6}$$

and

$$\frac{P_o}{P_v} = 1 + \frac{\gamma_v}{2} \mathrm{Ma}_v^2 \tag{4.7}$$

Combining the temperature and pressure ratios with the ideal gas law yields an expression for the density ratio that when combined with the two expressions given above yields an equation for the axial heat flux in terms of the physical properties, geometrical dimensions, and the Mach numbers

$$q = \frac{A_v \rho_o \lambda (\gamma_v R_v T_o)^{1/2} \mathrm{Ma}_v (1 + \frac{1}{2}(\nu_v - 1)\mathrm{Ma}_v^2)^{1/2}}{1 + (\nu_v/2)\mathrm{Ma}_v^2} \tag{4.8}$$

The recognition that the Mach number should be equal to unity at sonic velocities yields an expression for the maximum axial heat transport,

$$q_s = A_v \rho_o \lambda \left[\frac{\gamma_v R_v T_o}{2(\gamma_v + 1)} \right]^{1/2} \tag{4.9}$$

prior to the onset of choked flow. It is important to note that while this expression was initially intended for use at the exit to the vaporator, it is also valid for the condenser section. However, in this region the Mach number may be greater than unity.

An alternative approach has been presented by Busse (1973). Using one-dimensional laminar flow, the Navier–Stokes equations yield

$$\frac{dP}{dx} = \frac{d}{dx} \overline{\rho v^2} \tag{4.10}$$

Integrating this expression and applying a non-slip boundary condition yields

$$P_v - P = \overline{\rho v^2} = a \rho v^2 \tag{4.11}$$

where a is a coefficient that accounts for the deviation from a parabolic velocity profile. Combining this expression with the continuity equation and the constitutive relationships for an ideal gas yields an expression for the maximum heat transport capacity as a function of the thermophysical properties:

$$q = \lambda \left(\frac{\rho_v P_v}{A} \right)^{1/2} \left[\frac{P}{P_v} \left(1 - \frac{P}{P_v} \right) \right]^{1/2} \tag{4.12}$$

Setting the first derivative $dq/dp = 0$ yields a relationship for the sonic limit of

$$q_s = 0.474 \lambda A_v (\rho_v P_v)^{1/2} \tag{4.13}$$

where ρ_v and P_v are the vapor density and pressure at the end of the evaporator. The greatest difficulties in determining the heat flux level at which choked flow will occur is in determining these two quantities along with the inlet pressure to the condenser. As is apparent in equation 4.13, these values may have a significant effect on the vapor velocity and can cause large variations in the existence or location of the choked flow condition.

4.3.3 Transition from Viscous to Sonic Limits

The easiest way to avoid viscous limitations encountered during startup is to increase the heat flux applied at the evaporator. As this heat flux is increased, the vapor temperature and hence vapor pressure increase to a point where the pressure gradient along the vapor flow path exceeds the viscous pressure drop. The danger of this procedure is that at low vapor pressures, such as those encountered near the viscous limit, heat pipe operation may quickly transition from a viscous limit to a sonic limit.

Because both the viscous and sonic limitations in heat pipes occur at relatively low pressures and, hence, low temperatures, in many cases a transition from the viscous to the sonic limit can be observed. As discussed previously, Busse (1973) developed a one-dimensional hydrodynamic model of isothermal vapor flow assuming an ideal gas. The result was a relationship for the sonic limit of the form

$$q_s = \lambda\left(\frac{\rho_v P_v}{A}\right)^{1/2}\left[\frac{P}{P_v}\left(1-\frac{P}{P_v}\right)\right]^{1/2} \tag{4.14}$$

that assumes that the heat pipe was operating in the region where inertial effects were dominant. In the same analysis an expression for the viscous limit was developed as

$$q_v = d_v^2 \lambda A_v \frac{P_v \rho_v}{64\mu_v L_e} \tag{4.15}$$

These two equations illustrate the relationship that exists between the product $P_v\rho_v$.

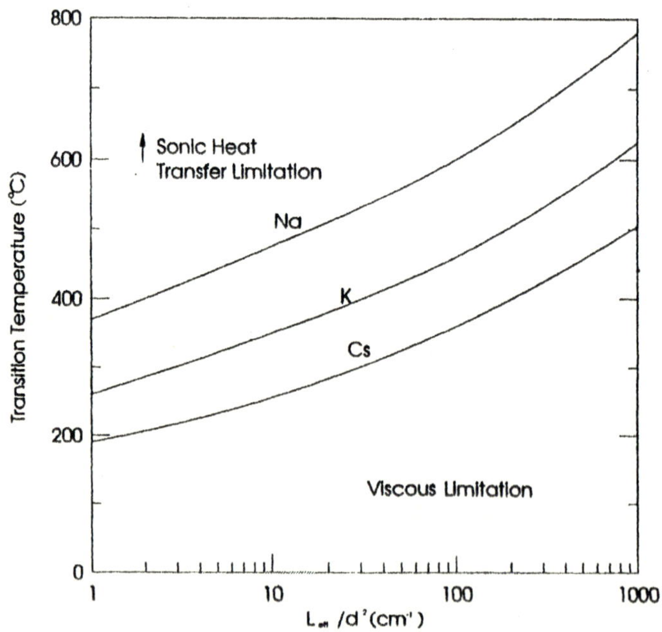

Figure 4.4 Transition from viscous to sonic limit (Busse, 1973).

The sonic limit varies as the product $(P_v \rho_v)^{1/2}$ while the viscous limit varies linearly with respect to $P_v \rho_v$. As a result, when this product is small, the transport capacity is typically limited by viscous effects, but when the product is large, inertial effects begin to dominate and a transition is made from the viscous to the sonic limit. The boundary between these two limits can be determined by setting these two equations equal to each other and solving for the combined terms as a function of temperature (Ivanovskii et al., 1982). Busse (1973) made just such a comparison for liquid metal heat pipes utilizing sodium, potassium, and cesium as the working fluids. The results are illustrated in Fig. 4.4, which indicates that the transition temperature is dependent upon the thermophysical properties of the working fluid, the geometry of the heat pipe, and the size of the evaporator and condenser regions. Further experimental work by Vinz and Busse (1973) experimentally verified that for heat pipes utilizing sodium as a working fluid, the transition temperature from the viscous to sonic limitation regions was in the range of 450–530°C. As shown in Fig. 4.4, this compares quite well with the predicted values for sodium. Other experimental and theoretical investigations of the sonic limitation have been performed by Kemme (1969) and Deverall et al. (1970).

4.4 ENTRAINMENT LIMITATION

In operating thermosyphons and heat pipes, the liquid and vapor flow in opposite directions. The interaction between this countercurrent liquid and vapor flow and the viscous shear forces occurring at the liquid–vapor interface may inhibit the return of liquid to the evaporator. When this occurs, the heat pipe is said to have reached the flooding limit. Further increases in the heat input result in increased vapor velocities, which may cause the liquid flow to become unstable. In the most severe cases, waves may form and the interfacial shear forces may become greater than the liquid surface tension forces, resulting in liquid droplets being picked up or entrained in the vapor flow and carried to the condenser. This entrainment of liquid droplets can limit the axial heat flux and is referred to as the entrainment limit. Although both flooding and entrainment result in excess liquid accumulation in the condenser and dryout of the evaporator wicking structure, they can be differentiated by the presence of liquid droplets entrained in the vapor stream (Peterson and Bage, 1991).

Entrainment in multicomponent annular pipe flow is a problem of fundamental importance and has been studied extensively. Early investigations of Kutateladze and Sorokin (1969) summarized the onset of the "critical" conditions occurring in several liquid vapor systems. The onset of droplet formation, the rates of entrainment, and the contribution to momentum transfer by the entrained droplets are some of the more important areas investigated (Lopes and Dukler, 1986; Langner and

Mayinger, 1979; Hewitt, 1979; Andreussi, 1980; Kataoka et al., 1983). However, the phenomenon of entrainment in multiphase, single component flow in the presence of wicking structures (i.e., heat pipes) has received only limited attention. Although more recent work by Prenger and Kemme (1981), Nguyen-Chi and Groll (1981), Prenger (1984), Weber-Carstanjen (1987), and Rice and Fulford (1987) has clarified several issues, the major problem of establishing criteria that can be used to predict the onset of entrainment remains unresolved. Shortly after the invention of the heat pipe by Grover et al. (1964), Cotter (1967) presented a method by which the entrainment limitation in an operating heat pipe could be determined. This method utilized the Weber number, defined as the ratio of the viscous shear force to the forces resulting from the surface tension, or

$$\mathrm{We} = \frac{2r_{h,w}\rho_v V_v^2}{\sigma} \tag{4.16}$$

By relating the vapor velocity and the heat transport capacity to the axial heat flux as

$$v_v = \frac{q}{A_v \rho_v \lambda} \tag{4.17}$$

and assuming that to prevent entrainment of liquid droplets in the vapor flow, the Weber number must be less than unity, a value for the maximum transport capacity based on entrainment can be written as

$$q_l = A_v \lambda \left(\frac{\sigma \rho_v}{2r_{h,w}} \right)^{1/2} \tag{4.18}$$

where $r_{h,w}$ is the hydraulic radius of the wick structure.

Since the initial development of this expression, numerous investigations of entrainment in both thermosyphons and heat pipes have been conducted. Presented here is a summary (Peterson and Bage, 1991) of these investigations and a comparison of the results as predicted by the various models.

4.4.1 Entrainment in Thermosyphons

There have been numerous investigations involving the entrainment limit in thermosyphons and several reasonably accurate prediction methods exist. Nguyen-Chi and Groll (1981) presented an experimental verification of three semiempirical entrainment correlations for thermosyphons. The first of these was a flooding correlation by Bezrodnyi (1978), which was based on the stability of the vapor–liquid interface.

$$q_m = C_B(\tfrac{1}{4}\pi d_i^2)\lambda\rho_v^{0.5}[\sigma g(\rho_l - \rho_v)]^{0.25}K_p^n \tag{4.19}$$

where

$$K_p = \sigma\{P/g[\sigma/(\rho_l - p_v)]\}^{1/4}$$

In this model, it was assumed that at the instant flooding occurred, the surface waves of the liquid film would increase in amplitude and, hence, the vapor flow pressure gradient would increase rapidly. It was also assumed that this would lead to a separation of liquid from the condensate film caused by the vapor–liquid pressure difference and liquid film wave profile.

The second correlation presented was based on the Wallis (1969) flooding criterion as modified for applications to thermosyphons by Sakhuja (1973). This model,

$$q_m = C_w^2\,(\tfrac{1}{4}\pi d_i^{2.5})\lambda[\rho_v g(\rho_l - \rho_v)]^{0.5}/[1 + (\rho_v/\rho)^{0.25}]^2 \tag{4.20}$$

was limited by the assumptions of incompressible vapor flow, constant liquid film thickness, and one-dimensional, steady-state, homogeneous, equilibrium two-phase, single-component flow.

The third and final thermosyphon entrainment/flooding correlation considered was one developed by Tien and Chung (1979),

$$q_m = C^2[\tfrac{1}{4}(\pi d_i^2)]\lambda(\rho_l^{-0.25} + \rho_v^{-0.25})^{-2}[\sigma g(\rho_l - \rho_v)]^{0.25} \tag{4.21}$$

This expression utilized a modification of the modified Kutateladze (1972) flooding correlation to include the effect of the tube diameter on the entrainment/flooding phenomenon by means of the Bond number $Bo = d_i[g/\sigma(\rho_l - \rho_v)]^{1/2}$ and defined the constant C_k to be $(3.2)^{0.5}\tanh(0.5Bo^{0.25})$ for water.

Nguyen-Chi and Groll (1981) were concerned with determining the influence of liquid fill charge, inclination angle, and operating temperature on the entrainment limit. Thus, in order to compare the data in terms of the three given models, the maximum heat fluxes (limited by entrainment or flooding) were reduced by the function

$$f(\beta) = \left[\frac{\beta}{180°} + \sqrt{\sin 2\beta}\right]^{0.65} \tag{4.22}$$

to eliminate the dependence on the inclination angle. As shown in Fig. 4.5, their data correlate well with the Wallis (1969) model, given in equation 4.20.

Fukano et al. (1983) conducted an experimental investigation to validate the Tien and Chung (1978) expression, given in equation 4.21. In addition

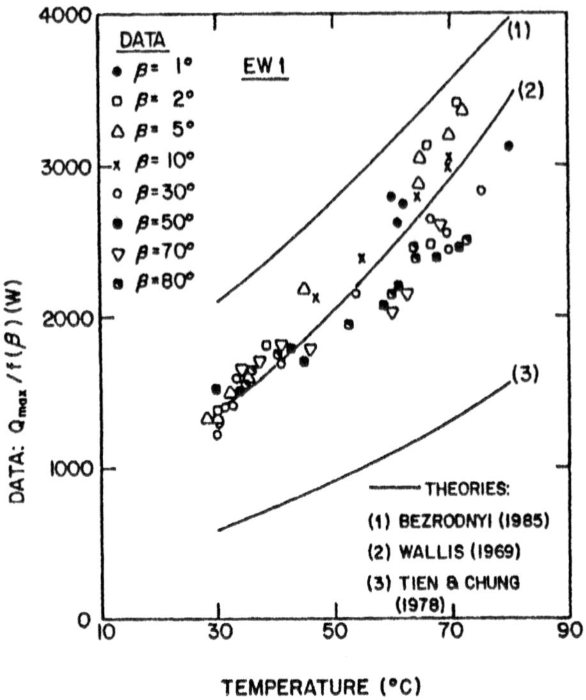

Figure 4.5 Comparison of reduced experimental data for inclined thermosyphons (Nguyen-Chi and Groll, 1981).

to observing dryout, evidence of both boiling and oscillation of the liquid–vapor interface were also observed. However, their experimental data indicated that the Tien and Chung (1978) model overpredicted the entrainment limit, while the Nguyen-Chi and Groll (1981) data indicated an underprediction. Faghri et al. (1988) have since modified and extended the Tien and Chung (1978) expression (equation 4.21) and defined C_k as $(3.2)^{0.5}$ tanh $\mathrm{Bo}^{0.25}$ for water and $C_k^2 = (P/P_v)^{0.14} \tanh^2 \mathrm{Bo}^{0.25}$ for other fluids.

Whereas Nguyen-Chi and Groll (1981) investigated the effects of inclination and fill charge on the entrainment limit of the thermosyphons, Shiraishi et al. (1984) investigated only the influence of fill charge on the operating limit of the thermosyphons using the fill ratio V_{fc}/V_e to differentiate between large and small fill charges. Two types of dryout phenomena were visually observed: One associated with the dryout of the falling film and the other with the flooding limit. For a small filling ratio Shiraishi et al. (1984) used the model proposed by Cohen and Bayley (1955) shown on the next page,

$$q_m = \frac{\pi g d_c \rho_l^2 \lambda}{3\mu}\left(\frac{V_e(V_{fc}/V_e - \rho_v/\rho_l)}{[\pi d_c(0.8L_c + L_{ac}) + (d_e/d_c)^{2/3}(L_{ac} + 0.75L_c)][1-(\rho_v/\rho_l)]}\right)^3 \tag{4.23}$$

For large filling ratios, the models developed by Nejat (1981)

$$q_m = \frac{0.09(d_e/L_e)^{0.9}\mathrm{Bo}^{0.5}\pi d_e L_e \rho_v^{0.5}\lambda[\sigma g(\rho_l - \rho_v)]^{0.25}}{[1+(\rho_v/\rho_l)^{1/4}]^2} \tag{4.24}$$

and Katto (1978)

$$q_m = \frac{0.01\pi d_e L_e \rho_v^{0.5}\lambda[\sigma g(\rho_l - \rho_v)]^{0.25}}{\left[1 + \dfrac{0.0491 L_e}{d_e \mathrm{Bo}^{0.3}}\right]} \tag{4.25}$$

were utilized.

Equation 4.23 correlates the dryout of the falling liquid film, while equations 4.24 and 4.25 correlate the flooding limit in a confirmed channel. As shown in Fig. 4.6, the experiments of Shiraishi et al. (1984) compared well with the models representing the large filling ratio (flooding correlation). However, for the small filling ratio model, the data followed the same trend but were a factor of approximately 5 smaller than the experimental data.

Another correlation that used a similar approach to that in equations 4.24 and 4.25 was developed by Imura et al. (1983). Their expression,

$$q_e = 0.64\left[\frac{\rho_l}{\rho_v}\right]\left[\frac{d_i}{4L_e}\right]\lambda[\sigma g \rho_v(\rho_l - \rho_v)] \tag{4.26}$$

accurately predicted the associated experimental data and also predicted the data of several other investigators to within ±30%. While the investigation of Shiraishi et al. (1984) was primarily concerned with the influence of fill charge on the operating limits of thermosyphons, the investigation conducted by Imura et al. (1983) focused on developing an expression for the optimum fill charge for thermosyphons.

In the experimental investigation of Shiraishi et al. (1984) only two types of dryout phenomena, the dryout of the falling liquid film and the flooding limit, were observed. Andros and Florschuetz (1978), however, visually observed three types of dryout, one due to insufficient fill charge and the other two due to expulsion of the liquid from the evaporator. Alternatively, Bezrodnyi (1985) identified five crisis phenomena associated with the flooding limit in thermosyphons. These are listed in Table 4.1, along with the corresponding correlations. It should be noted that this work was

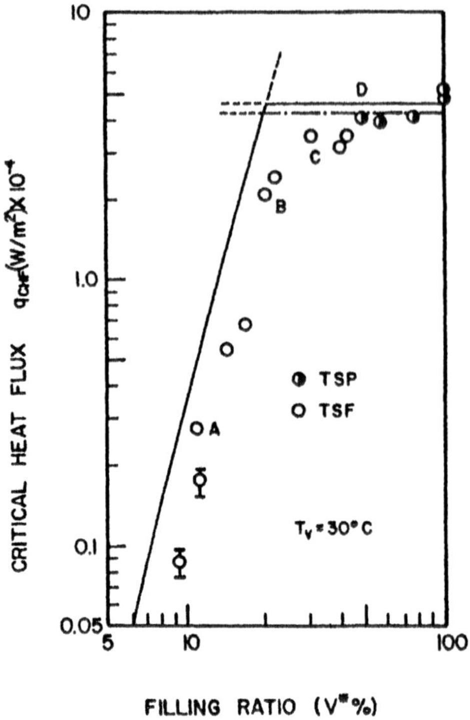

Figure 4.6 Critical heat flux as a function of charge (Shiraishi et al., 1984).

TABLE 4.1 Five Crisis Phenomena

Crisis	Stability criterion (K)	Restrictions
Limiting film thickness	$\dfrac{\pi Bo^{0.5}}{[986j^2(y_0/\delta)^2 + 0.962Bo]^{0.5}}$	$j = 3.6$ for $Y \le 0.016$, $j = 5710\ Y^{1.8}$ for $Y \ge 0.016$
Start of droplet	$\dfrac{\pi Bo^{0.5} f(Kp)}{[234j^2(y_0/\delta)^2 + 0.962Bo]^{0.5}}$	$j = 3.6$ for $Y/Ka^{1/12} \le 0.019$, $j = 0.5Y^{-0.5}Ka^{1/24}$ for $Y/Ka^{1/12} > 0.019$, $f(Kp) = 4.06Kp^{-0.17}$ for $Kp \le 40{,}00$, $f(Kp) = 0.67$ for $Kp \ge 40{,}000$
Film Suspension	$8.19Kp^{-0.17}$ 1.35	For $Kp \le 40{,}000$ For $Kp \ge 40{,}000$
Liquid expulsion from flow core	$10.24Kp^{-0.17}$ 1.7	For $Kp \le 40{,}000$ For $Kp \ge 40{,}000$
Flow inversion	3.2	None

an extension of an earlier investigation published by Bezrodnyi and Volkov (1981) in which only three crisis phenomena were identified.

It is significant to note that equations 4.19 through 4.21 and 4.23 through 4.26 are all based on either flooding correlations or dimensional analysis and require a large amount of data to properly determine the various constants for a given application.

In all of the previously cited investigations, the models were based upon the assumption of dry vapor flow. Although for some special cases, these models represent the data reasonably well, Vasil'ev et al. (1982) proposed that the vapor would contain microscopic drops of liquid. Several reasons were cited for this conclusion, the two most obvious being the interaction between the vapor and liquid flows and droplet formation due to condensation in the adiabatic zone of the thermosyphon. By simultaneously evaluating the mass flow rate and the vapor velocity, described as

$$\dot{m}_d = 18C\mu_v\left[\frac{\rho_v^3}{\sigma^3 g}\right]\left[\left(\frac{4z}{d_{vc}}\right)\sin\beta + \cos\beta\right]\left[\frac{q}{\rho_v r^*}\right] \tag{4.27a}$$

$$G_v(z) = \pi d_{vc}\frac{q}{r^*}z \tag{4.27b}$$

$$G_{lw}(z) = -G_W + \pi d_{vc}\frac{q}{R^*}(L_e - z) + 18\mu_v C\left[\frac{q}{r^*\rho v}\right]^4\left[\frac{\rho_v^3}{g\sigma^3}\right]$$

$$\times\left[\frac{2(L_e^2 - z^2)}{d_{vc}}\sin\beta + (L_e - z)\cos\beta\right]\pi d_{vc} \tag{4.27c}$$

$$G_d = 18\mu_v C\left(\frac{q}{r^*\rho_v}\right)\left(\frac{\rho_v^3}{g\sigma^3}\right)^{0.5}\left[\frac{2z^2}{d_{vc}}\sin\beta + z\cos\beta\right]\pi d_{vc} \tag{4.27d}$$

$$C = \frac{Y_f(L_e)}{1 - Y_f(L_e)}\,18\mu_v\left[\frac{q}{r^*\rho_v}\right]^3\left[\frac{\rho_v}{g\sigma^3}\right]^{0.5}\left[\frac{2L_e}{d_{vc}}\sin\beta + z\cos\beta\right] \tag{4.28}$$

an expression for the entrainment limit was obtained. The results of this process are shown in Fig. 4.7, where it is apparent that the amount of moisture contained in the liquid droplets is a function of the heat flux.

In all of the investigations cited thus far, the correlations model either flooding or dryout of the falling liquid film. Dobran (1985) developed a lumped parameter model for thermosyphons and concluded that the predicted maximum heat flux could be related to the entrainment or flooding limit. This model,

$$(Q_i^*)^2\left[\frac{2}{\pi\rho^*}\left(\frac{x_{20}^*}{x_{10}^*}\right)\right] - Q_i^*\left[\left(\frac{f_w}{f_{wL}}\right)\frac{(x_{10})^*}{4}\frac{(x_{10})^*}{(x_{20})^*} + \left(\frac{f_i}{f_{iL}}\right)\frac{\mu^*}{2}(x_{10})^*\left(1 + \frac{2}{P}\frac{(x_{20})^*}{(x_{10})^*}\right)\right]$$

$$+ (N_L)^2\frac{\pi}{256}(x_{10})^*\left(\frac{(x_{20})^*}{(x_{10})^*}\right) = 0 \tag{4.29}$$

is a steady-state model in which the input heat flux is a function of the film thickness, capillary number, Grashof number, liquid charge, thermosyphon

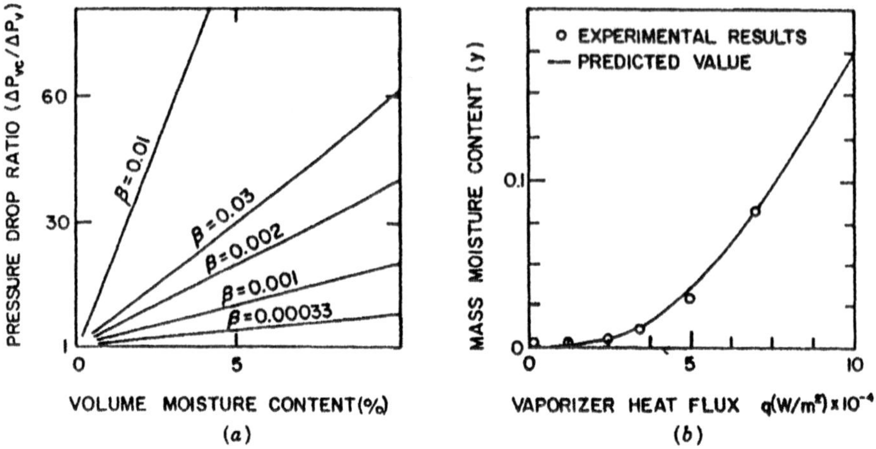

Figure 4.7 (*a*) Dependence of the ratio of pressure drops in the vapor channel of a heat pipe in a vapor drop flow and a flow of dry vapor for different volume moisture contents. (*b*) Dependence of mass moisture content of vapor in a heat pipe on the heat flux in the evaporator (Vasil'ev et al., 1982).

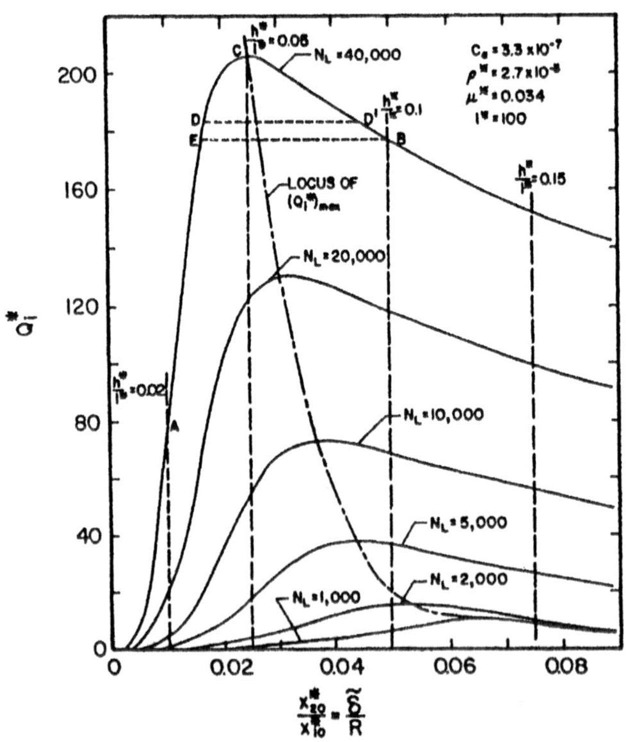

Figure 4.8 Steady-state distribution of input heat flux to the evaporator (Dobran, 1985).

length, density, and viscosity. The limits are plotted in Fig. 4.8 as a function of the heat supply parameter Q^* and the ratio of the average film thickness of the pipe radius.

4.4.2 Entrainment in Heat Pipes

In contrast to the many investigations of flooding/entrainment in thermosyphons, the study of these phenomena in heat pipes is somewhat limited. Correlations developed for thermosyphons have been expanded, while other studies have focused specifically on entrainment in heat pipes, but for the most part, as noted by Tien and Chung (1978), the determination of the entrainment limit in heat pipes is still rather speculative and is not well understood.

A majority of the work done to date has been on gravity-assisted heat pipes. Busse and Kemme (1980) have expressed doubt as to whether entrainment actually occurs in a capillary-driven heat pipe, since the capillary structure would most likely retard the growth of any surface waves. In a majority of cases studied thus far, the wicking structure of the heat pipe has become flooded, allowing entrainment to occur. Matveev et al. (1977) showed that by blowing air over a wire mesh, entrainment could only be induced when the mesh was fully saturated with liquid.

Tien and Chung (1978) presented correlations for thermosyphons and for horizontal, vertical, grooved, and rotating heat pipes. Of these, only the one for vertical (gravity-assisted) heat pipes is applicable here. This correlation (equation 4.21) was applied to data reported by Kemme (1976), who expanded the Weber number criterion suggested by Cotter (1967) to include the balancing force term of buoyancy, or

$$q_e = A_v \lambda \left[\left(\frac{\rho_v}{A^*} \right) \left(\frac{2\pi\sigma}{z} + \rho_l g d \right) \right]^{0.5} \tag{4.30}$$

A comparison of Kemme's data along with equations 4.21 and 4.30 is illustrated in Fig. 4.9.

Prenger (1984) developed a correlation for textured wall, gravity-assisted heat pipes. A model was presented that included both liquid and vapor inertia terms. It was found that for textured wall heat pipes, the liquid inertia term was dominant, because the liquid was partially shielded from the vapor flow. This fact allowed the vapor inertia term to be neglected and the model to be reduced to

$$q_e = 2 A_v \lambda \frac{d_{\text{wire}}}{d_{\text{pipe}}} \left(\frac{\rho_l \sigma}{\pi d_{\text{pipe}}} \right)^{0.5} \tag{4.31}$$

This expression correlated well with previous data taken by Prenger and Kemme (1981). While the model presented by Tien and Chung (1978) had

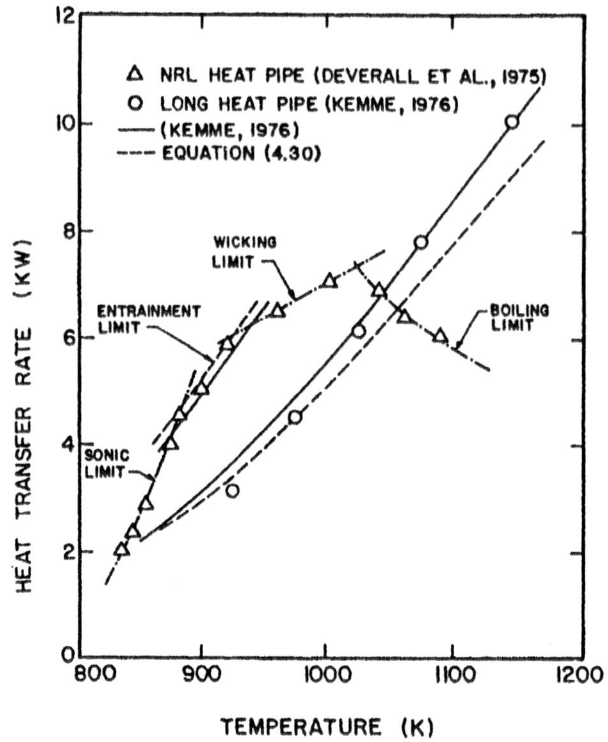

Figure 4.9 Maximum heat transfer rate in gravity-assisted wicked heat pipes (Tien and Chung, 1979).

the heat flux limited by entrainment or flooding as a function of the capillary structure, Prenger (1984) showed it to be a function of the depth of the liquid layer or flow channel.

Weber-Carstanjen (1987) presented a model for heat pipes with a textured surface, as did Prenger (1984). The model was derived from the Wallis flooding correlation based on the Kutateladze number, making it similar to the expression proposed by Tien and Chung (1978). The Weber-Carstanjen (1987) correlation

$$q_{\text{dry}} = A_\nu \lambda((\rho_l)^3 \sigma g)^{0.25} C\left[\left(\frac{\rho_l}{\rho_\nu}\right)^{0.25} + \left(\frac{A_\nu}{A_l}\right)^{0.25}\right]^n \tag{4.32}$$

is an empirical expression developed from data obtained by several different investigators. The values for the constants C and n were determined by best fit curves, thus identifying a need for much more experimentation to aid in the interpretation of these constants. The form of the expression indicates that the heat flux required for entrainment is a function of both the saturation temperature in the adiabatic zone of the heat pipe and the cross-sectional areas of the liquid and vapor passages.

In all the previously presented models except that of Prenger (1984) (equation 4.21) the authors presumed the wick to be flooded. However, classical heat pipe theory suggests that the wick is normally not flooded and the liquid is retained within the wicking structure. Rice and Fulford (1987) recognized this and suggested that the entrainment limited heat flux should be related to the energy required to draw liquid from the wick and not to the energy required to form capillary waves on a flooded wicking structure. With this and the assumption that the only pressure difference between the liquid in the mesh and the vapor is the kinetic head of the vapor flow, Rice and Fulford (1987) derived the expression

$$q_e = A_v \lambda \left(\frac{8 \sigma \rho_v}{\imath} \right)^{0.5} \tag{4.33}$$

Although this expression is similar to the Cotter (1967) model, which was based on the Weber number, the characteristic wick constant $\imath$ now has physical meaning in terms of the mesh size and is not arbitrarily set. This value can be approximated as twice the radius of curvature of the largest droplet that can be formed on the innermost mesh. It should be noted that this expression has not yet been experimentally verified.

In equations 4.30 and 4.31, the characteristic length of the wick was arbitrarily selected depending on the type of wick in the heat pipe. In equation 4.33, the characteristic length was physically determined for different wicks. In equation 4.32, like many of the thermosyphon correlations, it was based on a flooding correlation. Thus, once equation 4.33 has been experimentally verified, it may prove to be most useful.

4.4.3 Comparisons of Various Entrainment Models

In order to compare the various predicted entrainment limitations for thermosyphons and heat pipes, two separate comparisons have been conducted. Using a simple copper–water heat pipe with an inner diameter of 0.017 m, an evaporator length of 1 m, an adiabatic length of 0.5 m, and a condenser length of 1 m (see Table 4.2), the entrainment limitations for thermosyphons as predicted by the models proposed by Wallis (1969), Bezrodnyi (1978), Tien and Chung (1978), Katto (1978), Nejat (1981), Imura et al. (1983), and Tien and Chung (1978) with modification by Faghri et al. (1988) were computed for a temperature range of 20–200°C. Figure 4.10 illustrates a comparison of these results. As shown, the predicted values vary from approximately 0.254 to 1.377 kW at an operating temperature of 20°C and from 4.908 to 9.055 kW at an operating temperature of 200°C or a variation of from 100 to 500%. Because of the strong influence of fill charge,

TABLE 4.2 Thermosyphon and Heat Pipe Dimensions for Modeling Comparison

Parameter	Thermosyphon	Heat Pipe
Pipe inner diameter	0.017 m	0.010 m
Condenser length	1.000 m	0.250 m
Adiabatic length	0.500 m	0.250 m
Evaporator length	1.000 m	0.250 m
Mesh number	—	3937 m^{-1}
Wire diameter	—	0.114 mm

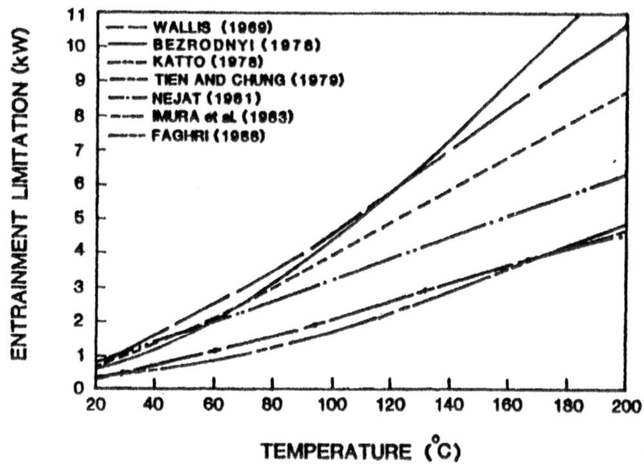

Figure 4.10 Comparison of techniques for predicting the entrainment limit in thermosyphons (Peterson and Bage, 1991).

the Cohen and Bayley (1955) expression (equation 4.23) was not included in this comparison.

Similarly, a comparison of the predicted entrainment limitation in a screen-wicked copper–water heat pipe 0.010 m in diameter with an evaporator length of 0.5 m, an adiabatic length of 0.25 m, a condenser length of 0.25 m, and a screen mesh number of 3937 m^{-1} (see Table 4.2) was made using the analytical methods developed by Cotter (1967), Kemme (1976), Prenger (1984), and Rice and Fulford (1987). Again, as shown in Fig. 4.11, significant variations in the predicted values were observed, with variations from 0.014 to 0.402 kW at 20°C and from 0.048 to 6.575 kW at 200°C. The model developed by Weber-Carstanjen (1987) was not included in this comparison since it required two dimensionless constants, C and n, not available for water. Because both the Cotter (1967) and Kemme (1976) expressions (equations 4.18 and 4.30, respectively) were derived from the Weber number concept, similar values were obtained. The Prenger (1984) and Rice and Fulford (1987) models (equations 4.31 and

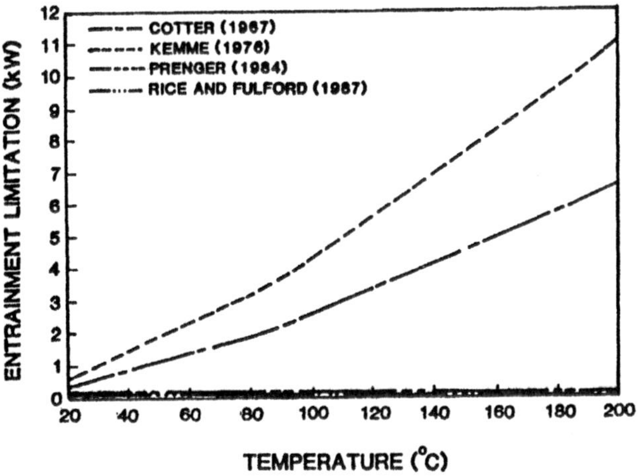

Figure 4.11 Comparison of techniques for predicting the entrainment limit in heat pipes (Peterson and Bage, 1991).

4.33) resulted in significantly different values, because these expressions considered the specific wicking structure. The four models resulted in values for the heat pipe entrainment limit that varied by as much as a factor of 27.

4.4.4 Entrainment Closure

Sixteen different correlations for predicting the onset of flooding and / or entrainment in thermosyphons and heat pipes have been reviewed and compared. Although many of these studies have helped investigators to develop a better understanding of the fundamental phenomena that govern this limit, these investigations have resulted in wide variations in the predicted level at which entrainment would occur. Using a simple copper–water thermosyphon and a similar screen-wicked heat pipe, comparisons of the entrainment limit as predicted from seven analytical models for thermosyphons and four analytical models for heat pipes were compared. These comparisons indicate variations ranging from a factor of 5 for thermosyphons to a factor of 27 for heat pipes.

4.5 BOILING LIMITATION AND REWETTING OF DRY SURFACES

When the input heat flux is sufficient, nucleate boiling may occur in the wicking structure and bubbles may become trapped in the wick, blocking the liquid return and resulting in premature evaporator dryout. This phenomenon, referred to as the heat flux or boiling limit, differs from the other limitations previously discussed in that it depends on the radial or circumferential heat flux applied to the evaporator as opposed to the axial heat flux or thermal power transported by the heat pipe.

4.5.1 Heat Flux Limit

Determination of the heat flux or boiling limit is based on nucleate boiling theory and is comprised of two separate phenomena—bubble formation and the subsequent growth or collapse of the bubbles (Chi, 1976). Bubble formation is governed by the number and size of nucleation sites on a solid surface and the temperature difference between the heat pipe wall and the working fluid. This temperature difference, termed the *superheat*, governs the formation of the bubbles and can typically be defined in terms of maximum heat flux as

$$q_m = \left(\frac{k_{\text{eff}}}{T_w}\right) \Delta T_{\text{cr}} \tag{4.34}$$

where k_{eff} is the effective thermal conductivity of the liquid–wick combination and ΔT_{cr} is the critical superheat, defined by Marcus (1972) as

$$\Delta T_{\text{cr}} = \left(\frac{T_{\text{sat}}}{\lambda \rho_v}\right)\left(\frac{2\sigma}{r_n} - \Delta P_{i,m}\right) \tag{4.35}$$

Here T_{sat} is the saturation temperature of the fluid and r_n is the critical nucleation site radius, which according to Dunn and Reay (1982) can be assumed to be from 2.54×10^{-5} to 2.54×10^{-7} m for conventional metallic heat pipe case materials.

As discussed by Brennan and Kroliczek (1979), this model yields a very conservative estimate of the amount of superheat required for bubble formation. This is true even when the lower bound for the critical nucleation site radius is used. Some researchers attribute this to the absence of adsorbed gases on the surface of the nucleation sites caused by the degassing and cleaning procedures used in the preparation and charging of heat pipes.

The growth or collapse of a given bubble once established on a flat or planar surface is dependent upon the liquid temperature and corresponding pressure difference across the liquid–vapor interface caused by the vapor pressure and surface tension of the liquid. By performing a pressure balance on any given bubble and using the Clausius–Clapeyron equation to relate the temperature and pressure, an expression for the heat flux beyond which bubble growth will occur may be developed (Chi, 1976). This expression, which is a function of the fluid properties, can be written as

TABLE 4.3 Effective Thermal Conductivity for Liquid-saturated Wick Structures

Wick Structures	k_e
Wick and liquid in series	$\dfrac{k_l k_2}{\epsilon k_w + k_l(1-\epsilon)}$
Wick and liquid in parallel	$\epsilon k_l + k_w(1-\epsilon)$
Wrapped screen	$\dfrac{k_l[(k_l + k_w) - (1-\epsilon)(k_l - k_w)]}{(k_l + k_w) + (1-\epsilon)(k_l - k_w)]}$
Packed spheres	$\dfrac{k_l[(2k_l + k_w) - 2(1-\epsilon)(k_l - k_w)]}{(2k_l + k_w) + (1-\epsilon)(k_l - k_w)}$
Rectangular grooves	$\dfrac{(w_f k_l k_w \delta) + w k_l(0.185 w_f k_w + \delta k_l)}{(w + w_f)(0.185 w_f k_w + \delta k_l)}$

Source: Chi, 1976.

$$q_{b,e} = \left(\frac{2\pi L_e k_{\text{eff}} T_v}{\lambda \rho_v \ln(r_i / r_v)}\right)\left(\frac{2\sigma}{r_n} - \Delta P_{c,m}\right) \tag{4.36}$$

where k_{eff} is the effective thermal conductivity of the liquid–wick combination (given in Table 4.3), r_i is the inner radius of the heat pipe wall, and r_n is again the nucleation site radius.

4.5.2 Recovery from Dryout of Circumferential Wicks

Equation 4.36 provides some insight into how to prevent the formation of nucleate bubbles in the evaporator wicking structure. However, it gives no indication of how a heat pipe or thermosyphon might recover from partial or complete dryout of the evaporator wicking structure. To do this, it is necessary to understand the rewetting phenomenon and the parameters that govern this phenomenon.

Under normal operating conditions, heat supplied to the evaporator is conducted around the circumference of the vapor channel and through the heat pipe walls to a circumferential wicking structure. This wicking structure typically consists of a porous material or circumferential grooves. However, during high thermal loads or adverse gravitational conditions the circumferential wall grooves may not be capable of supplying sufficient liquid, resulting in dryout. When this occurs, the wicking structure may become starved of liquid. In these situations, it is necessary to reduce the evaporator heat flux to allow the circumferential wall grooves to rewet. This rewetting process results from the high capillary pressure induced by the small characteristic radius of the wall wicking structure and is extremely complex. The fluid flow in the wall wick may be further complicated by the physical geometry of the heat pipes of interest.

Several researchers, including Elliott and Rose (1970), Simopoules et al. (1979), Stroes et al. (1990), Ueda et al. (1971, 1983), Elias and Yadigaroglu (1977), and Raj and Pate (1986), have investigated the rewetting characteristics of liquid films on heated or hot rods, tubes, or flat surfaces. Saeed and Peterson (1991) have reviewed the work of these and others and found that although these investigations have provided substantial experimental data and considerable insight into the behavior of these films on both circular tubes and flat plates, no general physical model exists that is capable of describing the governing phenomena or the behavior of the liquid in this region.

More recently, Peng and Peterson (1992), Peng et al. (1991a,b), and Peterson et al. (1992) have proposed several models for investigating the rewetting behavior of specific cases, including the wetting and rewetting of heated and unheated flat surfaces and the rewetting of porous cover layers. In all of these models, the vaporization of the liquid at the rewetting front was assumed to be an important aspect of rewetting. Consequently, the rewetting velocity was first related to the liquid flow velocity, and then exact analytical solutions were obtained to determine the rewetting velocity for a thin liquid film as a function of the bulk liquid flow velocity, the applied heat flux, and the thermophysical properties of the liquid and the surface over which the liquid was flowing. These theoretical analyses were then compared with several different sets of experimental data. These investigations have all focused on the rewetting behavior of thin liquid films on various types of surfaces and have utilized a similar approach to address the problem of rewetting of hot or heated surfaces in which the liquid is supplied primarily by capillary pumping. Because of the propensity of circumferential grooves in heat pipe applications, the analytical technique for this particular configuration described by Peterson et al. (1992) will be presented here.

Unheated Surfaces

If the entire liquid layer on the inner surface of the grooved heat pipe wall is taken as the control volume and the liquid mass is assumed to increase as the length of the liquid layer increases, the liquid layer length is fixed for a given time t and the liquid velocity is the same for any point in the control volume (by continuity). As a result, the liquid film advances at an average tangential velocity U that will vary with respect to the length of the liquid film. The physical model for this case is shown in Fig. 4.12. As illustrated, the flow is assumed to be one-dimensional curvilinear flow and the liquid film thickness is assumed to be constant along the groove, as are the thermophysical properties of both the liquid and the channel. In addition, the channel radius R is assumed to be very large compared with groove width w and the thickness of the channel wall, δ.

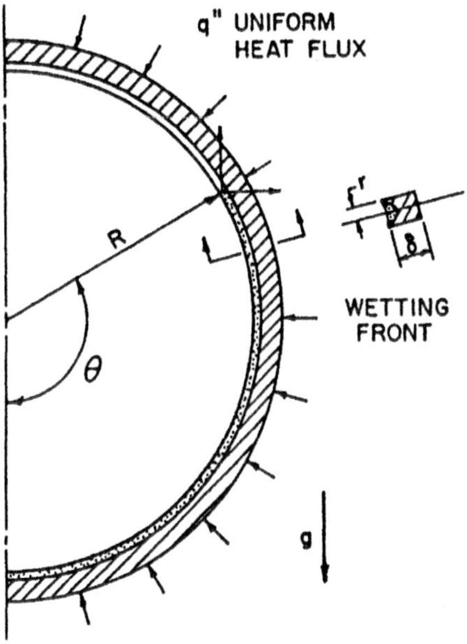

Figure 4.12 Physical model (Peng et al., 1992).

Based on the considerations above, Newton's law for the liquid film in the groove can be written as

$$F_s - F_g - F_f = \frac{d}{dt}(mU) \tag{4.37}$$

For a wetting fluid, the capillary driving force can be expressed as

$$F_s = \frac{2\sigma}{r}\frac{\omega^2}{4}\cot\frac{\gamma}{2} = \frac{\sigma\omega^2}{2r}\cot\frac{\gamma}{2} \tag{4.38}$$

The gravitational force is much more complex because the orientation of the groove with respect to the gravitational vector changes for different positions, but can be expressed as

$$F_g = \int_0^\theta \frac{\omega^2}{4}\rho_l \cot\frac{\gamma}{2}\sin\theta R d\theta = \tfrac{1}{4}\omega^2\rho_l g R\cot\frac{\gamma}{2}(1-\cos\theta) \tag{4.39}$$

For Newtonian fluids in laminar flow, the shear stress at the wall is

$$C_w = \mu_l \frac{\partial u}{\partial y}\bigg|_{\text{at wall}} \approx \frac{5\mu_l U}{\omega} \tag{4.40}$$

Using this value, an expression for the wall friction can be derived as

$$F_f = 2\omega\theta RT_w = 2\omega\theta R\frac{5\mu_l U}{w} = 10\mu_l RU \tag{4.41}$$

and because

$$m = \tfrac{1}{4}\omega^2 R\theta\rho_l \cot\frac{\gamma}{2} \tag{4.42}$$

$$d(mU) = \tfrac{1}{4}\omega^2 R\rho_l \cot\frac{\gamma}{2}\left(U\frac{d\theta}{dt} + \theta\frac{dU}{dt}\right) \tag{4.43}$$

The differential time dt in equations 4.37 and 4.43 can be replaced by

$$dt = \frac{Rd\theta}{U} \tag{4.44}$$

Substituting equations 4.38–4.44 into equation 4.37 yields

$$\frac{\sigma\omega^2}{2r}\cot\frac{\gamma}{2} - \tfrac{1}{4}\rho_l gR\cot\frac{\gamma}{2}(1-\cos\theta) - 10\mu_l\theta RU = \tfrac{1}{4}\omega^2\rho_l\cot\frac{\gamma}{2}\left(U^2 + \theta U\frac{dU}{d\theta}\right) \tag{4.45}$$

or

$$\frac{dU}{d\theta} = \frac{(\sigma\omega^2/2r)\cot(\gamma/2) - \frac{1}{4}\rho_l gR\cot(\gamma/2)(1-\cos\theta) - 10\mu_l\theta RU - \frac{1}{4}\omega^2\rho_l U^2\cot(\gamma/2)}{\frac{1}{4}\omega^2\rho_l U\theta\cot(\gamma/2)} \tag{4.46}$$

with boundary condition

$$\theta = 0 \qquad U = 0 \tag{4.47}$$

Equation 4.46 represents an expression for the rewetting velocity U as a function of the angular position of the wetting front, θ, and assumes that the surface is at ambient conditions with no heat addition. Because of the functional dependence of the velocity on position, a direct analytical solution is not possible, and hence a numerical technique must be employed.

Heated Surfaces

For the case of rewetting of a surface after dryout, the liquid front is assumed to be driven by capillary pressure and to advance along the grooves in the channel wall to which a uniform circumferential heat flux q'' has been applied, as shown in Fig. 4.12. Because the surface is at an elevated temperature (at or near the Leidenfrost temperature), some of the liquid at the leading edge of the advancing liquid front is vaporized. The remaining liquid advances with a velocity U_w, referred to as the wetting front velocity. The heat required to vaporize this liquid is supplied by conduction from the dry hot zone of the channel surface. When dryout occurs, the heat supplied to the wetting front exceeds that required to vaporize all of the liquid. For the case of no heat addition previously described, no vaporization occurs and the wetting front velocity and liquid velocity are equal. However, for a heated plate the liquid velocity is higher than the wetting front velocity, because some of the liquid is vaporized, and this reduces the liquid mass flow rate.

To determine the amount of heat absorbed by the vaporization process, the conduction equation for the channel wall can be transformed to a coordinate system moving with a velocity equal to that of the wetting front U_w. In this analysis, a curvilinear coordinate system was again employed and the conduction through the walls was assumed to be one dimensional. Several other fundamental assumptions were made and can be summarized as follows:

1. The grooves on the surface were assumed to be located immediately adjacent to one another and the size of the grooves was assumed to be much smaller than the thickness of the wall (i.e., $w \ll \delta$).
2. The liquid temperature at the rewetting front was assumed to be different from the rewetting temperature, which was assumed to be constant and equal to saturation temperature T_s.
3. The convective heat transfer between the plate and the vapor along with radiation from both the heated surface and the vapor was neglected.

Utilizing these assumptions, the conduction equation for the wall can be written as

$$\frac{k}{R^2}\frac{\partial^2 T}{\partial \theta'^2} + \frac{q}{\delta} = \rho c \frac{\partial T}{\partial t} \tag{4.48}$$

or

$$k\frac{d^2 T}{d\theta'^2} + \frac{qR^2}{\delta} = \rho c R U_w \frac{dT}{d\theta'} \tag{4.49}$$

The general solution of equation 4.48 is

$$T = C_1 + C_2 e^{[\rho c U_w R\theta'/k]} + \frac{q'' R\theta'}{\rho c U_w \delta} \tag{4.50}$$

If the circular grooves are assumed to have a finite length, the location and condition at the groove ends are difficult to determine, especially in a moving system. However, the problem of interest here is the prediction of the heat conduction at the wetting front, and it is therefore reasonable to develop a solution assuming infinite grooves.

Differentiating equation 4.50 with respect to θ' yields

$$\frac{dT}{d\theta'} = C_2\left(\frac{cU_w R}{k}\right)e^{[\rho c U_w R\theta'/k]} + \frac{q''R}{\rho c U_w \delta} \tag{4.51}$$

This expression indicates that $dT/d\theta'$ will approach infinity as $\theta' \to \infty$, which is not physically possible. Therefore, $C_2 = 0$ and at the rewetting front the plate temperature is at the Leidenfrost or rewetting temperature, that is, $\theta' = 0$, $T = T_w$. Using this condition, a solution for this expression can be found as

$$T(\theta') = T_w + \frac{q''R}{\rho c U_w \delta}\theta' \tag{4.52}$$

From this, the total heat conduction at $\theta' = 0$ can be found as

$$q = \left(\delta w - \tfrac{1}{4}\omega^2 \cot\frac{\delta}{2}\right)k\frac{dT}{Rd\theta'}\bigg|_{\theta'=0} \approx \frac{\omega q''\alpha}{U_w} \tag{4.53}$$

Utilizing an energy balance, the total heat conduction in the region of the wetting front is equal to the energy absorbed by the liquid evaporation, or

$$q = \tfrac{1}{4}(U - U_w)\rho_l \lambda w^2 \cot\frac{\gamma}{2} \tag{4.54}$$

where U is determined from equation 4.46. As noted by Peng et al. (1991a), at the rewetting front liquid sputtering may be significant and some liquid may leave the groove as droplets due to the explosive forces resulting from boiling. Therefore, in actuality, the amount of vaporized liquid predicted by equation 4.54 may be slightly higher than the amount actually vaporized at the wetting front. To compensate for this, an expression for the thermal boundary layer thickness for laminar flow presented by Kays and Crawford (1980),

$$\frac{\delta_t}{x} = 4.64\,\mathrm{Re}_x^{-1/2}\,\mathrm{Pr}_l^{-1/3} \tag{4.55}$$

was modified as

$$\frac{\delta_t}{w} = c'\,\mathrm{Re}_w^{-1/2}\,\mathrm{Pr}_l^{-1/3} \tag{4.56}$$

where δ_t represents the thickness of the liquid layer, c' is an experimentally determined constant, and the rewetting Reynolds number is defined as

$$\mathrm{Re}_w = \frac{\rho_l U w}{\mu_l} \tag{4.57}$$

Moreover, for a liquid layer of thickness, δ_t, equation 4.54 becomes

$$q = (U - U_w)\frac{w}{\sin(\gamma/2)}\delta_t \rho_l \lambda \tag{4.58}$$

and combining equations 4.54 and 4.58 yields a quadratic expression of the form

$$\frac{2(U - U_w)\delta_t \rho_l \lambda}{\sin(\gamma/2)} = \frac{q''\alpha}{U_w} \tag{4.59}$$

Substituting equation 4.55 into equation 4.54 and rearranging yield

$$U_w^2 - UU_w + \frac{q''\alpha \sin(\gamma/2)}{c'\rho_l \lambda w \,\mathrm{Re}_w^{-1/2}\,\mathrm{Pr}_l^{-1/3}} = 0 \tag{4.60}$$

and solving this expression for the rewetting velocity yields

$$U_w = \frac{1}{2}\left[U \pm \left(U^2 - \frac{4q''\alpha \sin(\gamma/2)}{c'\rho_l \lambda \omega \,\mathrm{Re}_w^{-1/2}\,\mathrm{Pr}_l^{-1/3}}\right)^{1/2}\right] \tag{4.61}$$

Previous analyses indicated that this solution is correct only when the second term on the right-hand side is positive. Hence, the rewetting velocity can be expressed as

$$U_w = \frac{1}{2}\left[U + \left(U^2 - \frac{4q''\alpha \sin(\gamma/2)}{c'\rho_l \lambda w \,\mathrm{Re}_w^{-1/2}\,\mathrm{Pr}_l^{-1/3}}\right)^{1/2}\right] \tag{4.62}$$

for rewetting to occur

$$U^2 - \frac{4q''\alpha \sin(\gamma/2)}{c'\rho_l \lambda \omega \,\mathrm{Re}_w^{-1/2}\,\mathrm{Pr}_l^{-1/3}} \geq 0 \tag{4.63}$$

must be true. Here the limiting conditions are

$$q_{\max} = \frac{c'(\rho_l \mu_l)^{1/2} w^{1/2} \lambda}{4\,\mathrm{Pr}_l^{1/3}\alpha \sin(\gamma/2)} U^{3/2} \tag{4.64}$$

This limiting condition indicates the maximum heat flux under which rewetting can occur. Because the liquid velocity corresponds to the angular position, the maximum heat flux is different for every angular location and the liquid wetting front will exhibit a maximum angular position or rise height for any specified heat flux q''

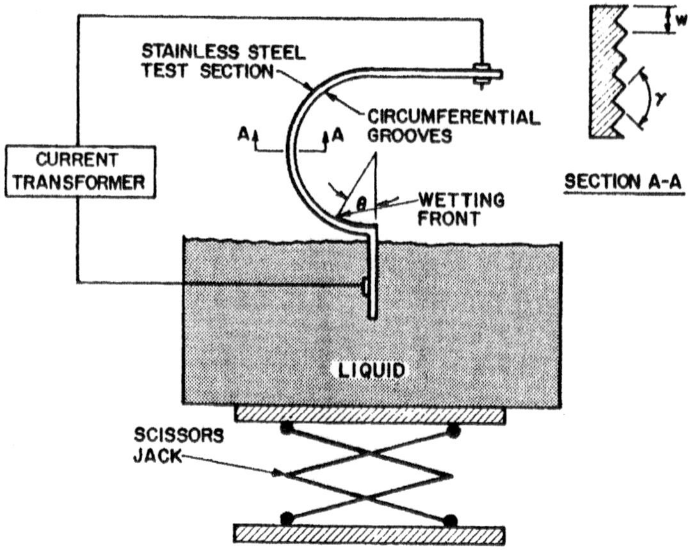

Figure 4.13 Experimental test facility (Peng et al., 1992).

To verify the analysis procedure, several experimental investigations have been conducted (Peterson et al., 1992). In this investigation, the test facility illustrated schematically in Fig. 4.13 was used to evaluate the rewetting for three circular test sections each 25.4 mm in diameter fabricated from 0.765-mm-thick stainless steel plate. A series of V-shaped grooves with an apex angle of grooves, α, of 60° was machined into the inner surface of each of the three plates. The widths of the three sets of grooves were 0.635, 0.381, and 0.177 mm, and the resulting physical characteristics associated with each are summarized in Table 4.4. Tests were conducted for three different working fluids: R11, methanol, and acetone.

The expression given in equation 4.46 with the boundary conditions shown in equation 4.47 was solved numerically for each of the three working fluids, and Figs. 4.14*a*–*c* illustrate the results for all three working

TABLE 4.4 Properties of Grooves Evaluated

Groove Number	Width (mm)	Cone Angle (degrees)	Capillary Radius (mm)	Deviation of Capillary Radius (%)
1	0.635	60	0.26	1
2	0.381	60	0.19	1.2
3	0.177	60	0.14	1.2

Source: Peterson et al., 1992.

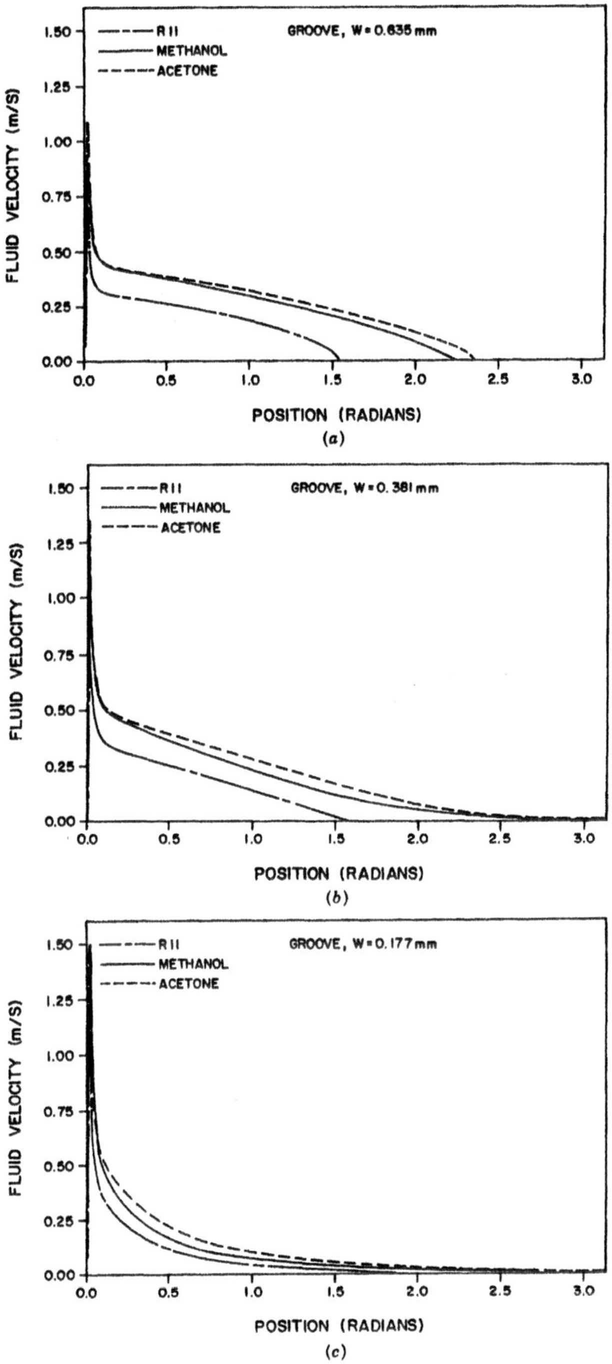

Figure 4.14 Fluid velocity as a function of radial position for an unheated plate with (*a*) 0.635-mm-wide grooves (Peterson et al., 1992); (*b*) 0.381-mm-wide grooves (Peterson et al., 1992); and (*c*) 0.177-mm-wide grooves (Peterson et al., 1992).

fluids and groove widths of 0.635, 0.381, and 0.177 mm, respectively. As illustrated, the liquid velocity as a function of the angular position is considerably different for the three fluids. In addition, comparison of these results for the three different sizes of grooves clearly indicates that the liquid velocity is strongly dependent upon the groove width.

Figures 4.15*a*–*c* illustrate the relationship that exists between the uniformly applied heat flux and the maximum radial position (or rise height) for the three different groove widths. Each figure illustrates the experimental results for the three working fluids evaluated and compares these results with the values predicted by the analytical model using a value of $c' = 0.00065$. The rewetting experimental data for the surface with 0.635-mm-wide grooves indicates that the rewetting front position for methanol is largest and that for R11 is smallest (see Fig. 4.15*a*) for a given applied heat flux. Similar results (Fig. 4.15*b*) are shown for the surface with 0.381-mm-wide grooves. However, for the surface with 0.177-mm-wide grooves, the rewetting front angles for methanol and acetone approach each other at a given heat flux, and that of R11 is only slightly smaller. The results of the numerical model given in Figs. 4.14*a*–*c* help to explain this phenomenon. When no heat is applied to the surface of the test section with 0.177-mm grooves (Fig. 4.14*c*), the liquid velocity and mass flux are quite small and indicate only a very slight difference for the three liquids. In the experimental tests, the liquid flowing along the grooves is not sufficient to cool the surface and replenish the liquid vaporized and/or sputtered at the rewetting front. As a result, the liquid rise angle is much smaller and appears to be approximately the same for all three liquids.

To better illustrate the effect of groove size, the experimental data for methanol, acetone, and R11 flowing in all three grooved surfaces are compared in Figs. 4.16*a*–*c*, respectively. The results indicate that there is not a significant difference between the data for surfaces with 0.635-mm-wide grooves and those with 0.381-mm-wide grooves. However, the liquid rise for the surface with 0.177-mm-wide grooves is significantly lower than for the other two surfaces. The experimental results for R11 (Fig. 4.16*c*) indicate that the liquid front position decreases as the groove width becomes smaller for a specified applied heat flux.

Based on the experimental data and the corresponding numerical results for liquid velocity given by equation 4.36, the coefficient c' in equation 4.64 was determined empirically and found to be 0.00065 for all cases investigated. As mentioned previously, the theoretical results predicted by equation 4.64 with $c' = 0.00065$ are also shown in Figs. 4.15 and 4.16 for the three different liquids and three different grooved surfaces. As illustrated, these predicted values are in reasonably good agreement with the measured data. Also as illustrated, the empirical constant c' appears to be independent of the liquid properties and groove size and can therefore be assumed to be a universal constant for this and other rewetting processes.

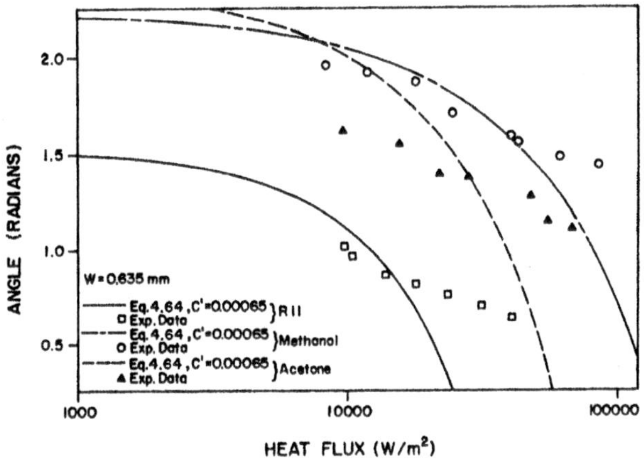

(a)

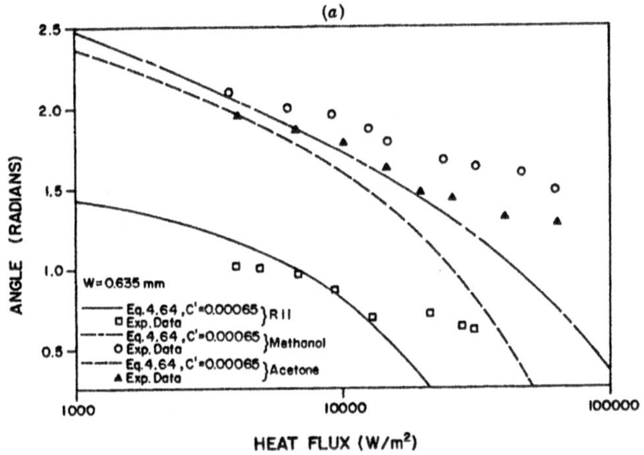

(b)

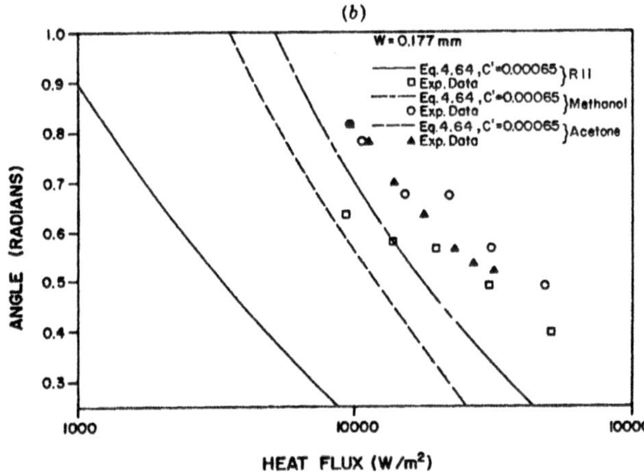

(c)

Figure 4.15 Maximum rewetting angle as a function of input heat flux for a plate with (*a*) 0.635-mm-wide grooves (Peterson et al., 1992); (*b*) 0.381-mm-wide grooves (Peterson et al., 1992); and (*c*) 0.177-mm-wide grooves (Peterson et al., 1992).

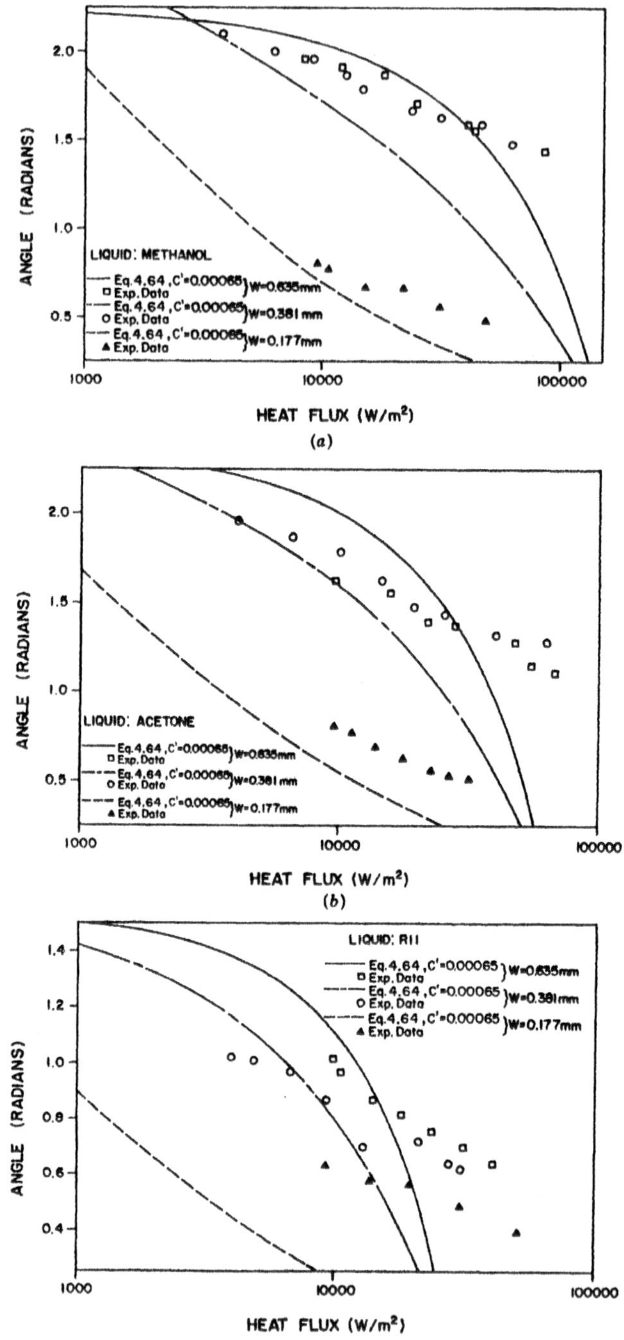

Figure 4.16 Maximum rewetting angle as a function of input heat flux for (*a*) methanol (Peterson et al., 1992); (*b*) acetone (Peterson et al., 1992); and (*c*) (Peterson et al., 1992).

4.5.3 Rewetting of Axial Grooves and Arteries

In addition to the rewetting of circumferential grooves, investigations have been conducted to determine the rewetting of axial grooves and liquid arteries. Unlike the case of circumferential grooves, dryout and depriming of axial grooves or liquid arteries are typically caused by a combination of evaporator heat flux and axial accelerations. These accelerations could result in redistribution of the working fluid and require a reduction in the evaporator heat flux to allow rewetting.

Recently Peng and Peterson (1992) conducted an analytical investigation to determine the effect of short term longitudinal accelerations on the liquid–vapor interface in two external artery heat pipes under consideration for use as radiator elements on Space Station Freedom (Alario et al., 1981; Ambrose and Holmes, 1991). Both of these utilize a liquid and vapor channel connected by a longitudinal slot to provide axial pumping (Fig. 4.17). Assuming the heat pipe is properly primed and operating normally in a zero-gravity environment, the most damaging accelerations will be those occurring along the longitudinal axis since these are most likely to deprime the liquid channel. When the magnitude and / or duration of the acceleration is sufficient, liquid will flow out of the liquid channel through the connecting slot and accumulate in the end of the liquid and vapor channels, opposite the direction of acceleration. Because this flow may occur quite rapidly, some of the liquid may remain in the slot or adhere to the surface of the liquid and vapor channels forming a thin liquid film flowing along the length of the heat pipe. Although the flow of liquid out of the liquid channel and into the vapor channel is the most significant, the flow of liquid in the slot and the flow of the thin film on the surface of the liquid and vapor channels must also be evaluated.

The inertial force resulting from the longitudinal acceleration along with the surface tension forces induced by the capillary radius existing in the liquid channel and the slot separating the two channels will determine the location and movement of the majority of the working fluid. The liquid configuration at some time τ after the onset of the acceleration is illustrated

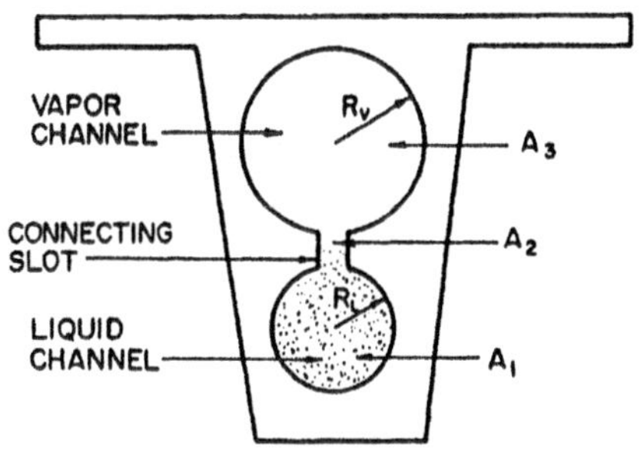

Figure 4.17 Typical external artery heat pipe configuration (Peng and Peterson, 1992).

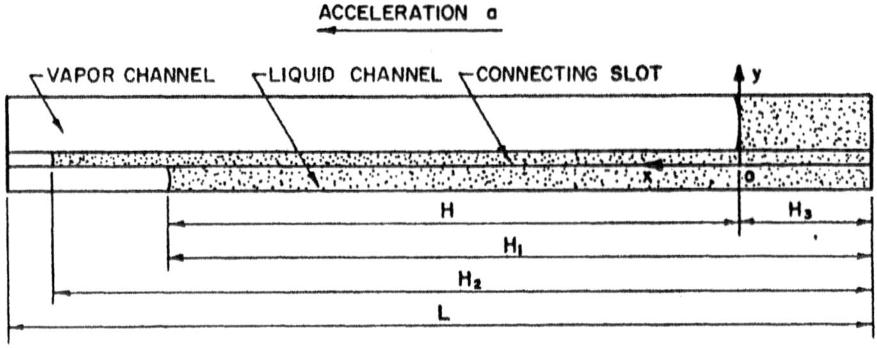

Figure 4.18 Location of liquid level in the liquid channel, vapor channel and connecting slot (Peng and Peterson, 1992).

in Fig. 4.18, where H_1 and H_3 are the length of the liquid column in the liquid and vapor channels, respectively. Although when properly charged and primed, H_3 would be zero, there may be some liquid located at the far end of the condenser region due to overcharging. In order to evaluate how the liquid configuration varies with time, a coordinate system was selected with origin at the location of the liquid level in the vapor channel. At any position x, an energy balance of the form

$$\rho_l a(H - x) - \frac{2\sigma}{R_1} - \frac{2\sigma}{R_2} = \tfrac{1}{2}\rho_l v_x^{\,2} \tag{4.65}$$

can be written, where v_x is the velocity of the liquid flowing out of the liquid channel and into the vapor channel and the velocity distribution for the flow can be written as

$$v_x = \left[2a(H - x) - \frac{4\sigma}{\rho_l}\left(\frac{1}{R_1} + \frac{1}{R_2}\right)\right]^{1/2} \tag{4.66}$$

Because of surface tension, there exists a position H' such that when $x \geq H'$, the velocity $v_x = 0$. From equation 4.66, it is apparent that this occurs when

$$H' = H - \frac{2\sigma}{\rho_l a}\left(\frac{1}{R_1} + \frac{1}{R_2}\right) \tag{4.67}$$

The volumetric flow rate at any time τ can be written as

$$\dot{V}(t) = \int_0^{H'} \delta v_x \, dx = \delta \int_0^{H'} \left[2a(H - x) - \frac{4\sigma}{\rho_l}\left(\frac{1}{R_1} + \frac{1}{R_2}\right)\right]^{1/2} dx \tag{4.68}$$

where δ is the width of the slot connecting the liquid and vapor channels. Integrating and rearranging yield

$$\dot{V}(t) = \frac{\delta}{3a}\left[2aH - \frac{4\sigma}{\rho_l}\left(\frac{1}{R_1} + \frac{1}{R_2}\right)\right]^{3/2} \tag{4.69}$$

Clearly, the change in volume of the liquid in the liquid channel can be expressed by continuity as

$$dV(t) = -A_1\, dH_1 = \dot{V}(t)\, dt \tag{4.70}$$

and also, by conservation of mass, as

$$A_1\, dH_1 = -A_3\, dH_3 \tag{4.71}$$

Defining H as the difference in the length of the liquid column in the liquid and vapor channel yields

$$H = H_1 - H_3 \quad \text{or} \quad dH = dH_1 - dH_3 \tag{4.72}$$

and combining equations 4.71 and 4.72 yields

$$dH = \left(1 + \frac{A_1}{A_3}\right) dH_1 \tag{4.73}$$

Substituting equations 4.70 and 4.73 into equations 4.69 results in an expression for H as a function of time:

$$\frac{-A_1 A_3}{A_1 + A_3}\, dH = \frac{\delta}{3a}\left[2aH - \frac{4\sigma}{\rho_l}\left(\frac{1}{R_1} + \frac{1}{R_2}\right)\right]^{3/2} dt \tag{4.74}$$

If the heat pipe is properly charged (i.e., no overcharge) $H = L$ at $t = 0$ and equation 4.72 becomes

$$H(t) = \frac{1}{2a}\left[\frac{\delta(A_1 + A_3)}{3A_1A_3}\, t + \left[2aL - \frac{4\sigma}{\rho_l}\left(\frac{1}{R_1} + \frac{1}{R_2}\right)\right]^{-1/2}\right]^{-2} + \frac{2\sigma}{\rho_l a}\left(\frac{1}{R_1} + \frac{1}{R_2}\right) \tag{4.75}$$

Differentiating yields

$$dH = -\frac{1}{a}\left\{\frac{\delta(A_1 + A_2)}{3A_1A_3}\, t + \left[2aL - \frac{4\sigma}{\rho_l}\left(\frac{1}{R_1} + \frac{1}{R_2}\right)\right]^{-1/2}\right\}^{-3} \times d\left[\frac{\delta(A_1 + A_3)}{3A_1A_3}\, t\right] \tag{4.76}$$

and substituting equation 4.76 into equation 4.74 yields an expression for the change in the length of the liquid column in the liquid channel:

$$dH_1 = \frac{-A_3}{a(A_1 + A_3)} \left\{ \frac{\delta(A_1 + A_3)}{3A_1A_3} t + \left[2aL - \frac{4\sigma}{\rho_l} \left(\frac{1}{R_1} + \frac{1}{R_2} \right) \right]^{-1/2} \right\}^{-3} d\left[\frac{\delta(A_1 + A_3)}{3A_1A_3} t \right] \tag{4.77}$$

Integrating yields an expression for the liquid column length as a function of time:

$$H_1 = L + \frac{A_3}{2a(A_1 + A_3)} \left\{ \frac{\delta(A_1 + A_3)}{3A_1A_3} t + \left[2aL - \frac{4\sigma}{\rho_l} \left(\frac{1}{R_1} + \frac{1}{R_2} \right) \right]^{-1/2} \right\}^{-2} - \frac{A_3}{2a(A_1 + A_3)} \left[2aL - \frac{4\sigma}{\rho_l} \left(\frac{1}{R_1} + \frac{1}{R_2} \right) \right] \tag{4.78}$$

Similarly, the length of the liquid column in the vapor channel can be expressed as

$$H_3 = L - \frac{A_1}{2a(A_1 + A_3)} \left\{ \frac{\delta(A_1 + A_3)}{3A_1A_3} t + \left[2aL - \frac{4\sigma}{\rho_l} \left(\frac{1}{R_1} + \frac{1}{R_2} \right) \right]^{-1/2} \right\}^{-2} - \frac{A_3}{2a(A_1 + A_3)} \left[2aL - \frac{4\sigma}{\rho_l} \left(\frac{1}{R_1} + \frac{1}{R_2} \right) \right] - \frac{2\sigma}{\rho_l a} \left(\frac{1}{R_1} + \frac{1}{R_2} \right) \tag{4.79}$$

As discussed previously, flow also occurs in the slot separating the liquid and vapor channels and in the thin film on the surface of the liquid and vapor channels. The liquid flow in the slot was assumed to behave as one-dimensional laminar flow between two parallel plates; hence the momentum equation can be written as

$$\mu_l \frac{d^2u}{dz^2} = a\rho_l \tag{4.80}$$

where $u = 0$ at $z = 0$ and $u = 0$ at $z = \delta$. Integrating yields an expression for the local velocity in the x direction in the slot, that is,

$$u = -\frac{\delta^2}{2\mu_l} (a\rho_l) \left[\frac{z}{\delta} - \left(\frac{z}{\delta} \right)^2 \right] \tag{4.81}$$

The average velocity in the slot can then be found by integrating with respect to z:

$$U_2 = \frac{1}{\delta} \int_0^\delta u\, dz = -\frac{\delta}{2\mu_l} (a\rho_l) \int_0^\delta \left[\frac{z}{\delta} - \left(\frac{z}{\delta} \right)^2 \right] dz = -\frac{\delta^2}{12\mu_l} (a\rho_l) \tag{4.82}$$

and the length of the liquid column in the slot, H_2, can be written as

$$H_2 = L + U_2 t = L - \frac{\delta^2}{12\mu_l}(a\rho_l)t \tag{4.83}$$

Because the liquid film on the surface of the liquid channel is thin compared to the liquid channel radius, the film may be assumed to behave as flow over a flat plate. The resulting force balance yields

$$\mu_l \frac{d^2u}{dy^2} = a\rho_l \tag{4.84}$$

where $u = 0$ at $y = 0$ and $du/dy = 0$ at $y = b$. Combining and integrating yield an expression for the velocity of the liquid film in the x direction of

$$u = \frac{b^2}{2\mu_l}(a\rho_l)\left[\left(\frac{y}{b}\right)^2 - 2\left(\frac{y}{b}\right)\right] \tag{4.85}$$

When the liquid film initially forms, the film region is stationary and only inertial and capillary forces act upon it. The thickness of this liquid film can be approximated by R^1. Integrating yields an expression for the average film velocity in the liquid channel of

$$U_1 = \frac{1}{2\pi R_l b}\int_0^b u 2\pi R_l\, dy = \frac{1}{b}\int_0^b \frac{b^2}{2\mu_l}(a\rho_l)\left[\left(\frac{y}{b}\right)^2 - 2\frac{y}{b}\right]dy = -\frac{b^2}{3\mu_l}(a\rho_l) \tag{4.86}$$

Substituting as before results in expressions for the length of the liquid film in the liquid and vapor channels of

$$H_1 = L + U_1 t = L - \frac{R_l^2}{3\mu_l}(a\rho_l)t \tag{4.87}$$

and

$$H_3 = L + U_3 t = L - \frac{R_3^2}{3\mu_l}(a\rho_l)t \tag{4.88}$$

respectively. Using equations 4.80, 4.81, 4.86, 4.88, and 4.89, the location of the liquid–vapor interface in the liquid channel, vapor channel, and connecting slot along with the interface for the wall films in both the liquid and vapor channels can be determined for a constant acceleration applied over some time interval.

Figure 4.19 illustrates the change of the liquid column lengths in the liquid and vapor channels, H_1 and H_3, respectively, along with the difference between the two, H, all as a function of time τ. As illustrated,

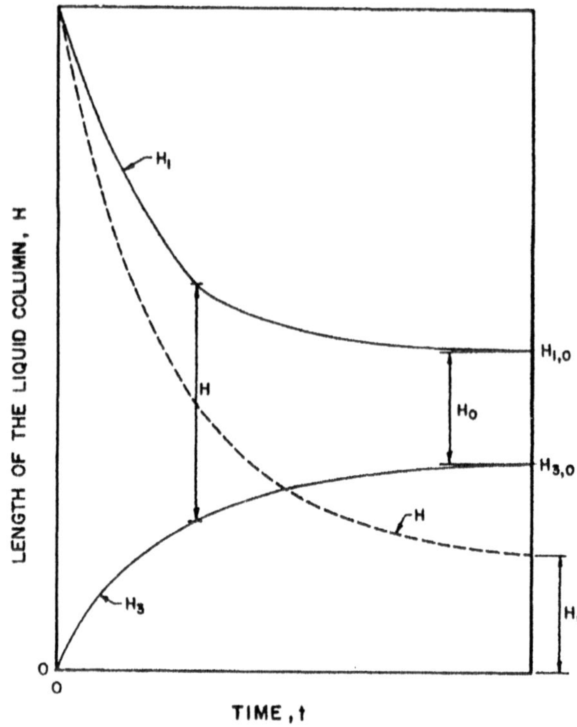

Figure 4.19 Liquid position as a function of time for the liquid channel, the vapor channel, and the connecting slot (Peng and Peterson, 1992).

when H_1 decreases, H_3 experiences a corresponding increase. This results in the accumulation of the liquid in the end of the heat pipe opposite the direction of acceleration. If a constant acceleration is maintained for a sufficient period of time, both H_1 and H_3 will approach a constant value. If equations 4.76, 4.70, and 4.81 are evaluated as the time over which the acceleration is applied approaches infinity (i.e., $t \rightarrow \infty$), the following relationships result:

$$H_{1,0} = L - \frac{A_3}{2a(A_1 + A_2)}\left[2aL - \frac{4\sigma}{\rho_l}\left(\frac{1}{R_1} + \frac{1}{R_2}\right)\right] \quad (4.89)$$

$$H_{3,0} = L - \frac{A_3}{2a(A_1 + A_2)}\left[2aL - \frac{4\sigma}{\rho_l}\left(\frac{1}{R_1} + \frac{1}{R_2}\right)\right] - \frac{2\sigma}{\rho_l a}\left(\frac{1}{R_1} + \frac{1}{R_2}\right) \quad (4.90)$$

and

$$H_0 = \frac{2\sigma}{\rho_l a}\left(\frac{1}{R_1} + \frac{1}{R_2}\right) \quad (4.91)$$

Equations 4.90 and 4.91 illustrate that the final length of the liquid column in both the vapor and liquid channels depends not only on the time and magnitude of the acceleration and properties of the working fluid but also on the geometric parameters of the heat pipe.

NOMENCLATURE

a	acceleration, coefficient to account for deviation from parabolic velocity profile (m/s^2)
A	area (m^2)
A^*	Dimensionless constant, $\simeq \frac{1}{8}\pi^2$ for $P_{e,exit}/P_{e,top} \simeq 1$; $\simeq 2.2$ for turbulent flow
b	Thickness (m)
Bo	Bond number, $= d_i(g/\sigma(\rho_l - \rho_v))^{0.5}$
c_p	Specific heat $(J/kg \cdot K)$
c'	Experimental constant
C	Constant
CHF	Critical heat flux (W/m^2)
d	Diameter (m)
f	Interfacial friction coefficient
F	Force (N)
g	Gravitational acceleration (m/s^2)
G	Mass velocity $(kg/m^2 \cdot s)$
h	Heat transfer coefficient, length of liquid film, liquid rise height $(W/m^2 \cdot K)$, (m)
H	Length of liquid column (m)
k	Thermal conductivity $(W/m \cdot K)$
Kp	Kutateladze pressure criterion
L	Length
m	Mass
$\dot{m}$	Mass flow rate
Ma	Mach number
N_L	Two-phase Grashof number, $= (gD^3\rho_l/\mu_l^2)^{5/2}$
P	Pressure (N/m^2)
Pr	Prandtl number
q	Heat transfer (W)
$\dot{Q}_i^*$	Heat supply parameter, $= \dot{Q}/8\mu_l Rh$
q''	heat flux (W/m^2)
r	Capillary radius, radial coordinate, radius of curvature (m), (m^{-1})
R	Gas constant
Re	Reynolds number
t	Time (s)
T	Temperature (K)
u	x Component of velocity (m/s)
v	Velocity (m/s)

U Rewetting velocity (m / s)
V Volume (m^3)
$\dot{V}$ Volume flow rate (m^3 / s)
We Weber number
x Length coordinate
$(x_{10})^*$ Steady-state nondimensional liquid film length, $= u'/R$
$(x_{20})^*$ Steady-state nondimensional film cross-sectional area, $= \delta u'/R^2$
Y δ/d_i
z Axial position coordinate (m)

Greek Symbols

α Thermal diffusivity (m^2 / s)
β Angle of inclination, contact angle (deg)
γ Cone angle of groove, ratio of specific heat (deg)
ΔT Wall superheat at rewet point, average film thickness = λ (K), (m)
δ Thickness, tube diameter factor, equations 4.59 and 4.60 (m), groove depth (m)
θ Angle (deg)
θ' Angle in moving coordinate frame (deg)
ι Characteristic length of wick (m)
λ Latent heat of vaporization (J / kg)
μ Absolute viscosity ($N \cdot s / m^2$)
ρ Density (kg / m^3)
σ Surface tension (N / m)
ω Groove width (m)

Subscripts

a Adiabatic
b Boiling
c Condenser, capillary
cr Critical
d Droplet, dry, dryout
e Evaporator
eff effective
f Frictional, fin
fc Fill change
g Gravitational
h Hydraulic
i Inner, interface
L Leidenfrost
l Liquid
m Maximum

n	Nucleation
o	Outer
s	Surface
t	Total
v	Vapor
w	Wick, wetting, wall
0	Location of thermocouple
1	Liquid flow channel
2	Slot
3	Vapor flow channel

REFERENCES

Alario, J., Haslett, R., and Kossan, R., 1981, "The Monogroove High Performance Heat Pipe," Paper No. AIAA-81-1156, 16th AIAA Thermophysics Conference, Palo Alto, CA, American Institute of Aeronautics and Astronautics, June.

Ambrose, J., and Holmes, H. R., 1991, "Development of the Graded Groove High Performance Heat Pipe," Paper No. AIAA-91-0366, January, American Institute of Aeronautics and Astronautics.

Andreussi, P., 1980, "The Onset of Droplet Entrainment in Annular Downward Flows," *Canadian Journal of Chemical Engineering*, Vol. 58, pp. 267–270.

Andros, F. E., and Florschuetz, L. W., 1978, "The Two-Phase Closed Thermosyphon: An Experimental Study with Flow Visualization," *Two Phase Transfer and Reactor Safety*, Hemisphere, Washington, D.C.

Anonymous, 1979, "Heat Pipes—Performance of Capillary Driven Designs," data item No. 79012, Engineering Sciences Data Unit, London, England.

Bankston, C. A., and Smith, J. H., 1971, "Incompressible Laminar Vapor Flow in Cyclindrical Heat Pipes," Paper No. 71-WA/HT-15, American Society of Mechanical Engineering, New York.

Bezrodnyi, M. K., 1978, "The Upper Limit of Maximum Heat Transfer Capacity of Evaporative Thermosyphons," *Teploenergetika*, Vol. 25, No. 8, pp. 63–66.

Bezrodnyi, M. K., 1985, "Flooding of Liquid–Vapor Countercurrent Flow in Closed Thermosyphons," *Heat Transfer—Soviet Research*, Vol. 17, No. 2, pp. 71–76.

Bezrodnyi, M. K., and Volkov, S. S., 1981, "Study of Hydrodynamic Characteristics of Two-Phase Flow in Closed Thermosyphons," *Advances in Heat Pipe Technology*, Vol. 1, pp. 115–123.

Brennan, P. J., and Kroliczek, E. J., 1979, *Heat Pipe Design Handbook*, Contract Report No. NAS5-23406, National Aeronautics and Space Administration, Washington, D.C.

Busse, C. A., 1973, "Theory of the Ultimate Heat Transfer of Cylindrical Heat Pipes," *International Journal of Heat Mass Transfer*, Vol. 16, pp. 169.

Busse, C. A., and Kemme, J. E., 1980, "Dry-out Phenomena in Gravity-Assist Heat Pipes with Capillary Flow," *International Journal of Heat Mass Transfer*, Vol. 23, pp. 643–654.

Chi, S. W., 1976, *Heat Pipe Theory and Practice*, McGraw-Hill, New York.

Cohen, H., and Bayley, F. J., 1955, "Heat Transfer Problems of Liquid Cooled Gas Turbine Blades," *Proceedings of the Institute of Mechanical Engineers*, Vol. 169, pp. 1063–1074.

Cotter, T. P., 1967, "Heat Pipe Startup Dynamics," Proc. SAE Thermionic Conversion Specialist Conference, Palo Alto, CA.

Deverall, J. E., Kemme, J. E., and Florschuetz, L. W., 1970, "Sonic Limitations and Startup Problems of Heat Pipes," Report No. LA-4578, Los Alamos Scientific Laboratory, Los Alamos, NM.

Dobran, F., 1985, "Steady-State Characteristics and Stability Thresholds of a Closed Two-Phase Thermosyphon," *International Journal of Heat Mass Transfer*, Vol. 28, No. 5, pp. 949–957.

Dunn, P. D., and Reay, D. A., 1982, *Heat Pipes*, 3rd ed., Pergamon, New York.

Elias, E., and Yadigaroglu, G., 1977, "A General One-Dimensional Model for Conduction-controlled Rewetting of a Surface," *Nuclear Engineering and Design*, Vol. 42, pp. 185–186.

Elliott, D. F., and Rose, P. W., 1970, "The Quench of a Heated Surface by a Film of Water in a Steam Environment at Pressure up to 53 Bar," Report No. AEIW-M976, Atomic Energy Establishment, Winfrith, England.

Faghri, A., Chen, M. M., and Morgan, M., 1988, "Heat Transfer Characteristics in Two-Phase Closed Conventional and Concentric Annular Thermosyphons," Proc. ASME Nat'l Heat Transfer Conference, pp. 291–303.

Fukano, T., Chen, S. J., and Tien, C. L., 1983, "Operating Limits of the Closed Two-Phase Thermosyphon," ASME / JSME Thermal Engineering Joint Conference, pp. 95–101.

Grover, G. M., Cotter, T. P., and Erikson, G. F., 1964, "Structures of Very High Thermal Conductivity," *Journal of Applied Physics*, Vol. 218, pp. 1190–1191.

Grover, G.M., Kemme, J. E., and Keddy, E. S., 1968, "Advances in Heat Pipe Technology," 2nd Int. Symposium on Thermionic Electrical Power Generation Proc., Stresa, Italy.

Hewitt, G. F., 1979, "Liquid Mass Transport in Annular Two-Phase Flow," *Two-Phase Momentum, Heat and Mass Transfer in Chemical, Process and Energy Engineering Systems*, Vol. 1, Hemisphere, New York.

Imura, H., Sasaguchi, K., and Kozai, H., 1983, "Critical Heat Flux in a Closed Two-Phase Thermosyphon," *International Journal of Heat Mass Transfer*, Vol. 26, No. 8, pp. 1181–1188.

Ivanovskii, M. N., Sorokin, V. P., and Yagodkin, I. V., 1982, *The Physical Principles of Heat Pipes*, Clarendon, Oxford.

Kataoka, I., Ishii, M., and Mishima, K., 1983, "Generation and Size Distribution of Droplets in Annular Two-Phase Flow," *Journal of Fluids Engineering*, Vol. 105, pp. 230–238.

Katto, Y., 1978, "Generalized Correlation for Critical Heat Flux at the Natural Convection Boiling in Confined Channels," *Transactions of the JSME*, Vol. 44, pp. 3908–3911.

Kays, W. M., and Crawford, M. E., 1980, *Convective Heat and Mass Transfer*, McGraw-Hill, New York.

Kemme, J. E., 1969, "Ultimate Heat Pipe Performance," *IEEE Transactions on Electron Devices*, Vol. ed-16, pp. 717-723.

Kemme, J.E., 1976, "Vapor Flow Consideration in Conventional and Gravity-Assist Heat Pipes," Proc. 2nd Int. Heat Pipe Conference, Palo Alto, CA, pp. 11–21.

Kutateladze, S. S., 1972, "Elements of Hydrodynamics of Gas–Liquid Systems," *Fluid Mechanics—Soviet Research*, Vol. 1, pp. 29–50.

Kutateladze, S.S., and Sorokin, Y. L., 1969, "The Hydrodynamic Stability of Vapor–Liquid Systems," *Problems of Heat Transfer and Hydraulics of Two-Phase Media*, Pergamon, New York.

Langner, H., and Mayinger, F., 1979, "Entrainment in Annular Two-Phase Flow under Steady and Transient Flow Conditions," *Two-Phase Momentum, Heat and Mass Transfer in Chemical, Process, and Energy Engineering Systems*, Vol. 2, Hemisphere, New York, NY.

Levy, E. K., 1968, "Theoretical Investigation of Heat Pipes Operating at Low Vapor Pressure," *J. Engineering for Industry*, Vol. 90, pp. 547–552.

Lopes, J. C. B., and Dukler, A. E. 1986, "Droplet Entrainment in Vertical Annular Flow and Its Contribution to Momentum Transfer," *AIChE Journal*, Vol. 32, No. 9, pp. 1500–1515.

Marcus, B. D., 1972, "Theory and Design of Variable Conductance Heat Pipes," Report No. NASA CR, 2018, National Aeronautics and Space Administration. Washington, D.C. April.

Matveev, V. M., Filippov, Y. N., Dyuzhev, V. I., and Okhapkin, E. V., 1977, *Journal of Engineering Physics*, Vol. 33, No. 3, pp. 1008–1012.

Nejat, Z., 1981, "Effect of Density Ratio on Critical Heat Flux in Closed End Vertical Tubes," *International Journal of Multiphase Flow*, Vol. 7, pp. 321–327.

Nguyen-Chi, H., and Groll, M., 1981, "Entrainment or Flooding Limit in a Closed Two-Phase Thermosyphon," Proc. 4th Int. Heat Pipe Conference, Tsukuba, Japan, pp. 147–162.

Peng, X. F., and Peterson, G. P., 1992, "Acceleration Induced Depriming of External Artery Heat Pipes," *AIAA Journal of Thermophysics and Heat Transfer*, Vol. 6, No. 3, July, pp. 546–549.

Peng, X. F., Peterson, G. P., and Wang, B. X., 1991a, "Capillary Induced Rewetting in a Flat Porous Cover Layer," *International Journal Heat Mass Transfer*, Vol. 34, No. 10.

Peng, X. F., Peterson, G. P., and Wang, B. X., 1991b, "The Effect of Plate Temperature on the Onset of Wetting," *International Journal Heat Mass Transfer*, Vol. 35, No. 11, pp. 3085–3093.

Peterson, G. P., and Bage, B., 1991, "Entrainment Limitations in Thermosyphons and Heat Pipes," *ASME Journal of Energy Resources Technology*, Vol. 113, No. 3, September, pp. 147–154.

Peterson, G. P., Peng, X. F., Lu, X. J., and Wang, B.X., 1992, "Analytical and Experimental Investigation of the Rewetting of Circular Channels with Internal V-grooves," *International Journal of Heat Mass Transfer*, Vol. 35, No. 11, November, pp. 3085–3094.

Prenger, F. C., 1984, "Performance Limits of Gravity-Assist Heat Pipes," Proc. 5th Int. Heat Pipe Conference, pp. 1–5.

Prenger, F. C., and Kemme, J. E., 1981, "Performance Limits of Gravity-Assist Heat Pipes with Simple Wick Structures," Proc. 4th Int. Heat Pipe Conference, pp. 137–146.

Raj, V. V., and Pate, A. W., 1986, "Analysis of Conduction Controlled Rewetting of Hot Surfaces Based on Two-Region Model," Paper No. IP-20, Proc. 8th Int. Heat Transfer Conf., Vol. 4, 1987–1992.

Rice, G., and Fulford, D., 1987, "Influence of a Fine Mesh Screen on Entrainment in Heat Pipes," Proc. 6th Int. Heat Pipe Conference, pp. 168–172.

Rohani, A. R., and Tien, C. L., 1974, "Analysis of the Effects of Vapor Pressure Drop on Heat Pipe Performance," *International Journal of Heat Mass Transfer*, Vol. 17, pp. 61–67.

Saeed, B., and Peterson, G. P., 1991, "A Review of Rewetting of Hot Surfaces," 32*nd Heat Transfer and Fluid Mechanics Inst.*, Vol. 32, June, pp. 203–237.

Sakhuja, R. S., 1973, "Flooding Constraint in Wickless Heat Pipes," Paper No. 73-WA / HT-7, American Society of Mechanical Engineers, New York.

Shiraishi, M., Yoneya, M., and Yabe, A. 1984, "Visual Study of Operating Limit in the Two-phase Closed Thermosyphon," Proc. 5th Int. Heat Pipe Conference, Grenoble, France, pp. 10–17.

Simopoules, S. E., El-Shiribini, A. A., and Murgatroyd, W., 1979, "Experimental Investigation of the Rewetting Process in a Freon-113 Vapor Environment," *Nuclear Engineering and Design*, Vol. 55, pp. 17–24.

Stroes, G., Fricker, D., Issacci, F., and Catton, I., 1990, "Heat Flux Induced Dryout and Rewet in Thin Films," Proc. 9th Int. Heat Transfer Conference, Vol. 6, pp. 358–364.

Tien, C. L., and Chung, K. S., 1978, "Entrainment Limits in Heat Pipes," *AIAA Journal*, Vol. 17, No. 6, p. 643.

Ueda, T., Inoue, M., Iwata, Y., and Sogawa, Y., 1971, "Rewetting of a Hot Surface by a Falling Liquid Film," *International Journal of Heat Mass Transfer*, Vol. 14, No. 3, pp. 601–607.

Ueda, T., Tsunenori, S., and Koyanagi, M., 1983, "An Investigation of Critical Heat Flux and Surface Rewet in Flow Boiling Systems," *International Journal of Heat Mass Transfer*, Vol. 26, No. 8, pp. 1189–1195.

Vasiliev, L. L., Konev, S. V., and Khrolenok, V. V., 1982, "Study of Entrainment of Drops of Coolant by Wet Vapor in a Heat Pipe Vaporizer," *Journal of Engineering Physics*, Vol. 42, No. 1, pp. 7–14.

Vinz, P., and Busse, C. A., 1973, "Axial Heat Transfer Limits of Cylindrical Sodium Heat Pipes between 25 W-cm^{-2} and 15.5 kW cm^{-2}," Paper No. 2-1, Proc. 1st Int. Heat Pipe Conf., Stuttgart, Germany.

Wallis, G. B., 1969, *One-dimensional Two-Phase Flow*, McGraw-Hill, New York.

Weber-Carstanjen, C. T., 1987, "Flooding in a Gravity-assisted Visualization Heat Pipe," Proc. 6th Int. Heat Pipe Conference, Minsk, Russia, pp. 112–116.

Chapter

5

MODELING HEAT PIPE PERFORMANCE

5.1 INTRODUCTION

The modeling of heat pipes is interdisciplinary in nature, with even the simplest heat pipe configurations presenting two-dimensional axisymmetric geometries, involving two-phase constitutive relationships. The vapor flow may be compressible at high Mach numbers, and the model must compensate for the effects of vaporization and condensation at the liquid–vapor and solid–vapor interfaces, respectively (Hendrix, 1989). High vapor flows may result in some interfacial instability (Cassel, 1991) and flow through the wicking structures requires knowledge of liquid flow in porous media. Finally, other problems that may be unanticipated, such as chemical reactions between the working fluid, wick, and container material (Ingram et al. 1984), may result in noncondensible gas generation, fluid property degradation, freezing (Wells et al., 1985) or corrosion (Modlin and Colwell, 1992). All of these factors serve to complicate even the most simple steady-state models.

5.2 STEADY-STATE MODELING

The methods and procedures for calculating the various operational limits of heat pipes and thermosyphons have been presented in Chapters 3 and 4. These procedures can be utilized to determine the performance limits and operational characteristics for a specific heat pipe as a function of the mean adiabatic or operating temperature.

Typically this procedure begins with estimation of the capillary limit. Because the equations used to evaluate both the Reynolds number and the

Mach number are functions of the heat transport capacity, the solution procedure most commonly used assumes laminar, incompressible vapor flow. Using these assumptions, the maximum heat transport capacity $q_{c,m}$ can be determined by substituting the values of the individual pressure drops into equation 3.1 and solving for the axial heat transfer. Once this value has been obtained, the axial heat transfer can then be substituted into the expressions for the vapor Reynolds number and the Mach number to determine the accuracy of the original assumptions. Using this iterative approach, which is covered in more detail in Chi (1976), accurate values for the capillary limitation as a function of the operating temperature can be determined in units of watt-meters or watts for qL and q, respectively. This calculation can then be followed by determination of the other transport limitations, that is, the viscous, sonic, entrainment, and boiling limitations.

Using the analytical techniques presented in Chapters 3 and 4, the heat pipe designer can calculate or compute the value of each of the operational limits as a function of temperature. This procedure will yield a heat pipe performance map similar to that shown in Fig. 5.1. This map illustrates the individual performance limits defined and described in Chapters 3 and 4. As shown, these five separate performance limits define an operational range represented by the region bounded by the combination of the individual limits. In effect, this operational range defines the region or combination of temperatures and axial transport capacities at which the heat pipe will function effectively. If at a given transport power the

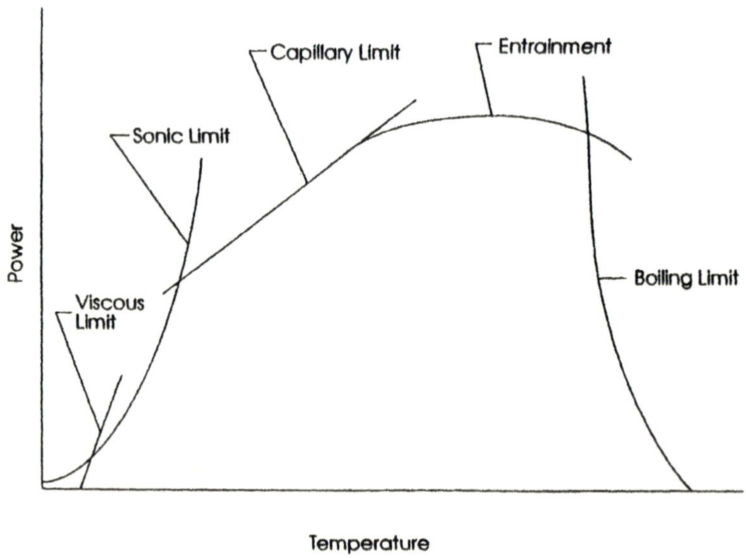

Figure 5.1 Heat pipe performance map.

adiabatic vapor temperature is such that the power lies outside of this operational range, failure due to one of the limiting conditions will result.

The following example illustrates the techniques outlined in Chapter 3 and demonstrates how this information can be used to determine the capillary limit and develop a preliminary design for a given application.

Example 5.1 A simple horizontal copper–water heat pipe is to be constructed from a 1.5-cm-internal-diameter, 0.75-m-long tube to cool an enclosed electrical cabinet as shown below. The evaporator and condenser lengths of the heat pipe are 0.25 m each and the wicking structure consists of two layers of 100-mesh copper screen. The maximum heat transport capacity of the heat pipe is estimated to be 20 W and to occur at an adiabatic vapor temperature of 30°C. If the working fluid is water and is assumed to fully wet the wicking structure, will this heat pipe be adequate? What happens to the maximum transport capacity if the wetting angle is increased to 45° due to poor cleaning?

SOLUTION First it is necessary to summarize the physical parameters and known information for this application:

Wick Geometry
$N = 100\ \text{in.}^{-1} = 3937\ \text{m}^{-1}$ (mesh number)
$d_\omega = 0.0045\ \text{in.} = 1.143 \times 10^{-4}$ m (wire diameter)
Spacing $= d_\omega = 1.143 \times 10^{-4}$ m (assume)

Fluid Properties at 30°C
$\lambda = 2425 \times 10^3\ \text{J/kg}$
$\rho_\ell = 995.3\ \text{kg/m}^3$
$\rho_v = 0.035\ \text{kg/m}^3$
$\mu_\ell = 769 \times 10^{-6}\ \text{N} \cdot \text{s/m}^2$
$\mu_v = 9.29 \times 10^{-6}\ \text{N} \cdot \text{s/m}^2$
$\sigma = 70.9 \times 10^{-3}\ \text{N/m}$

Next, calculate the vapor diameter:

$$\begin{aligned} d_v &= d - 2\ (\text{two layers} + \text{two spaces}) \\ &= 0.015 - 2[4(1.143 \times 10^{-4})] \\ &= 0.0141\ \text{m} \end{aligned}$$

To evaluate the maximum heat transport capacity, the capillary limit must be evaluated. This is represented by:

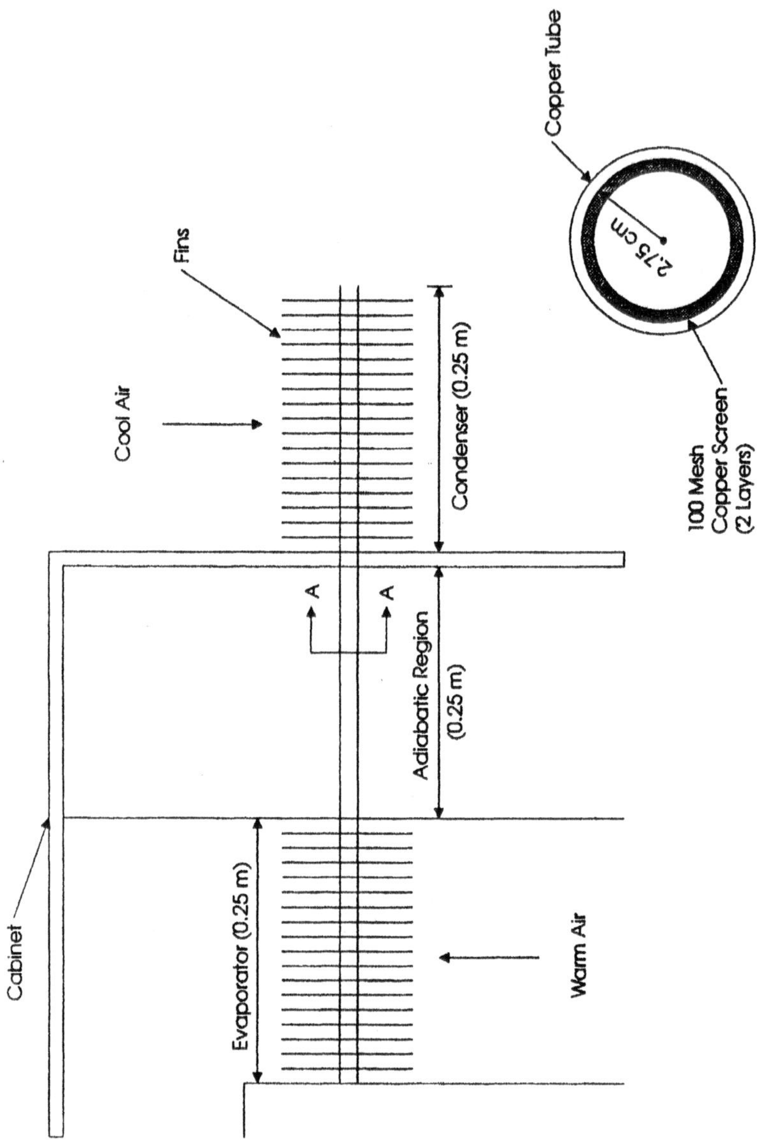
Cabinet
Evaporator (0.25 m)
Warm Air
Adiabatic Region
(0.25 m)
A
A
Condenser (0.25 m)
Cool Air
Fins
Copper Tube
2.75 cm
100 Mesh
Copper Screen
(2 Layers)

$$(\Delta P_c)_m \geq \int_{L_{\text{eff}}} \frac{dP_v}{dx}\,dx + \int_{L_{\text{eff}}} \frac{dP_l}{dx}\,dx + \Delta P_{\text{ph},e} + \Delta P_{\text{ph},c} + \Delta P_+ + \Delta P_\|$$

Assuming one-dimensional flow and that the wet point is at the end of the condenser yields

$$\int_{L_{\text{eff}}} \frac{dP_v}{dx}\,dx = \Delta P_v = \left(\frac{C(f_v\,\text{Re}_v)\mu_v}{2(r_{h,v})^2 A_v \rho_v \lambda}\right) L_{\text{eff}} q$$

$$\int_{L_{\text{eff}}} \frac{dP_l}{dx}\,dx = \Delta P_l = \left(\frac{\mu_l}{KA_\omega \lambda \rho_l}\right) L_{\text{eff}} q$$

$$\Delta P_+ = \rho_l g d_v \cos\psi = \rho_l g d_v \quad \text{since } \psi = 0°$$

and

$$(\Delta P_L)_m = \frac{2\sigma\cos\theta}{r_c} \quad \text{where } r_c = \frac{1}{2N}$$

Thus, equation 3.33 takes the form

$$\frac{2\sigma\cos\theta}{r_c} = \left(\frac{C(f_v\,\text{Re}_v)\mu_v}{2(r_{h,v})^2 A_v \rho_v \lambda}\right) L_{\text{eff}} q + \left(\frac{\mu_l}{KA_w \lambda \rho_l}\right) L_{\text{eff}} q + \rho_l g d_v$$

Next it is necessary to find the capillary radius r_c (from Table 3.1),

$$r_c = \frac{1}{2N} = \frac{1}{2(3937)} = \underline{1.27 \times 10^{-4}\ \text{m}}$$

the vapor space area A_v,

$$A_v = \tfrac{1}{4}\pi(d_v)^2 = \tfrac{1}{4}\pi(0.0141\ \text{m})^2 = \underline{1.56 \times 10^{-4}\text{m}^2}$$

the liquid flow area A_ω,

$$A_\omega = \tfrac{1}{4}\pi(d^2 - d_v^2) = \tfrac{1}{4}\pi(0.015^2 - 0.0141^2) = \underline{1.057 \times 10^{-5}\ \text{m}^2}$$

and the wick permeability K (from Table 3.2),

$$K = \frac{d_\omega^2 \epsilon^3}{122(1-\epsilon)^2} \quad \text{where } \epsilon = 1 - \frac{1.05\pi N d_\omega}{4}$$

$$\epsilon = 1 - \frac{1.05\pi N d_\omega}{4} = 1 - \frac{1.05\pi(3937)(1.143 \times 10^{-4})}{4} = 0.629$$

$$K = \frac{d_w^2 \epsilon^2}{122(1-\epsilon)^2} = \frac{(1.143 \times 10^{-4})^2 (0.629)^3}{122(1-0.629)^2} = \underline{1.94 \times 10^{-10}\ \mathrm{m}^2}$$

Because, at this point, it is not known if the vapor flow is laminar or turbulent, or compressible or incompressible, it is necessary to, as a first approximation, assume laminar, incompressible flow, that is, $f_v \mathrm{Re}_v = 16$ and $C = 1.0$. Substituting these and the other values from above into equation 3.33 yields

$$\frac{2(70.9 \times 10^{-3})\cos \theta^\circ}{1.27 \times 10^{-4}} = \frac{1.0(16)(9.29 \times 10^{-6})(0.50)q}{2(0.00705)^2 (1.56 \times 10^{-4})(0.035)(2425 \times 10^3)}$$
$$+ \frac{(769 \times 10^{-6})(0.50)q}{(1.94 \times 10^{-10})(2.057 \times 10^{-5})(2425 \times 10^3)(995.3)}$$
$$+ 995.3(9.81)(0.0141)$$

or

$$1116.5 = 0.0565q + 25.1q + 137.7$$

Solving for q yields

$$q = \frac{1116.5 - 137.7}{0.0565 + 39.9}$$

or

$$\underline{q_m = 24.5\ \mathrm{W}}$$

This value represents the maximum axial heat transfer the heat pipe can transport prior to reaching the capillary limit. Next, the assumption of laminar, incompressible flow must be verified. This can be done by evaluating the Reynolds number:

$$\mathrm{Re} = \frac{4\dot{m}}{\pi d_v \mu} = \frac{4q}{\pi d_v \mu \lambda} = \frac{4(24.5)}{\pi(0.0141)(9.29 \times 10^{-6})(2425 \times 10^3)}$$
$$\Rightarrow \mathrm{Re} = 97.9 \rightarrow \text{laminar flow assumption is valid}$$

For uniform mass addition (vaporization) and uniform mass removal (condensation)

$$L_{eff} = 0.5L_e + L_a + 0.5L_c$$

Finally, the Mach number must be calculated to verify the assumption of incompressible flow:

$$\text{Ma} = \frac{v_m}{c} = \frac{\dot{m}/A_v}{\sqrt{\gamma R T_v}} = \frac{4q/(\lambda \pi d_v^2)}{\sqrt{\gamma R T_v}}$$

$$= \frac{4(38.91)/((2425\times10^3)\pi(0.0141)^2)}{[(1.33)(461.89)(30+273)]^{1/2}}$$

$$= \frac{0.1028}{431.4} = 2.38\times10^{-4} \ll 0.3 \Rightarrow \underline{\text{incompressible}}$$

Because the original assumptions are valid, the maximum heat transport capacity is approximately 24.5 W.

If $\theta \to 45°$ due to poor cleaning, then the maximum capillary pressure becomes

$$(\Delta P_c)_m = \frac{2\sigma\cos\theta}{r_c} = \frac{2(70.9\times10^{-3})\cos 45°}{1.27\times10^{-4}} = 789.5\ \text{Pa}$$

and $q_m = 15.98$ W or 66% of q_m with $\theta = 0°$.

The preceding example illustrates the procedure for finding and estimating the maximum transport capacity as determined by the capillary limit. In the example, the heat pipe is assumed to be horizontal, which may not be the case. The next example illustrates how the effects of tilt angle will affect the capillary limit.

Example 5.2 For the heat pipe in Example 5.1, determine the effects of tilt angle (evaporator elevated above the condenser) on the heat pipe performance. What is the maximum tilt angle at which the heat pipe will still operate?

SOLUTION If $\Delta P_{\|}$ is no longer equal to zero, then the capillary limit is represented by

$$(\Delta P_{c,m}) = \int_{L_{\text{eff}}} \frac{dP_v}{dx}\,dx + \int_{L_{\text{eff}}} \frac{dP_l}{dx}\,dx + \Delta P_+ + \Delta P_{\|}$$

at 30°C these terms are equal to

$$1116.5 = 0.0565q + 39.9q + 137.7 + \rho_l g L \sin\psi$$

or

$$978.8 = 39.96q + 995.3(9.81)(0.75)\sin\psi$$

$$\Rightarrow q_m = \frac{978.8 - 7323\sin\psi}{39.96}$$

ψ (degrees)	$h = L \sin \psi$ (cm)	q_m (W)
0	0	24.5
1	1.31	21.28
2	2.62	18.18
3	3.92	14.89
4	5.23	11.70
5	6.54	8.52
6	7.83	5.33
7	9.14	2.16
8	10.43	—

These data can be illustrated graphically, as shown below.

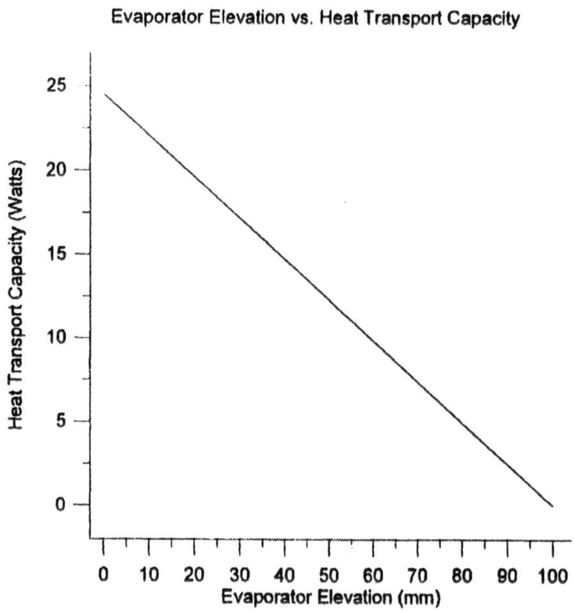

The maximum angle at which the heat pipe will still operate occurs when $q_m = 0$ or $\psi = 7.68°$ ($h = 10.02$ cm). However, to still transfer the required thermal load of 20 W, the tilt angle must be less than approximately 1.4°. Note that ΔP_+ will not change for small angles of ψ since $\cos \theta \simeq 1.0$ (e.g., $\cos 7° = 0.993$).

In addition to determining the effect of tilt angle on the transport capacity, variations in the mean operating temperature may also have a significant impact on the transport capacity. While in practice it is difficult to estimate what this value will be, the capillary transport limit can be estimated for a reasonable temperature range, allowing the designer to determine if the design is appropriate.

Example 5.3 For the heat pipe described in Example 5.1, determine the effect on the capillary limit of varying the adiabatic vapor temperature over a range of 10–120°C.

SOLUTION Variations in the adiabatic vapor temperature will cause corresponding changes in the properties of the working fluid. This in turn will affect the performance and the capillary limit. The basic equation is

$$\frac{2\sigma\cos\theta}{r_c} = \left(\frac{C(f_v\,\mathrm{Re}_v)\mu_v}{2(r_{h,v})^2 A_v \rho_v \lambda}\right)L_{\mathrm{eff}}q + \left(\frac{\mu_l}{KA_w\lambda\rho_l}\right)L_{\mathrm{eff}}q + \rho_l g d_v$$

which in terms of the fluid properties reduces to

$$\frac{2\sigma}{1.27\times10^{-4}} = \frac{1(16)\mu_v(0.50)q}{2(0.0075)^2(1.56\times10^{-4})\rho_v\lambda} + \frac{\mu_l(0.50)}{(1.94\times10^{-10})(2.057\times10^{-5})\lambda\rho_l} + \rho_l(9.81)(0.0141)$$

or

$$1.575\times10^4\sigma = \left(4.56\times10^8\frac{\mu_v}{\rho_v\lambda} + 1.25\times10^{14}\frac{\mu_l}{\lambda\rho_l}\right)q + \rho_l(0.138)$$

Simplifying yields

$$q = \frac{1.575\times10^4\sigma - 0.138\rho_l}{(4.56\times10^8)(\mu_v/\rho_v\lambda) + (1.25\times10^{14})(\mu_l/\lambda\rho_l)}$$

The results as a function of temperature can be calculated and tabulated as follows:

T	λ (kJ/kg)	σ (N/m)	ρ_v (kg/m^3)	ρ_l (kg/m^3)	μ_v (N·s/m^3)	μ_l (N·s/m^2)	ΔP_c (Pa)	$\Delta P_l/q$ (Pa/W)	$\Delta P_v/q$ (Pa/W)	ΔP_+ (Pa)	q (W)
10	2478.0	0.075	0.008	1000.0	82.9×10^{-7}	14.2×10^{-4}	1181.3	71.6	0.191	138.0	14.5
20	2453.8	0.073	0.0173	999.0	88.5×10^{-7}	10.0×10^{-4}	1149.8	51.0	0.095	137.86	19.8
40	2406.5	0.069	0.051	993.1	96.6×10^{-7}	6.51×10^{-4}	1086.8	34.1	0.036	137.05	27.8
60	2358.4	0.066	0.130	983.3	105.0×10^{-7}	4.63×10^{-4}	1039.5	25.0	0.016	135.7	36.0
80	2308.9	0.063	0.293	971.8	113.0×10^{-7}	3.51×10^{-4}	992.3	19.55	0.008	134.1	43.9
100	2251.2	0.059	0.597	958.8	121.0×10^{-7}	2.79×10^{-4}	929.3	16.2	0.004	132.3	49.2
120	2202.2	0.055	1.121	943.4	128.0×10^{-7}	2.3×10^{-4}	866.3	13.8	0.002	130.2	53.3

or shown graphically as

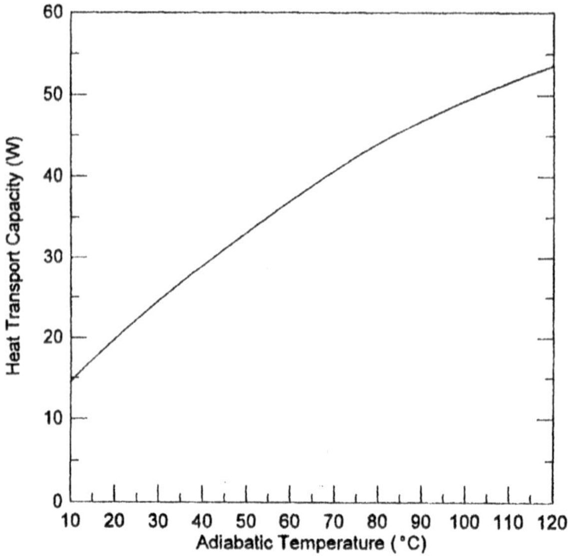

The preceding examples illustrate the effects the gravational environment and operating temperature can have on the capillary limit of heat pipes, but as mentioned previously, this is only one of several limits encountered during the design and operation of heat pipes. The following example illustrates the modeling procedures for finding the other limitations outlined in Chapter 4.

Example 5.4 In addition to the capillary limit, it is also necessary in many applications to determine the capillary, sonic, boiling, and entrainment limits as a function of mean operating temperature and tilt angle. Assume a round copper-water heat pipe with an overall length of 25.4 mm, a finned condenser section 9.39 mm long, an evaporator section 11.81 mm long, constructed from 3.2-mm-diameter copper tubing with a wall thickness of 0.9 mm, a wicking structure constructed from phosphor bronze wire mesh (No. 325) with a wire diameter of 0.0355 mm, and a condenser section with a series of 10 fins approximately 6 mm square, 0.2 mm thick, at a spacing of 1 mm. These limits can be found as follows:

SOLUTION The thermophysical properties of the working fluid are summarized below.

Operating Temp. (K)	ρ_l	ρ_v	μ_l	μ_v	k_{eff}	σ	λ
298.15	996.92	0.024	9.47×10^{-4}	1.03×10^{-5}	0.605	7.29×10^{-2}	2.347
323.15	996.92	0.083	5.5×10^{-4}	1.116×10^{-5}	0.640	6.93×10^{-2}	2.324
348.15	974.50	0.247	3.93×10^{-4}	1.119×10^{-5}	0.657	6.20×10^{-2}	2.254
373.15	960.72	0.580	2.82×10^{-4}	1.28×10^{-5}	0.680	5.84×10^{-2}	2.1

where ρ_l = liquid density $(\mathrm{kg/m^3})$
ρ_v = vapor density $(\mathrm{kg/m^3})$
μ_l = liquid viscosity $(\mathrm{kg/m \cdot s})$
k_{eff} = thermal conductivity $(\mathrm{W/m \cdot K})$
σ = surface tension $(\mathrm{N/m})$
μ_v = vapor viscosity $(\mathrm{kg/m \cdot s})$

For screen mesh, the capillary radius can be found from Table 3.1 as

$$r_c = \frac{1}{2N} = \frac{1}{2(12{,}795.25)} = 3.91 \times 10^{-5}\ \mathrm{m}$$

the maximum capillary pressure can be found from equation 3.4 as

$$P_{c,m} = 2\frac{\sigma}{r_c} \quad (\mathrm{N/m^2})$$

the normal hydrostatic pressure can be found from equation 3.52 as

$$\Delta P_+ = \rho_l g d_v \cos\psi \quad (\mathrm{N/m^2})$$

the axial hydrostatic pressure can be found from equation 3.53 as

$$\Delta P_\parallel = \rho_l g L \sin\psi \quad (\mathrm{N/m^2})$$

and the maximum effective pumping pressure can be expressed as

$$P_{p,m} = P_{c,m} - \Delta P_+ - \Delta P_\parallel$$

Using these expressions, the contributions of each of the pressure terms can be summarized in tubular form as shown below

Operating Temperature (K)	σ	$P_{c,m}$	ΔP_+	$\Delta P_\parallel$	$P_{p,m}$
298.15	7.29×10^{-2}	3,728.90	12.21	0	3716.70
232.15	6.93×10^{-2}	3,544.76	12.10	0	3532.66
348.15	6.20×10^{-2}	3,171.35	11.94	0	3159.41
373.15	5.84×10^{-2}	2,987.21	11.77	0	2975.44

Other intermediate values that must be evaluated are as follows:

Wick cross-sectional area: $A_\omega = \frac{1}{4}\pi(d_i^2 - d_v^2) = 2.9 \times 10^{-7}\ \mathrm{m^2}$
Wick porosity, $\epsilon = 1 - \frac{1}{4}\pi SNd = 0.625$ (Table 3.2, equation 3.39, where a wick crimping factor $S = 1.05$ is used)
Wick permeability, $K = d^2[t^3/122(1-t)^2] = 2 \times 10^{-11}\ \mathrm{m^2}$ (Table 3.2, equation 3.38)

Liquid friction coefficient, $F_e = \mu_e / K A_w P_e \lambda$ (equation 3.42)
Vapor core cross-sectional area, $A_v = \frac{1}{4}\pi d_v^2 = 1.227 \times 10^{-6}\ \text{m}^2$
Vapor core hydraulic radius, $r_{h,v} = \frac{1}{2} d_v = 0.000625\ \text{m}$
Drag coefficient, $(f_v\ \text{Re}_v) = 16$ (circular vapor flow passage)
Vapor frictional coefficient, $F_v = (f_v\ \text{Re}_v)[\mu_v / 2A_v r_v h^2 P_v \lambda]$

Assuming

$$\Delta P_{\text{ph}} \sim 0$$

the governing equation becomes

$$P_c = F_e L_{\text{eff}} q + F_v L_{\text{eff}} q + \Delta P_+ + \Delta P_\parallel + \Delta P_{\text{ph}}$$

The friction factors can be summarized as follows:

Temperature (K)	F_e	F_v	$(qL)_{c,m}$	$q_{c,m}$
298.15	69,782.83	3,052.10	0.0510	3.45
323.15	41,287.91	965.68	0.0836	5.65
348.15	30,848.13	335.50	0.1013	6.85
373.15	23,182.43	168.66	0.1275	8.61

Computing the effective length as

$$L_{\text{eff}} = 0.5L_c + L_a + 0.5L_e = 0.0148\ \text{m}$$

the transport capacity can be found as a function of length or in terms of the total power:

$$(q_{c,m}L) = \frac{P_{pm}}{F_e + F_v} \quad (\text{W} - \text{m}) \quad \text{or } q_{c,m} = \frac{q_{c,m}L}{L_{\text{eff}}}$$

Using a similar approach, the individual pressure terms and transport capacities can be determined for tilt angles of 15° as follows:

Temperature (K)	$P_{c,m}$	ΔP_+	$\Delta P_\parallel$	$P_{p,m}$
298.15	3728.90	11.79	63.26	3653.85
323.15	3544.76	11.69	62.71	3470.36
348.15	3171.35	11.53	31.84	3097.98
373.15	2987.21	11.38	60.96	2914.87

Temperature (K)	$q_{c,m}L$	$q_{c,m}$
298.15	0.0502	3.39
323.15	0.0821	5.55
348.15	0.0993	6.71
373.15	0.1249	8.44

and for tilt angles of 45 degrees as follows:

Temperature (K)	$q_{c,m}L$	ΔP_+	$\Delta P_\|$	$P_{p,m}$
298.15	3728.90	8.64	172.64	3547.62
323.15	3544.76	8.56	171.14	3365.06
348.15	3171.35	8.44	168.75	2994.16
378.15	2987.21	8.32	166.37	2812.52

Temperature (K)	$q_{c,m}L$	$q_{c,m}$
298.15	0.0487	3.29
323.15	0.0796	5.38
348.15.	0.0960	6.48
373.15	0.1205	8.14

SONIC LIMIT
The sonic limit can be found from equation 4.9:

$$q_{s,m} = A_v \rho_o \lambda \left[\frac{\gamma_v R_v T_o}{2(\gamma_v + 1)} \right]^{1/2} \quad \text{(W)}$$

Vapor molecular weight $M = 18$
Vapor specific heat ratio $\gamma_v = 1.33$
Universal gas constant, $\bar{R} = 8.314 \times 10^3 \text{ J/kg} \cdot \text{mol} \cdot \text{K}$
Vapor constant $R_v = \dfrac{8.314 \times 10^3}{18} = 462 \text{ J/kg} \cdot \text{K}$

Temperature (K)	$q_{s,m}$
298.15	13.70
323.15	48.85
348.15	146.36
373.15	344.76

BOILING LIMIT
The boiling limit can be found from equation 4.36:

$$q_{b,e} = \frac{2\pi L_e k_{eff} T_v}{\lambda \rho_v \ln(r_i / r_v)} \left[\frac{2\sigma}{r_n} - P_{c,m} \right]$$

where $P_{c,m}$ is the capillary pressure in the wicking structure or, if $P_c < P_{c,m}$ the maximum capillary pressure found earlier, and the nucleation radius r_n is in the range 2.54×10^{-5}–2.54×10^{-7}. The effective conductivity of the saturated wick, k_{eff}, can be found from

$$k_{eff} = \frac{k_l(k_l + k_\omega - (1 - \epsilon)(k_l - k_\omega))}{k_l + k_\omega + (1 - \epsilon)(k_l - k_\omega)}$$

where $L_e = 0.0118$ m
$r_n = 2.54 \times 10^{-7}$ m

or

$$k_\omega = 402 \text{ W/m} \cdot \text{K}$$

Temperature (K)	k_{eff}	$q_{b,m}$
298.15	1.327	2797.53
323.15	1.404	890.54
348.15	1.441	305.24
373.15	1.491	140.35

ENTRAINMENT LIMIT
The entrainment limit can be estimated using equation 4.18 as

$$q_{e,m} = A_v \lambda \left[\frac{\sigma \rho_v}{2 r_{h,w}} \right]^{1/2}$$

where the wick surface pore hydraulic radius is

$$r_{h,w} = \frac{1}{2N} - \frac{d}{2} = 2.13 \times 10^{-5} \text{ m}$$

Temperature (K)	$q_{e,m}$
298.15	18.45
323.15	33.13
348.15	52.44
373.15	75.56

VISCOUS LIMIT
Finally the viscous limit can be estimated from equation 4.15 as

$$q_v = d_v^2 \lambda A_v \frac{P_v \rho_v}{4 f_v \text{Re}_v L_e \mu_v}$$

where

$$f_v \, \mathrm{Re}_v = 16$$

Temperature	P_v (N / m^2)	$q_{v,m}$
298.15	3,293	42.93
323.15	12,349	219.25
348.15	37,290	760.72
373.15	101,350	2095.25

The five limits can be represented graphically as a function of the mean adiabatic or operating temperature as shown below.

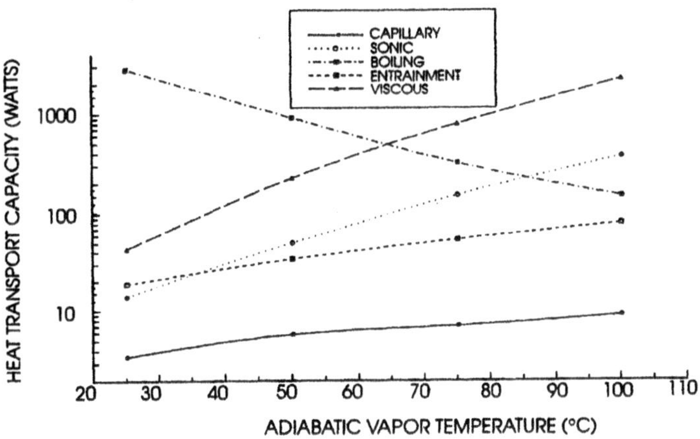

As shown above, this configuration is capillary limited over the entire temperature range.

It is important to note that in order to determine the actual heat transport capacity in the preceding example, the mean operating temperature or adiabatic vapor temperature must be known, a priori, which is not usually the case. The following section describes a method by which this value can be estimated.

Example 5.5 A heat pipe for an electronics application uses ethanol as the working fluid. The evaporator is 2 cm long, the condenser is 3 cm long, and the heat pipe has no adiabatic section. The diameter of the vapor space is 3 mm, and the wicking structure consists of three layers of #500 mesh 304 stainless steel screen (wire diameter = 0.00085 in.) If the heat pipe operates at 30°C in a horizontal position, determine the capillary, sonic, entrainment and boiling limits.

SOLUTION The relevant thermophysical properties of enthanol at 30°C are

Saturation Pressure:	$P_v = 10{,}000$ Pa
Latent Heat of Vaporization:	$\lambda = 888{,}600$ J/kg
Liquid Density:	$\rho_l = 781.0$ kg/m^3
Vapor Density:	$\rho_v = 0.38$ kg/m^3
Liquid Thermal Conductivity:	$k_l = 0.168$ W/m · K
Liquid Viscosity:	$\mu_l = 1.02 \times 10^{-3}$ kg/m · s
Vapor Viscosity:	$\mu_v = 0.91 \times 10^{-5}$ kg/m · s
Surface Tension:	$\sigma = 2.44 \times 10^{-2}$ N/m

The thermal conductivity of 304 stainless steel at 30°C is $k_w = 14.9$ W/m · K.

Neglecting the axial hydrostatic pressure drop, the inertial pressure gradient in the evaporator, and the pressure gradients across the phase transitions, equation (3.1) can be written as

$$\Delta P_{c,m} = \Delta P_v + \Delta P_l + \Delta P_+$$

The maximum capillary pressure for a heat pipe operating at steady state can be expressed as a function of the effective capillary radius of the evaporator wick:

$$\Delta P_{c,m} = \left(\frac{2\sigma}{r_{c,e}}\right)$$

From Table 3.1, for a wire mesh wick structure the effective capillary radius is:

$$r_c = \frac{1}{2 \cdot N} = \frac{1}{2 \cdot 500 \text{ mesh/in.}} = 0.001 \text{ in.} = 2.54 \times 10^{-5} \text{ m}$$

Substituting the values of the surface tension and the capillary radius into equation (3.4) yields the maximum capillary pressure:

$$\Delta P_{c,m} = \frac{2 \cdot 2.44 \times 10^{-2}}{2.54 \times 10^{-5}} = 1921 \text{ Pa}$$

Assuming one-dimensional laminar incompressible vapor flow, the expression for the vapor pressure drop, equation (3.33), reduces to:

$$\Delta P_v = \frac{16\mu_v L_{eff} q}{2r_{h,v}^2 A_v \rho_v \lambda}$$

The effective length is

$$L_{\text{eff}} = L_a + \frac{L_e + L_c}{2} = 0 + \frac{0.02\ \text{m} + 0.03\ \text{m}}{2} = 0.025\ \text{m}$$

The cross-sectional area of the vapor space is:

$$A_v = \frac{\pi d_v^2}{4} = \frac{\pi (0.003\ \text{m})^2}{4} = 7.069 \times 10^{-6}\ \text{m}^2$$

Substituting these known dimensions and the known thermophysical properties into the expression for the vapor pressure drop yields

$$\Delta P_v = \frac{16\,(0.91 \times 10^{-5})\,0.025 \cdot q}{2(0.0015)^2\,7.069 \times 10^{-6}\,(888{,}600)} = 0.1288 \cdot q$$

where ΔP_v is in pascals, and q is in watts.

The liquid pressure drop can be calculated by the following equation:

$$\Delta P_l = \frac{\mu_l L_{\text{eff}} q}{K A_w \lambda \rho_l}$$

where the permeability, K of a wrapped screen wick is given by

$$K = \frac{d^2 \epsilon^3}{122(1 - \epsilon)^2}$$

(note that the wire diameter d in this equation must be in inches), and the wick porosity ϵ can be determined as

$$\epsilon = 1 - \frac{\pi S N d}{4} \tag{3.51}$$

Substituting the known physical dimensions of the screen into equation (3.51) one obtains

$$\epsilon = 1 - \frac{\pi (1.05) 500 (0.00085)}{4} = 0.6495$$

Equation (3.50) then yields

$$K = \frac{(0.00085)^2 (0.6495)^3}{122(1 - 0.6495)^2} = 1.321 \times 10^{-8}$$

The cross-sectional area of the wick is based on the diameter of the vapor space and the outside diameter of the wicking structure:

$$A_w = \frac{\pi (d_i^2 - d_v^2)}{4}$$

The outside diameter of the wicking structure (inside diameter of the tubular heat pipe case) can be calculated assuming that the thickness of each layer of screen is equal to twice the wire diameter:

$$d_i = d_w + 12 \cdot d_\omega = 0.003\ \text{m} + 12(2.159 \times 10^{-5}\ \text{m}) = 0.003259\ \text{m}$$

The wick cross-sectional area is then

$$A_w = \frac{\pi[(0.003259)^2 - (0.003)^2]}{4} = 1.273 \times 10^{-6}\ \text{m}^2$$

Substituting these values and the known thermophysical properties into equation (3.54) yields

$$\Delta P_l = \frac{(1.02 \times 10^{-3})0.025 \cdot q}{(1.321 \times 10^{-8})(1.273 \times 10^{-6})888{,}600(781.0)} = 2.185 \cdot q$$

For a horizontal heat pipe, the expression for the normal hydrostatic pressure drop, equation (3.64), reduces to

$$\Delta P_+ = \rho_l g d_v$$

Inserting the known values into this equation gives

$$\Delta P_+ = 781.0\ \text{kg/m}^3\ (9.81\ \text{m/s}^2)\ 0.003\ \text{m} = 23.0\ \text{Pa}$$

Substituting yields an equation which can be solved for the limiting heat transport rate q:

$$1921\ \text{Pa} = 0.1288 \cdot q + 2.185 \cdot q + 23\ \text{Pa}$$

$$q = 820\ \text{W}$$

This result is based on the assumption of laminar vapor flow. To check this assumption, the vapor Reynolds number must be calculated using a slightly modified form of equation (3.34):

$$\text{Re}_v = \frac{2(r_{h,v})q}{A_v \mu_v \lambda} = \frac{4q}{\pi d_v \mu_v \lambda} = \frac{4(820)}{\pi(0.003)(0.91 \times 10^{-5})888{,}600} = 43{,}000$$

The vapor Reynolds number is greater than 2300, so the assumption of laminar vapor flow is not valid. The vapor pressure drop must be calculated using equation (3.38) substituted into equation (3.33):

$$\Delta P_v = 0.038\left(\frac{d_v q}{A_v \mu_v \lambda}\right)^{3/4} \cdot \frac{2\mu_v L_{\text{eff}} q}{d_v^2 A_v \rho_v \lambda}$$

$$= 0.038\left[\frac{0.003 \cdot q}{(7.069 \times 10^{-6})(0.91 \times 10^{-5})888{,}600}\right]^{3/4}$$

$$\times \frac{2(0.91 \times 10^{-5})0.025 \cdot q}{(0.003)^2(7.069 \times 10^{-6})0.38(888{,}600)}$$

$$AP_v = 0.0157 \cdot q^{7/4}$$

Substituting this result into equation (E1.1) one obtains:

$$1921 = 0.0157 \cdot q^{7/4} + 2.185 \cdot q + 23$$

$$q = 495 \text{ W}$$

The assumption that the inertial pressure gradient is negligible can be checked by using equation (3.40) to calculate the inertial pressure gradient of the vapor in the evaporator and comparing it to the maximum capillary pressure.

$$\Delta P_i = \frac{\rho_v \bar{V}_v^2}{2g}\left[\frac{28}{9} - \frac{0.68 \text{ Re}}{(29L_e/d_v) + \text{Re}} e^{(-60L_a/\text{Re}\, d_v)}\right]$$

For this heat pipe, $L_a = 0$, so equation (3.40) reduces to

$$\Delta P_i = \frac{\rho_v \bar{V}_v^2}{2g}\left[\frac{28}{9} - \frac{0.68 \text{ Re}}{(29L_e/d_v) + \text{Re}}\right]$$

The mean vapor velocity and the vapor Reynolds number can both be calculated from equation (3.29) using the revised value of $q = 495$ W:

$$\bar{V}_v = \frac{q}{\rho_v A_v \lambda} = \frac{495}{0.38(7.069 \times 10^{-6})888{,}600} = 207 \text{ m/s}$$

$$\text{Re}_v = \frac{\rho_v \bar{V}_v d_v}{\mu_v} = \frac{0.38(207)0.003}{0.91 \times 10^{-5}} = 25{,}900$$

The inertial pressure gradient can then be calculated as

$$\Delta P_i = \frac{0.38(207)^2}{2(9.81)}\left[\frac{28}{9} - \frac{0.68(25{,}900)}{[29(0.02)/0.003] + 25{,}900}\right] = 2020 \text{ Pa}$$

This is larger than the maximum capillary pressure, so it is evident that the inertial pressure gradient can not be neglected. If one assumes that $\text{Re}_v \gg$

$(29L_e/d_v) = 190$, then Re can be eliminated, and by substituting the expression for the vapor velocity from equation (3.29) into this equation, a simplified conservative estimate of the inertial pressure gradient can be obtained:

$$\Delta P_i = \frac{1.22 \cdot q^2}{g\rho_v A_v^2 \lambda^2}$$

$$= \frac{1.22 \cdot q^2}{9.81(0.38)(7.069 \times 10^{-6})(888{,}600)^2} = 0.0083 \cdot q^2$$

When the inertial contribution to the vapor pressure drop is used to calculate the capillary limit, it is common to assume that 80% of the pressure drop is recovered in the condenser. The net pressure drop is then 20% of the inertial pressure drop in the evaporator, so the equation above becomes:

$$\Delta P_i = 20\% \cdot (0.0083 \cdot q^2) = 0.00166 \cdot q^2$$

Assuming turbulent, incompressible vapor flow and including the inertial limit in the estimate of the capillary limit yields

$$1921 = 0.0157 \cdot q^{7/4} + 2.185 \cdot q + 23 + 0.00166 \cdot q^2$$

$$q = 432 \text{ W}$$

The sonic limit can be obtained using equation (4.13) with the assumption that the vapor pressure at the end of the evaporator is equal to the saturation pressure corresponding to the 30°C operating temperature of the heat pipe:

$$q_s = 0.474\lambda A_v(\rho_v P_v)^{1/2}$$

$$= 0.474(888{,}600)(7.069 \times 10^{-6})[0.38(10{,}000)]^{1/2} = 184 \text{ W}$$

The entrainment limit can be calculated directly from equation (4.18):

$$q_e = A_v \lambda \left(\frac{\sigma \rho_v}{2r_{h,w}}\right)^{1/2}$$

where the hydraulic radius of the wicking structure is equal to the cross-sectional area of the wicking structure divided by the total wetted perimeter of the wick:

$$r_{h,w} = \frac{A_w}{\pi(d_v + d_i)} = \frac{1.273 \times 10^{-6}}{\pi(0.003 + 0.003259)} = 6.474 \times 10^{-5} \text{ m}$$

Substituting this into equation (4.18) one obtains

$$q_e = (7.069 \times 10^{-6})888{,}600\left[\frac{(2.44 \times 10^{-2})0.38}{2(6.474 \times 10^{-5})}\right]^{1/2} = 53.2\ \text{W}$$

The boiling limit can be found from equation (4.36):

$$q_b = \frac{2\pi L_e k_{\text{eff}} T_v}{\lambda \rho_v \ln(r_i / r_v)}\left(\frac{2\sigma}{r_n} - \Delta P_{c,m}\right)$$

According to Table 4.3, the effective thermal conductivity, k_{eff}, of liquid in a wrapped screen wick can be determined from

$$k_{\text{eff}} = \frac{k_l[k_l + k_w - (1-\epsilon)(k_l - k_w)]}{k_l + k_w + (1-\epsilon)(k_l - k_w)}$$

Substituting the known values of the wick porosity and the thermal conductivities of the liquid and screen wick material, one obtains:

$$k_{\text{eff}} = \frac{0.168[0.168 + 14.9 - 0.3505(0.168 - 14.9)]}{0.168 + 14.9 + 0.3505(0.168 - 14.9)} = 0.343\ \text{W/m} \cdot \text{K}$$

The operating temperature $T_v = 30°\text{C} = 303\ \text{K}$, and r_n is the nucleation site radius, which is assumed to be 2.54×10^{-7} m, the lower bound of the conventional range. Also

$$\ln\left(\frac{r_i}{r_v}\right) = \ln\left(\frac{d_i}{d_v}\right) = \ln\left(\frac{0.003259}{0.003}\right) = 0.0828$$

Using the result from equation (E1.2), the boiling limit, equation (4.36), can be evaluated as

$$q_b = \frac{2\pi \cdot 0.02(0.343)303}{888{,}600(0.38)0.0828}\left[\frac{2(2.44 \times 10^{-2})}{2.54 \times 10^{-7}} - 1921\right] = 89\ \text{W}$$

The heat transport limits for this heat pipe can be summarized as follows:

Capillary Limit:	432 W
Sonic Limit:	184 W
Entrainment Limit:	53.2 W
Boiling Limit:	89 W

The dominant effects which limit the transport capacity of this heat pipe are the entrainment limit and the boiling limit. Both of these limits could be increased by slightly increasing the diameter of the vapor space.

Example 5.6 A heat pipe for an electronics application uses methanol as the working fluid. The evaporator, condenser and adiabatic sections are each 2 cm long. The diameter of the vapor space is 4 mm, and the wall is constructed of 1 mm thick 2024-T6 aluminum with ten equally spaced 0.5 mm square axial grooves forming the wicking structure. The heat pipe operates at 30°C with an adverse tilt angle of 5 degrees. Determine the capillary, sonic, entrainment and boiling limits for this heat pipe.

SOLUTION The thermal conductivity of 2024-T6 aluminum at 30°C is $k_w = 177\ \mathrm{W/m \cdot K}$

The relevant thermophysical properties of methanol at 30°C are:

Latent Heat of Vaporization:	$\lambda = 1{,}155{,}600\ \mathrm{J/kg}$
Liquid Density:	$\rho_l = 782.0\ \mathrm{kg/m^3}$
Vapor Density:	$\rho_v = 0.31\ \mathrm{kg/m^3}$
Liquid Thermal Conductivity:	$k_l = 0.203\ \mathrm{W/m \cdot K}$
Liquid Viscosity:	$\mu_l = 5.21 \times 10^{-4}\ \mathrm{kg/m \cdot s}$
Vapor Viscosity:	$\mu_v = 9.8 \times 10^{-5}\ \mathrm{kg/m \cdot s}$
Saturation Pressures	25,000 Pa
Surface Tension:	$\sigma = 0.0218\ \mathrm{N/m}$

Neglecting the pressure drop across the phase transitions and the inertial pressure gradient, the pressure balance associated with the capillary limitation can be expressed as

$$\Delta P_{c,m} = \Delta P_v + \Delta P_l + \Delta P_+ + \Delta P_{\|}$$

From equation (3.4) and Table 3.1, the maximum capillary pressure can be calculated as

$$\Delta P_{c,m} = \frac{2\sigma}{w} = \frac{2(0.0218\ \mathrm{N/m})}{0.0005\ \mathrm{m}} = 87.2\ \mathrm{Pa}$$

Assuming laminar incompressible vapor flow, the vapor pressure drop can be calculated using equation (3.33) with $C = 1$ and $(f_v\ \mathrm{Re}_v) = 16$:

$$\Delta P_v = \frac{16\mu_v L_{\mathrm{eff}} q}{2(r_{h,v})^2 A_v \rho_v \lambda}$$

The effective length of this heat pipe is 4.0 cm, the hydraulic radius of the vapor space is 2.0 mm, and the cross-sectional area of the vapor space can be calculated as

$$A_v = \frac{\pi d_v^2}{4} = \frac{\pi (0.004\ \mathrm{m})^2}{4} = 1.257 \times 10^{-5}\ \mathrm{m^2}$$

Substituting these values into equation (E2.3), one obtains

$$\Delta P_v = \frac{16(9.8 \times 10^{-6})(0.04) \cdot q}{2(0.002)^2(1.257 \times 10^{-5})0.31(1{,}155{,}000)} = 0.1742 \cdot q$$

where ΔP_v is in pascals and q is in watts.

The liquid pressure drop can be calculated from equation (3.54):

$$\Delta P_l = \frac{\mu_l L_{\text{eff}} q}{K A_w \lambda \rho_l}$$

where K is the permeability as expressed in equation (3.48):

$$K = \frac{2\epsilon (r_{h,l})^2}{f_l \,\text{Re}_l}$$

From Fig. 3.5*a*, for a square groove, $f \cdot \text{Re} = 14.3$. The porosity is equal to unity, and the hydraulic radius is defined as twice the cross-sectional area of an individual liquid passage divided by its wetted perimeter:

$$r_{h,l} = \frac{2w^2}{4w} = \frac{w}{2} = \frac{0.0005\,\text{m}}{2} = 0.00025\,\text{m}$$

Substituting this into equation (3.48), the permeability can be calculated:

$$K = \frac{2(0.00025\,\text{m})^2}{14.3} = 8.741 \times 10^{-9}\,\text{m}^2$$

The total cross-sectional area of the wick, A_w, can be calculated by multiplying the number of grooves by the cross-sectional area of each groove:

$$A_w = N \cdot w^2 = 10(0.0005\,\text{m})^2 = 2.5 \times 10^{-6}\,\text{m}^2$$

The liquid pressure drop can then be determined by substituting yields

$$\Delta P_l = \frac{(5.21 \times 10^{-4})(0.04) \cdot q}{(8.741 \times 10^{-9})(2.5 \times 10^{-6})1{,}155{,}000(782.0)} = 1.056 \cdot q$$

The normal hydrostatic pressure drop can be expressed as:

$$\Delta P_+ = \rho_l g d_v \cos \Psi = 782.0(9.81)(0.004)\cos(5°) = 30.6\,\text{Pa}$$

and the axial hydrostatic pressure drop is:

$$\Delta P_{\parallel} = \rho_l g L \sin \Psi = 782.0(9.81)(0.06)\sin(5°) = 40\,\text{Pa}$$

Substituting the values yields:

$$87.2 = 0.1742 \cdot q + 1.056 \cdot q + 30.6 + 40$$

This equation can be solved to obtain the capillary heat transport limitation:

$$q = \frac{87.2 - 30.6 - 40}{0.1742 + 1.056} = 13.5\ \mathrm{W}$$

To check the assumption of laminar flow, the vapor Reynolds number can be determined using equation (3.34):

$$\mathrm{Re}_v = \frac{d_v q}{A_v \mu_v \lambda} = \frac{(0.004)13.5}{(1.257 \times 10^{-5})(9.8 \times 10^{-6})1{,}155{,}000} = 380$$

The Reynolds number is less than 2300, so the flow is indeed laminar.

The sonic limit can be determined from equation (4.13):

$$\begin{aligned} q_s &= 0.474 \lambda A_v (\rho_v P_v)^{1/2} \\ &= 0.474(1{,}155{,}000)(1.257 \times 10^{-5})[0.31(25{,}000)]^{1/2} = 606\ \mathrm{W} \end{aligned}$$

The entrainment limit can be calculated using equation (4.18) and the previously calculated value of the hydraulic radius of the wick:

$$q_e = A_v \lambda \left(\frac{\sigma \rho_v}{2 r_{h,w}} \right)^{1/2} = (1.257 \times 10^{-5})1{,}155{,}000 \left(\frac{0.0218(0.31)}{2(0.00025)} \right)^{1/2} = 53.4\ \mathrm{W}$$

The boiling limit is given by equation (4.36):

$$q_b = \frac{2\pi L_e k_{\mathrm{eff}} T_v}{\lambda \rho_v \ln(r_i / r_v)} \left(\frac{2\sigma}{r_n} - \Delta P_{c,m} \right)$$

where the effective thermal conductivity of the liquid and wick combination, k_{eff}, can be calculated from the relation for rectangular gooves from Table 4.3:

$$k_{\mathrm{eff}} = \frac{(w_f k_l k_w \delta) + w k_l (0.185 w_f k_w + \delta k_l)}{(w + w_f)(0.185 w_f k_w + \delta k_l)}$$

where $w_f \equiv$ width of aluminum between grooves $= 0.000757$ m
$\delta \equiv$ depth of groove $= 0.0005$ m
$w \equiv$ width of groove $= 0.0005$ m

Substituting these values into the equation above, one obtains: $k_{\mathrm{eff}} = 0.515\ \mathrm{W/m \cdot K}$. The temperature of the saturated vapor is 303 K, the nucleation site radius is assumed to be 2.54×10^{-7} m, and $\ln(r_i / r_v) = \ln(2.5 \text{ mm} / 2.0 \text{ mm}) = 0.2231$, so equation (4.36) can be evaluated as

$$q_b = \frac{2\pi(0.04)(0.515)303}{1{,}155{,}000(0.31)0.2231}\left[\frac{2(0.0218)}{2.54 \times 10^{-7}} - 87.2\right] = 84.2 \text{ W}$$

The operational limits for this heat pipe can be summarized as follows:

Capillary Limit:	13.5 W
Sonic Limit:	606 W
Entrainment Limit:	53.4 W
Boiling Limit:	84.2 W

The capillary and entrainment limits could be improved by designing the heat pipe with narrower grooves.

5.2.1 Heat Pipe Thermal Resistance

The temperature drop between the evaporator and condenser of a heat pipe is of particular interest to the designer of heat pipe thermal control systems for electronics and is most readily found by utilizing an analogous electrothermal network. Figure 5.2 illustrates the electrothermal analog for a simple cylindrical heat pipe. As shown, the overall thermal resistance is comprised of nine different resistances arranged in a series–parallel combination. These nine resistances can be summarized as follows:

$R_{p,e}$: Radial resistance of the pipe wall at the evaporator

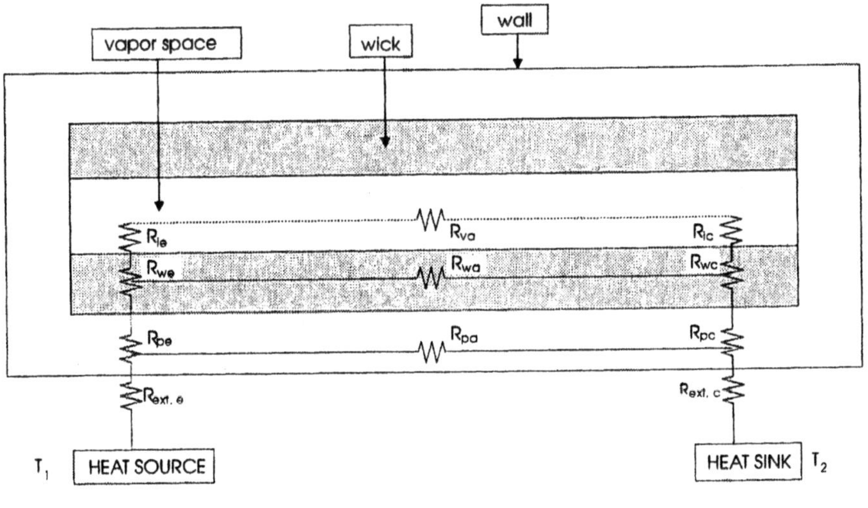

Figure 5.2 Equivalent thermal resistance of a heat pipe.

$R_{\nu,e}$: Resistance of the liquid–wick combination at the evaporator
$R_{i,e}$: Resistance of the liquid–vapor interface at the evaporator
$R_{\nu,a}$: Resistance of the adiabatic vapor section
$R_{p,a}$: Axial resistance of the pipe wall
$R_{\nu,a}$: Axial resistance of the liquid / wick combination
$R_{i,c}$: Resistance of the liquid / vapor interface at the condenser
$R_{\nu,c}$: Resistance of the liquid / wick combination at the condenser
$R_{p,c}$: Radial resistance of the pipe wall at the condenser

There are two other resistances shown in Fig. 5.2 that play a very important role in the design of heat pipes for electronic applications. These are the external, or contact, resistances occurring between the heat source and the heat pipe evaporator, $R_{\text{ext},e}$, and the contact resistance occurring between the heat sink and the condenser, $R_{\text{ext},c}$. In many applications, the combination of these two resistances will be of the same order of magnitude as the overall heat pipe resistance. An in-depth discussion of the factors that affect these resistances is beyond the scope of this book, but numerous investigations have been conducted and a considerable amount of information is available in the literature.

Asselman and Green (1973) give estimates of the order of magnitude for each of these resistances (Table 5.1) for a typical application. In comparing these values, it is apparent that several simplifications can be made. First, because of the comparative magnitudes of the resistance of the vapor space and the axial resistances of the pipe wall and liquid–wick combinations, the axial resistance of both the pipe wall and the liquid–wick combination may be treated as open circuits and neglected. Second, again because of the comparative resistances, the liquid–vapor interface resistances and the axial vapor resistance can, in most situations, be assumed to be negligible. This leaves only the pipe wall radial resistances and the liquid–wick resistances at both the evaporator and condenser.

The radial resistances at the pipe wall can be computed from Fourier's law as

TABLE 5.1 Comparative Values for Heat Pipe Resistances

Resistance	°C / W
$R_{p,e}$ and $R_{p,c}$	10^{-1}
$R_{\omega,e}$ and $R_{\omega,c}$	10^{+1}
$R_{i,e}$ and $R_{i,c}$	10^{-5}
$R_{v,a}$	10^{-8}
$R_{p,a}$	10^{+2}
$R_{\omega,a}$	10^{+4}

Source: Asselman and Green, 1973.

$$R_{p,e} = \frac{\delta}{k_p A_e} \tag{5.1}$$

for flat plates, where δ is the plate thickness and A_e is the evaporator area, or

$$R_{p,e} = \frac{\ln (d_o / d_i)}{2\pi L_e k_p} \tag{5.2}$$

for cylindrical pipes, where L_e is the evaporator length.

The most complicated and perhaps least understood of the resistances is the resistance of the liquid–wick combination. An expression for the equivalent thermal resistance in the evaporator of a circular pipe is

$$R_{\omega,e} = \frac{\ln (d_o / d_i)}{2\pi L_e k_{\text{eff}}} \tag{5.3}$$

where values for the effective conductivity k_{eff} can be found in Table 5.2. Expressions for the condenser would be similar. For sintered metal materials, an investigation performed by Peterson and Fletcher (1987) found that methods presented by Alexander (1972) and later by Ferrell et al. (1972) for determining the effective thermal conductivity of saturated sintered materials was the most accurate over the widest range of conditions.

The adiabatic vapor resistance, although usually negligible, can be found as

TABLE 5.2 Effective Thermal Conductivity for Liquid-Saturated Wick Structures

Wick Structures	k_{eff}
Wick and liquid in series	$\dfrac{k_l k_w}{\epsilon k_w + k_l(1-\epsilon)}$
Wick and liquid in parallel	$\epsilon k_l + k_\omega(1-\epsilon)$
Wrapped screen	$\dfrac{k_l[(k_l + k_\omega) - (1-\epsilon)(k_l - k_\omega)]}{(k_l + k_\omega)(1-\epsilon)(k_l - k_\omega)}$
Packed spheres	$\dfrac{k_l[(2k_l + k_\omega) - 2(1-\epsilon)(k_l - k_\omega)]}{(2k_l + k_\omega)(1-\epsilon)(k_l - k_\omega)}$
Rectangular grooves	$\dfrac{\omega_f k_l k_\omega \delta) + \omega k_l(0.185\omega_f k_\omega + \delta k_l)}{(\omega + \omega_f)(0.185\omega_f k_l + \delta k_f)}$

Source: Chi, 1976

$$R_{v,a} = \frac{T_v(P_{v,e} - P_{v,c})}{\rho_v \lambda q} \tag{5.4}$$

where $P_{v,e}$ and $P_{v,c}$ are the vapor pressures at the evaporator and condenser.

Combining these individual resistances provides a mechanism by which the overall thermal resistance can be computed and hence the temperature drop associated with various axial fluxes can be determined. In addition, this electrical analog provides a means by which the mean operating temperature or adiabatic vapor temperature can be estimated. This value can in turn be used in conjunction with the thermal performance map for a given heat pipe, similar to the one shown in Example 5.4, to determine the suitability of a particular heat pipe for a given application.

Example 5.7 For the heat pipe described in Example 5.4, determine the overall thermal resistance of the heat pipe. Neglect the convective heat transfer at the finned condenser end and the contact resistance external to the case in the evaporator.

SOLUTION The overall heat transfer coefficient is comprised of the resistance of the pipe wall at the evaporator, the resistance of the liquid–wick combination in the evaporator, the resistance of the vapor in the adiabatic section, the resistance of the liquid–wick combination in the condenser, and the resistance of the pipe wall at the condenser. This combination of resistances can be expressed as

$$U_p = \frac{1}{R_{p,e} + R_{\omega,e} + R_v + R_{\omega,c} + R_{p,c}}$$

Assuming a temperature of 373 K, the individual resistance can be found as follows:

the thermal resistance due to the pipe wall in the evaporator, $R_{p,e}$,

$$R_{p,e} = \frac{r_0 \delta_p}{2L_e k_p} = \frac{(1.6 \times 10^{-3}\ \text{m})(8.89 \times 10^{-4})}{2(0.01181)(379)} = 1.59 \times 10^{-7}\ \text{m}^2 \cdot \text{K/W}$$

the thermal resistance due to the liquid–wick combination in the evaporator, $R_{\omega e}$,

$$R_{\omega,e} = r_0^2 \frac{\delta_\omega}{2} L_e r_i k_{\text{eff}} = \frac{(1.6 \times 1.0^{-3}\ \text{m})^2 (7.10 \times 10^{-5})}{2(0.01181)(0.000695)(1.491)}$$

$$= 7.43 \times 10^{-6}\ \text{m}^2 \cdot \text{K/W}$$

the thermal resistance of the vapor flow, R_v,

$$R_v = \frac{\pi r_0^2 T_v F_v(\frac{1}{6}L_e + L_a + \frac{1}{6}L_c)}{P_v \lambda}$$

$$= \frac{(3.14)(1.6 \times 10^{-3})^2(373.15)(168.66)(0.0077)}{(0.580)(2.184 \times 10^6)(1)}$$

$$= 3.08 \times 10^{-9}\ \text{m}^2 \cdot \text{K/W}$$

the thermal resistance due to the liquid–wick combination in the condenser, $R_{\omega,c}$

$$R_{\omega,c} = \frac{r_0 \epsilon_p}{2L_c r_i k_{\text{eff}}} = \frac{(1.6 \times 10^{-3}\ \text{m})^2(7.10 \times 10^{-5})}{2(0.00939)(0.000695)(1.491)} = 9.34 \times 10^{-6}\ \text{m}^2 \cdot \text{K/W}$$

and the thermal resistance due to the pipe wall in the condenser, $R_{p,c}$,

$$R_{p,c} = \frac{r_0 \epsilon_p}{2L_c k_p} = \frac{(1.6 \times 10^{-3}\ \text{m})(8.89 \times 10^{-4}\ \text{m})}{2(0.00939)(379)} = 1.99 \times 10^{-7}\ \text{m}^2 \cdot \text{K/W}$$

These values can be summarized as follows:

$R_{p,e}$	$R_{\omega,e}$	R_v	$R_{\omega,c}$	$R_{p,c}$
1.59×10^{-7}	7.43×10^{-6}	3.0×10^{-6}	9.34×10^{-9}	1.99×10^{-7}

Substitution yields an overall heat transfer coefficient of

$$U_p = 5.84 \times 10^5\ \text{W/m}^2 \cdot \text{K}$$

Example 5.8 (*a*) Calculate the thermal resistance of the heat pipe in example 5.1. (*b*) The evaporator of the heat pipe in Example 2 is enclosed in a silicon casing, and the contact resistance between the heat pipe and its casing is $2.0 \times 10^{-5}\ \text{m}^2 \cdot \text{K/W}$. The condenser is immersed in a stream of liquid, and the convective heat transfer coefficient at the condenser is $500\ \text{W/m}^2 \cdot \text{K}$. If the liquid is at 10°C, and the heat pipe is operating at a heat transport rate of 2.0 W, what is the temperature of the silicon?

SOLUTION (*a*) Neglecting the liquid / vapor interface resistances and the axial resistances, the total thermal resistance of the heat pipe can be expressed as

$$R_{\text{tot}} = R_{p,e} + R_{p,c} + R_{w,e} + R_{w,c}$$

The evaporator and condenser are physically identical, so the above equation can be simplified to

$$R_{\text{tot}} = 2(R_{p,e} + R_{w,e})$$

The radial resistance of the pipe wall can be calculated from equation (5.2):

$$R_{p,e} = \frac{\ln(d_o / d_i)}{2\pi L_e k_p} = \frac{\ln(0.006\ \text{m} / 0.005\ \text{m})}{2\pi(0{,}02\ \text{m})(177\ \text{W/m}\cdot\text{K})} = 0.0082\ \text{K/W}$$

The resistance of the liquid / wick combination can be determined from equation (5.3):

$$R_{w,e} = \frac{\ln(d_o / d_i)}{2\pi L_e k_{\text{eff}}} = \frac{\ln(0.005\ \text{m} / 0.004\ \text{m})}{2\pi(0.02\ \text{m})(0.515\ \text{W/m}\cdot\text{K})} = 3.4\ \text{K/W}$$

Substitution of these two results into equation (E3.1) yields:

$$R_{\text{tot}} = 2(0.0082\ \text{K/W} + 3.4\ \text{K/W}) = 6.8\ \text{K/W}$$

(*b*) The change in temperature is equal to the heat transport rate multiplied by the sum of the series resistance:

$$T = T_\infty + q\left(\frac{R''_{t,c}}{\pi d_o L_c} + R_{\text{tot}} + \frac{1}{h\pi d_o L_c}\right)$$

Substitution of the known values into this equation yields

$$T = 10 + 2.0\left[\frac{2.0 \times 10^{-5}}{\pi(0.006)(0.02)} + 6.8 + \frac{1}{(500)\pi(0.006)(0.02)}\right]$$

which can be reduced to:

$$T = 10°\text{C} + 2.0\ \text{W}(0.053\ \text{K/W} + 6.8\ \text{K/W} + 5.3\ \text{K/W}) = 34°\text{C}$$

5.2.2 Temperature Profiles

Another operational characteristic of interest to the heat pipe designer is the axial temperature distribution along the length of the heat pipe. One method for determining this distribution utilizes the expressions for heat flux as a function of Mach number, equation 4.8,

$$q = \frac{A_v \rho_o \lambda(\gamma_v R_v T_o)^{1/2}\,\text{Ma}_v[1 + \frac{1}{2}(\gamma_v - 1)\,\text{Ma}_v^2]^{1/2}}{1 + (\nu_v / 2)\text{Ma}_v^2} \tag{5.5}$$

and the expression for maximum heat transport at the sonic limit, equation 4.9,

$$q_{s,m} = A_v \rho_o \gamma\left(\frac{\gamma_v R_v T_o}{2(\gamma_v + 1)}\right)^{1/2} \tag{5.6}$$

Combining these two equations, an expression for the axial distribution of the vapor Mach number can be obtained as

$$\frac{q}{q_{s,m}} = \frac{\mathrm{Ma}_v\{2(\gamma_v + 1)[1 + \frac{1}{2}(\nu_v - 1)\,\mathrm{Ma}_v^2]\}^{1/2}}{1 + \nu_v\,\mathrm{Ma}_v^2} \tag{5.7}$$

Utilizing this expression at a given heat flux, the local vapor Mach number can be established and the resulting vapor temperature can be calculated from

$$\frac{T_o}{T_v} = 1 + \frac{\gamma_v - 1}{2}\,\mathrm{Ma}_v^2 \tag{5.8}$$

where T_o represents the stagnation temperature and T_v represents the vapor temperature.

This technique will yield reasonably accurate estimations of the liquid–vapor interface temperature distribution along the entire length of the heat pipe, particularly in the evaporator section. It should be noted, however, that while the vapor mach number Ma is less than unity in the evaporator section, it is possible for the vapor Mach number to be greater than unity in the condenser section, which would result in liquid-vapor interface temperatures in this region that could be significantly different from the actual values.

It should not be surprising that the axial vapor temperature distribution in heat pipes is extremely small, since any small variation in vapor temperature will result in a pressure gradient and hence vapor flow. It is this very small temperature gradient in the vapor region of heat pipes that has given them the reputation of being "isothermal" devices. When initially considering heat pipes for a particular application, however, designers are sometimes surprised to find that there is actually a noticeable and sometimes significant temperature drop from the heat source to the heat sink. This temperature drop does not result from the temperature distribution of the liquid–vapor interface, but from the combined resistance of the liquid–wick combination and the wall resistance of the pipe or container.

The overall temperature drop or the external end-to-end temperature drop is further influenced by the direction of the heat flow and the point at which the temperature measurement is made. At the evaporator, the external wall temperature is higher than the vapor temperature due to the direction of heat flow, and in the condenser the external wall temperature is lower. As mentioned previously, the combination of these factors may result in overall temperature gradients that are substantial, particularly in situations where the effective thermal conductivity of the liquid–wick combination is low.

Figure 5.3 illustrates the temperature variation one might expect to find

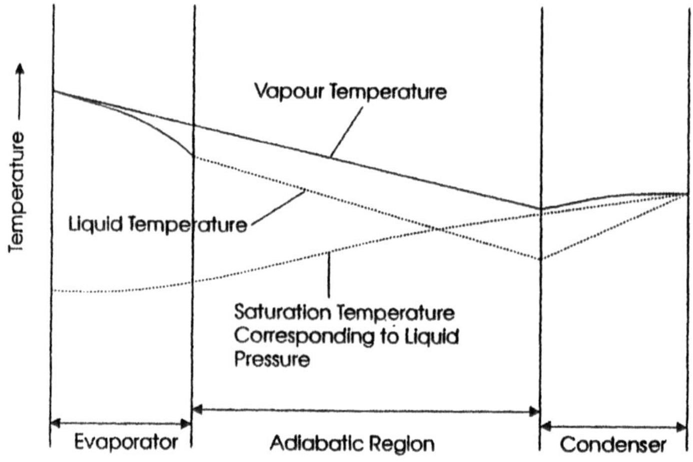

Figure 5.3 Axial temperature variation of the liquid and vapor regions in operating heat pipes.

in the vapor and liquid phases of an operating heat pipe. Beginning at the condenser and moving in the direction of fluid flow, it is apparent that the liquid temperature drops as the adiabatic section is approached, at which time a slight temperature rise occurs. This increase in temperature in the adiabatic section is often attributed to insufficient insulation. However, even in the ideal case where the adiabatic section is perfectly insulated, some increase in temperature will be observed due to heat transfer from the flowing vapor to the liquid flowing in the opposite direction. Once the liquid reaches the evaporator section, a rapid increase in temperature is observed due to heat addition through the wall. Note that in the evaporator section the vapor temperature slightly exceeds that of the liquid, that is, the liquid is slightly subcooled. While initially this may be counterintuitive, it is important to remember that the pressure of the liquid in the evaporator is slightly below that of the vapor due to capillary forces, and hence the liquid can evaporate into the vapor region.

Because slight variations in temperature result in corresponding variations in the vapor pressure, the temperature gradients in the vapor region are somewhat smaller than those occurring in the liquid. As shown in Fig. 5.3, the vapor temperature decreases steadily through the evaporator region and on into the adiabatic section, due to slight decreases in the vapor pressure and some condensation of the vapor on the liquid surface. Upon reaching the inlet to the condenser, a slight temperature increase in the vapor is possible due to pressure recovery.

An alternative method for evaluating the axial temperature distribution in heat pipes is to utilize the methods in Chapter 3 to determine the pressure distribution in the vapor phase. Then using the Clapeyron equation, estimations of the axial temperature variation occurring in the vapor space can be made. Recently, experimental investigations have demonstrated that this technique for evaluating axial temperature variations in the vapor space can result in surprisingly accurate estimations of the actual vapor temperature.

5.3 TRANSIENT MODELING

Colwell and Modlin (1992) have presented an excellent overview of transient thermal models for heat pipes and associated systems and have outlined a series of issues that should be considered prior to developing or selecting a particular model. These issues include the following:

- Type of model, that is, analytical or numerical
- Computational hardware available
- Minimum acceptable level of detail
- Methods by which the model will be coupled with the environment
- Nonstandard operational issues such as transition from free molecular to continuum flow, choked flow, freezing, thawing, and boiling

Because of the high effective thermal conductivity and small mass, heat pipes have a relatively low thermal capacitance and a low overall thermal resistance when compared to many of the other components in the thermal system. In these types of situations, the heat pipe can typically be modeled using a lumped capacitance method assuming uniform temperature at any given point in time. This approach may also be applicable in situations where free convection is used as the heat addition or heat rejection mechanism due to the large convective resistance occurring between the heat pipe and either the hot or cold flow stream.

Frequently, however, it may be necessary to know or accurately predict the precise axial temperature distribution occurring in the heat pipe. For these situations, numerical techniques are normally utilized. These numerical techniques include finite difference models (FDMs) or finite element models (FEMs). Finite difference modeling techniques are generally more straightforward and require less computational time; however, finite element techniques are more appropriate for situations where the shape is extremely complex or where the resulting temperature distributions may be used as input to other computational schemes (Colwell and Modlin, 1992).

NOMENCLATURE

A Area (m^2)
c Speed of sound (m / s)

C Constant
d Diameter (m)
f Friction factor
F Molecular free energy, correction factor in equation 3.16, functional coefficient (J)
g Gravitational acceleration ($\mathrm{m/s^2}$)
j Molecular flux $\left(\frac{\mathrm{mol}}{\mathrm{m^2 \cdot s}}\right)$
k Thermal conductivity $\left(\frac{\mathrm{W}}{\mathrm{m \cdot K}}\right)$
K Wick permeability ($\mathrm{m^2}$)
L Length (m)
M Molecular weight
Ma Mach number
$\bar{m}$ Mass flow rate $\left(\frac{\mathrm{k_g}}{\mathrm{s}}\right)$
N Screen mesh number ($\mathrm{m^{-1}}$)
P Pressure ($\mathrm{N/m^2}$)
q Heat transfer, heat transport (W)
R Thermal resistance, or universal gas constant $\left(\frac{\mathrm{m^2K}}{\mathrm{W}}\right)\left(\frac{\mathrm{J}}{\mathrm{mol \cdot K}}\right)$
Re Reynolds number
r Radius, radial coordinate (m)
S Crimping factor
T Temperature (K)
t Time (s)
v Axial component of velocity (m/s)
w Groove width, wire spacing (m)
We Weber number
x Axial position, coordinate (m)

Greek Symbols

γ Ratio of specific heats
δ Groove depth (m)
ϵ Wick porosity
λ Latent heat of vaporization (J/kg)
μ Absolute viscosity $\left(\frac{\mathrm{N \cdot s}}{\mathrm{m^2}}\right)$
ρ Density ($\mathrm{kg/m^3}$)
θ Wetting angle (deg)
σ Surface tension (N/m)
Ψ Angle with respect to horizontal (deg)
ν Vapor

Subscripts

a	Adiabatic section, air
b	boiling
c	Capillary, capillary limitation, condenser
e	Entrainment, evaporator section
eff	effective
ext	external
f	Frictional, fin
h	Hydraulic
i	Inner, inertial
l	Liquid
m	Maximum
n	Nucleation
o	Outer
p	Pipe
ph	Phase transition
s	Sonic or sphere
v	vapor
ω	Wire spacing, wick
$\parallel$	Axial hydrostatic pressure
$+$	Normal hydrostatic pressure

REFERENCES

Alexander, E. G. Jr., 1972, "Structure-Property Relationships in Heat Pipe Wicking Materials," Ph.D. Thesis, North Carolina State University, Dept. of Chemical Engineering.

Asselman, G. A. A., and Green, D. B., 1973, "Heat Pipes," Phillips Technical Review, Vol. 16, pp. 169–186.

Cassel, S. D., 1991, "The Effect of Increasing Length on the Overall Conductance and Capacitance of Long Heat Pipes," Ph.D. Dissertation, Georgia Institute of Technology.

Chi, S. W., 1976, *Heat Pipe Theory and Practice*, McGraw-Hill New York.

Colwell, G. T., and Modlin, J. M., 1992, "Mathematical Heat Pipe Models," *Proc. 8th Int. Heat Pipe Conf.*, Vol. 1, pp. 162–166.

Ferrell, J. K., Alexander, E. G., and Piver, W. T., 1972, "Vaporization Heat Transfer in Heat Pipe Wick Materials," AIAA Paper No. 72-256.

Hendrix, W. A., 1989, "An Analysis of Body Force Effects on Transient and Steady-State Performance of Heat Pipes," Ph.D. Dissertation, Georgia Institute of Technology.

Ingram, T. J., Haman, L. L., Andes, G. M., Colwell, G. T., and Wepfer, W. J., 1984, "Non-Metallic Heat Pipes for Flue Gas Reheat," Report No. 84-JPGC-APC-7, American Society of Mechanical Engineers, New York.

Modlin, J. M., and Colwell, G. T., 1992, "Surface Cooling of Scramjet Engine Inlets Using Heat Pipe, Transpiration, and Film Cooling," *AIAA Journal of Thermophysics and Heat Transfer*, Vol. 6, No. 2.

Peterson, G. P., and Fletcher, L.S., 1987, "Effective Thermal Conductivity of Sintered Heat Pipe Wicks," *AIAA J. Thermophysics and Heat Transfer*, Vol. 1, No. 4, pp. 343–347.

Wells, K. J., Colwell, G. T., and Berry, J. T., 1985, "Two-Dimensional Numerical Simulation of Casting Solidification with Heat Pipe Controlled Boundary Conditions," *American Foundryman's Society Transactions*, pp. 84–95.

Chapter

6

TYPES AND CLASSIFICATION OF HEAT PIPES

6.1 INTRODUCTION

Heat pipes can be classified in several different ways: by operating temperature range (cryogenic, ambient, or liquid metal), by wicking structure (arterial or composite), or by function (rotating / revolving heat pipes, micro-heat pipes, variable conductance heat pipes, or thermal diodes) (Cotter, 1965). The following summary utilizes this latter approach and classifies heat pipes by function. The purpose of each of the following sections is to provide adequate background information for the engineer to understand the principal issues involved in the use of these different types of heat pipes. In this way, significant time and effort can be saved by eliminating or at least highlighting the potential problems involved in the various types or classifications of heat pipes used in the thermal control of electronic equipment, systems, and individual devices.

6.2 ROTATING AND REVOLVING HEAT PIPES

An extensive review of rotating and revolving heat pipes has been conducted by Wu and Peterson (1991b). This review begins by first defining the difference between rotating and revolving heat pipes, which are sometimes used interchangeably. For the purpose of this discussion, a rotating heat pipe will be defined as one that rotates longitudinally around its own central axis, as illustrated in Fig. 6.1*a*, such as the shaft of an electric motor. The heat pipe case may be uniform in cross section or may

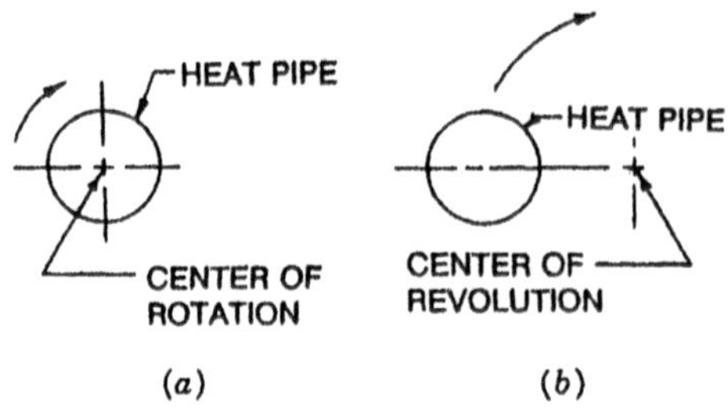

Figure 6.1 Rotating and revolving heat pipes.

be tapered to promote return of the working fluid to the evaporator. By comparison, a revolving heat pipe is one that rotates around an axis located some distance from and parallel to the central axis of the heat pipe, as illustrated in Fig. 6.1b, similar to the type that might be used to equalize the temperature in a rotating print head drum. The resulting performance characteristics of these two types of heat pipes are quite different and therefore are addressed independently in the following.

6.2.1 Rotating Heat Pipes with an Internal Taper

Most of the research performed in the 1970s focused on rotating heat pipes with an internal taper. Numerous papers were published by Daniels and his fellow workers at the University College of Swansea, Wales (Daniels and Al-Baharnah, 1979; Daniels and Al-Jumaily, 1972; Daniels and Williams, 1976–1979). Other related publications from this group include Ph.D. theses by Morris (1964), Al-Jumaily (1973), and Williams (1976). A detailed report was also prepared by Marto (1973a) for NASA, which summarized the research work done at the Naval Postgraduate School in Monterey, California. Other references include theses by Ballback (1969), Daley (1970), Newton (1971), Woodard (1972), Tucker (1974), Loynes (1976), and Wagenseil (1976) and papers by Marto et al. (1970), Marto (1973, 1976, 1984), Marto and Wagenseil (1979a,b), Marto and Wanniarachchi (1987), and Salinas and Marto (1991). Of these references, those of Daniels and Al-Jumaily (1972) and Marto (1973a) appear to be the most comprehensive. These two references both present theoretical investigations of rotating heat pipes with an internal taper starting with the momentum equation for the condensate and assuming

1. filmwise condensation,
2. laminar condensate film flow,
3. linear temperature distributions within the film,
4. constant fluid properties,
5. a fluid density much greater than that of the vapor, and

6. velocity gradients in the circumferential directions small enough to be negligible.

Utilizing these assumptions, a static force balance on a differential liquid element in the x direction, parallel to the tapered surface, yields an expression of the form

$$0 = \frac{\partial \tau}{\partial y} - \frac{\partial p}{\partial x} + \rho_l \omega^2 r \sin \alpha \tag{6.1}$$

and in the y direction, perpendicular to the tapered surface, yields

$$0 = -\frac{\partial p}{\partial y} - \rho_l \omega^2 r \cos \alpha \tag{6.2}$$

Although the two analyses are similar to this point, Daniels and Al-Jumaily (1972) further assumed constant vapor pressure and included the effect of gravity. In addition, the condenser film thickness was expressed as a fourth-order polynomial in the following form:

$$\mathrm{Sh}_x \left(\frac{\delta}{x}\right)^4 - \tfrac{1}{3}\,\mathrm{Dr}_x \left(\frac{\delta}{x}\right)^3 - \tfrac{1}{4}\,\mathrm{Re}_{vx} \left(\frac{\delta}{x}\right)^2 - 1 = 0 \tag{6.3}$$

$$\mathrm{Sh}_x = \frac{\rho_l^2(\omega^2 r - g)\lambda x^3 \sin \alpha}{4\mu_l k_l\,\Delta T}$$

$$\mathrm{Dr}_x = \frac{\rho_l \tau_v \lambda x^2 \cos \alpha}{\mu_l k_l\,\Delta T}$$

$$\mathrm{Re}_{vx} = \frac{\rho_l \bar{v} x \cos \alpha}{\mu_l}$$

The resulting equation was solved numerically, and the local heat flux was obtained by assuming one-dimensional Fourier heat transfer,

$$q = \frac{k_f\,\Delta T}{\delta} \tag{6.4}$$

The performance predictions obtained using this method related rates of heat transfer to the rotational speed, the temperature difference across the condensate film, the fluid properties, and the heat pipe geometry. However, because the radial force caused by the rotation and the gravitational force vary as a function of the rotational position (i.e., at the top of the rotational path, the radial force vector is upward and the gravitational vector is down while at the bottom of the rotational path the radial force vector and gravitational vector are both down), the gravitational effect indicated by this analysis may not be valid.

Marto (1973) neglected gravity and assumed one-dimensional isothermal turbulent vapor flow. The thermal resistance of the condenser wall was also considered. This resulted in a system of two first-order, nonlinear differential equations with unknowns of vapor velocity and condensate film thickness:

$$\frac{dV}{dx} = \frac{2(r_e + x \sin \alpha)(T_s - T_\infty)}{\rho_v(\delta/k_l + \delta/k_w + 1/h)\lambda r^2} - \frac{2V}{r}\left(\sin \alpha - \cos \alpha \frac{d\delta}{dx}\right) \tag{6.5}$$

and

$$\frac{d\delta}{dx} = \tan \alpha - \frac{\rho_v f V^2(P_1(x)/r + P_2(x)/2) + \rho_v V(\mu_l r^2/2\rho_l - 2P_1(x)\,(dV/dx))}{P_1(x) \cos \alpha(\rho_l \omega^2 r + 2\rho_v V^2/r)} \tag{6.6}$$

where

$$P_1(x) = (r_e + x \sin \alpha)\, \tfrac{1}{3}\delta^3 - \tfrac{5}{24}\, \delta^4 \cos \alpha$$

and

$$P_2(x) = (r_e + x \sin \alpha)\, \tfrac{1}{2}\delta^2 - \tfrac{1}{3}\delta^3 \cos \alpha$$

A Runge–Kutta numerical integration technique was used to solve these two equations simultaneously.

In addition to the analytical portion of the investigations, experiments were conducted by both research groups. Daniels and Al-Jumaily (1972) constructed a copper heat pipe with an overall length of 325 mm. The diameter of the evaporator and condenser ends were 55 and 30 mm, respectively, and the taper angle used was 2°. The heat pipe was tested at rotational speeds ranging from 600 to 1200 rpm and Arcton 113, Arcton 21, and water were all used as working fluids. Although, as illustrated in Fig. 6.2, the test results for Arcton 113 and Arcton 21 agree reasonably well with theoretical predictions, the experimental results using water as the working fluid were not in good agreement.

Marto (1973) constructed and tested a stainless steel heat pipe with a 1° internal taper. Water, ethyl alcohol, and Freon-113 were used as the working fluids and two rotational speeds were evaluated, 700 and 2800 rpm. As shown in Fig. 6.3, the theoretical results for water were from 10 to 20% higher than the experimental data, but for ethyl alcohol and Freon-113, shown in Figs. 6.4 and 6.5, respectively, the theoretical results were considerably lower than the experimental data. In general, this model was able to predict tapered rotating heat pipe performance to within ±20% for several working fluids. Combining the information presented in these and other investigations, several general conclusions can be made. These can be summarized as follows:

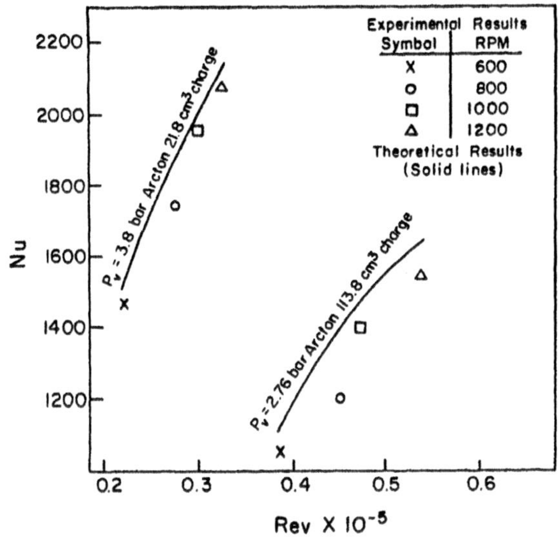

Figure 6.2 Nusselt–Reynolds relationship for different working fluids (Daniels and Al-Jumaily, 1992).

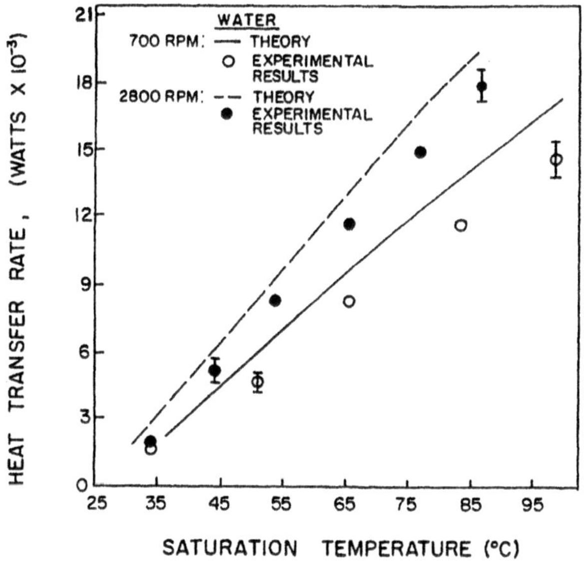

Figure 6.3 Comparison of condenser experimental results with theoretical prediction for water (Marto, 1973).

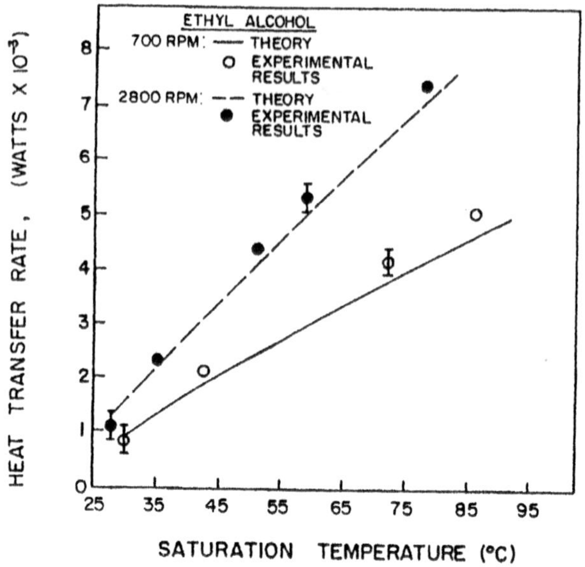

Figure 6.4 Comparison of condenser experimental results with the theoretical prediction for alcohol (Marto, 1973).

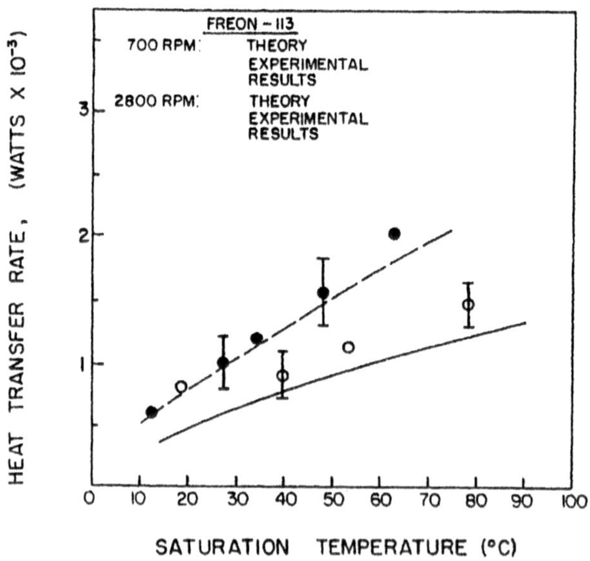

Figure 6.5 Comparison of condenser experimental results with the theoretical prediction for Freon-113 (Marto, 1973).

1. The rotational speed enhances the heat transfer capability of the heat pipe due to the decrease in the thermal resistance of the condensate film brought about by the reduction in the film thickness. The analyses of Ballback (1969) and Newton (1971) indicate that the heat pipe response approaches a theoretical relationship in which the heat transfer is proportional to the square root of the rotational speed.
2. The rate of heat transfer can be improved by increasing the length of the condenser, the taper angle, or the average radius of the heat pipe.
3. The optimum amount of working fluid depends on the fluid properties, heat flux, and rotational speed.
4. Water is the optimum working fluid in the temperature range of 20–100°C.
5. Dropwise condensation can improve the heat transfer performance of the condenser.
6. The presence of noncondensible gases can seriously impair the heat transfer performance of rotating heat pipes with an internal taper due to condenser blockage.

6.2.2 Rotating Heat Pipes without an Internal Taper

A majority of the literature found related to rotating heat pipes without an internal taper was the result of investigations conducted in the 1980s. The results of experimental investigations by Giessler et al. (1987), Katsuta et al. (1984, 1987), Krivosheev (1983), Krivosheev et al. (1979, 1980), Kukharskii et al. (1977), Lou and Chen (1990), Nakayama et al. (1984ab), Reddy (1989), Reddy et al. (1987, 1990), Shimizu and Yamazaki (1987), Vasiliev et al. (1984), and Vasiliev and Khrolenok (1983, 1990) indicate that the performance of rotating heat pipes without an internal taper is principally affected by rotational speed, working fluid fill ratio, properties of the working fluid, working temperature, inclination angle, and wicking structure. As was the case for rotating heat pipes with an internal taper, the rotational speed appears to be the most significant parameter.

Several researchers, including Lou and Chen (1990) and Katsuta et al. (1984), classified the flow inside a rotating heat pipe into one of four distinct patterns based on the rotational speed. These flow patterns are as follows:

1. *Pool Flow.* At low rotational speeds, most of the condensate remains at the bottom of the pipe to form a liquid pool. The remainder forms a very thin film around the upper part of the heat pipe. This flow pattern, shown in Fig. 6.6*a*, is dominated by gravity.
2. *Partial Annular Flow.* At moderate rotational speeds, the liquid pool is still primarily located at the bottom of the pipe but the upper liquid film becomes thicker as the rotational speed increases. Although both gravity and centrifugal acceleration are important in this flow

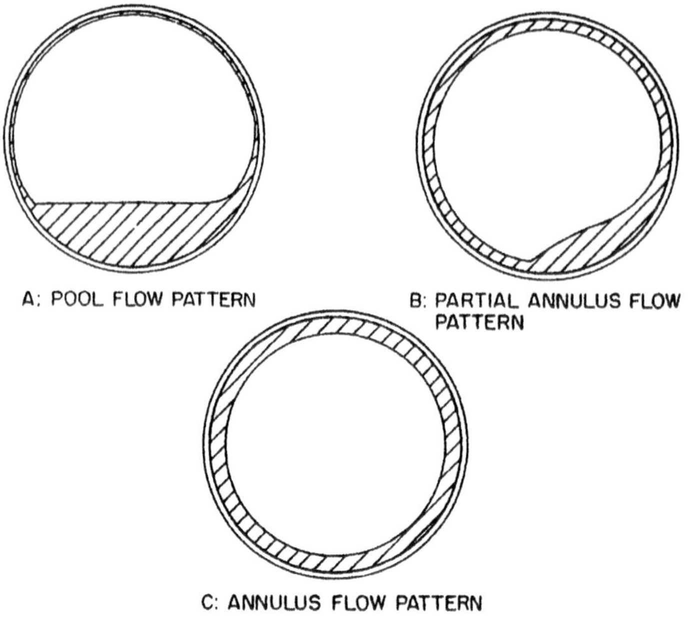

Figure 6.6 Schematic diagram of condensate flow pattern (Wu and Peterson, 1991b).

pattern, shown in Fig. 6.6*b*, the heat transfer coefficient is quite high and does not change significantly with changing rotational speed.

3. *Annular Flow.* At high rotational speeds, a point is reached where the centrifugal acceleration is much larger than the gravitational acceleration. At this "critical" point, the entire inside wall of the pipe is covered with a liquid layer and a stable liquid annulus is formed. This region is dominated by the centrifugal acceleration as shown in Fig. 6.6*c*, and once the critical rotational speed has been reached, the heat transfer coefficient drops rapidly to a level well below that occurring in the previous flow patterns.
4. *Hysteresis Phenomenon.* If the rotational speed is gradually decreased after the annular flow pattern has been established, the liquid annulus can be maintained at speeds much lower than the critical rotational speed. The annular flow pattern disappears when the rotational speed reaches a speed referred to as the "destruction" rotational speed.

A transparent experimental apparatus was constructed and used by Katsuta et al. (1984) to observe and classify the different flow patterns. The results verified the presence of four distinct patterns that were primarily dependent on the rotational radius and speed. Using the rotational radius and speeds given by several different authors, the critical rotational acceleration and the destruction acceleration have been calculated and are listed in Table 6.1. From this comparison, it appears as though the critical speed occurs at that point where the centrifugal acceleration is approximately 20 g. The destruction speed occurs at that point where the centrifugal acceleration is approximately 10 g. In addition, for all investigations, both the critical speed and the destruction speed increased with increasing temperature and decreasing tilt angle.

Most of the experimental results indicate that the heat transfer coefficient is at the highest level when the rotational speed is low. As the rotational speed increases, the flow passes through partial annular flow and the heat transfer coefficient decreases slowly, as shown in Fig. 6.7. When the rotational speed reaches the critical rotational speed, the flow pattern changes to annular flow and the heat transfer coefficient decreases rapidly to a very low level. Further increasing rotational speed does not improve the heat transfer performance of the heat pipe. Once the rotational speed surpasses the critical point, the annular flow pattern can be sustained (even at rotational speeds resulting in values well below the critical acceleration) and the heat transfer coefficient remains relatively

TABLE 6.1 Critical and Destructive Rotational Accelerations of Water–Copper Heat Pipe

Reference	Size $d(m)$	Inclined Angle $\alpha(°)$	Fill Ratio (%)	Temperature °C	Critical[a] Acceleration $\times g$	Destructive[a] Acceleration $\times g$
Katsuta et al. (1984)	0.038	0	20	100	13.6–30.6	6.4
	0.038	0	12	100	18.0–23.4	4.3
Lou and Chen (1990)	0.038	0	30	40	17.2–35.9	
	0.038	3	30	40	17.2–30.6	
	0.038	6	30	40	17.2–25.7	
	0.038	9	30	40	17.2–21.3	
	0.038	6	30	50	23.4–25.7	9.3
	0.038	9	30	50	18.0–25.7	99.0
Nakayama et al. (1984a)	0.037	0	12	–	16.0–20.7	
	0.037	0	25	–	16.8–35.0	
	0.037	0	17	–	16.8–35.0	7.5
Reddy et al. (1987)	0.0254	0	30	–	20.5–46.0	10.3[b]
	0.0254	0	30	–	36.4–65.7	9.1[c]
Reddy et al. (1990)	0.0254	0	30	100	17.2–46.0	10.3
	0.0254	2	30	100	20.5–46.0	
	0.0254	5	30	100	14.2–46.0	14.2
	0.0254	10	30	100	20.5–36.4	

[a] Calculated from information given in the original reference

[b] Self cooling

[c] Fan cooling

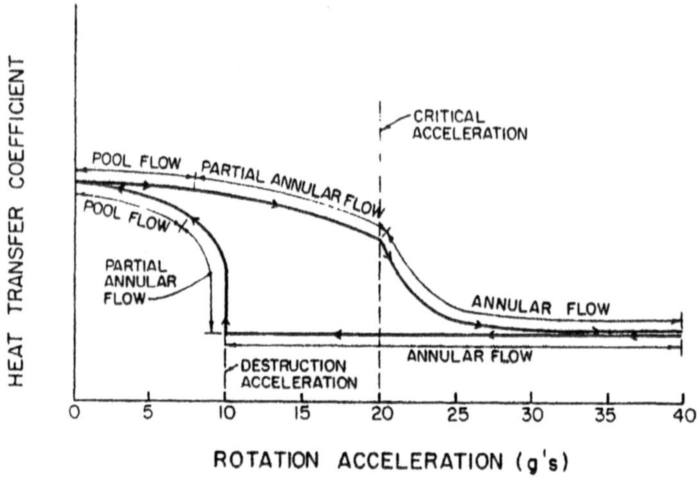

Figure 6.7 Effect of speed of revolution on the evaporation and condensation coefficient in horizontal rotating heat pipes. (Wu and Peterson, 1991b).

constant. Only when the rotational speed reaches a level sufficient to cause a rotational acceleration below the destructive acceleration will the flow return to the partial annular pattern.

In a majority of the past investigations a 30% by volume liquid fill ratio was recommended. However, comparisons have been given for fill ratios above and below 30% by Katsuta et al. (1984), Nakayama et al. (1984a), and Shimizu and Yamazaki (1987). Higher fill ratios further decrease the heat transfer coefficient in the annular flow pattern due to increasing liquid film thickness. Lower fill ratios result in partial evaporator dryout, which also decreases the heat transfer coefficient.

The effects of inclination angle relative to the gravitational field were investigated by Lou and Chen (1990) and Reddy et al. (1990). These results indicate that both critical rotational speed and destruction rotational speed decrease as the inclination angle increases. In general, inclined operation of a rotating heat pipe, either positive or negative, produces adverse effects on the heat transfer performance.

Copper–water rotating heat pipes have been evaluated over a wide range of operating temperatures by Katsuta et al. (1984) and Reddy et al. (1987). In the temperature range of 42–100°C, both investigations indicate that the heat transfer coefficient is not very sensitive to temperature variations. Other investigations include rotating heat pipes with axial grooves (Giessler et al., 1987; Vasiliev and Khrolenok, 1990), wicking structures (Nakayama et al., 1984*a*), a helical guide (Shimizu and Yamazaki, 1987), and some specific applications investigated by Jakes et al. (1984) and Quao (1990).

Independent theoretical investigations of rotating heat pipes without an internal taper have been conducted by Kamal et al. (1988), Semena and

Khmelev (1982, 1983), and Vasiliev and Khrolenok (1990). Katsuta et al. (1984) and Krivosheev et al. (1980) calculated the condensate liquid film thickness using a Nusselt-type analysis. The former corrected the theoretical correlation proposed by Leppert and Nimmo (1968) and used a semiempirical expression to calculate the filmwise condensation Nusselt number. Krivosheev et al. (1980) modified the Landau–Levich equation to compute the liquid film thickness.

Kamal et al. (1988) solved the equations of continuity, momentum, and energy numerically for both the vapor and porous regions of a cylindrical heat pipe with a porous wick using the SIMPLE method of Patankar (1980). Detailed vapor and liquid velocity distributions and pressure variations were plotted as a function of position. However, no experimental results are available to verify these theoretical predictions because a majority of the existing experimental data are for rotating heat pipes without any internal wicking structure. In this investigation, it was also found that the porous region was less affected by rotational speed and inclination than was the vapor region. While, normally, one would expect that the vapor flow field would not be affected by the gravitational force, it may be that the liquid, although not showing any significant changes with inclination, may influence the vapor behavior due to interactions occurring at the liquid–vapor interface.

Semena and Khmelev (1982, 1983) analyzed the hydrodynamic regimes of smooth-walled rotating heat pipes and assumed that the liquid charge was uniformly distributed over the internal surface of the heat pipe. The equations of continuity and conservation of momentum were written in cylindrical coordinates using an approximation for the boundary layer and were written with respect to a new variable $\epsilon = R - r$,

$$\frac{1}{R-\epsilon}\frac{\partial V_\phi}{\partial \phi} - \frac{\partial V_\epsilon}{\partial \epsilon} = 0 \tag{6.7}$$

$$\frac{1}{\rho(r-\epsilon)} V_\phi \frac{\partial V_\phi}{\partial \phi} = -V\left[\frac{\partial^2 V_\phi}{\partial \epsilon^2} - \frac{1}{r-\epsilon}\frac{\partial V_\phi}{\partial \epsilon} - \frac{V_\phi}{(r-\epsilon)^2}\right] - g \sin\phi \tag{6.8}$$

However, at steady state one would expect the azimuthal component of the velocity to be nearly equal to zero with the primary momentum transport resulting from the axial velocity, raising some question about the validity of these expressions.

Four operational regions were defined based on the Reynolds number ($\mathrm{Re} = (\omega\delta^2/V)$ and the Froude number ($\mathrm{Fr}_c = \omega^2 r/g$), namely, the region of rotation of a solid, the region of an entrained thin layer, the region of viscous flow, and the region of inertial flow. The relationships between the four regions and the Reynolds and Froude numbers are illustrated in Fig. 6.8. Different simplifications were applied for each of the corresponding flow regions and Equations 6.7 and 6.8 were solved for the liquid film flow

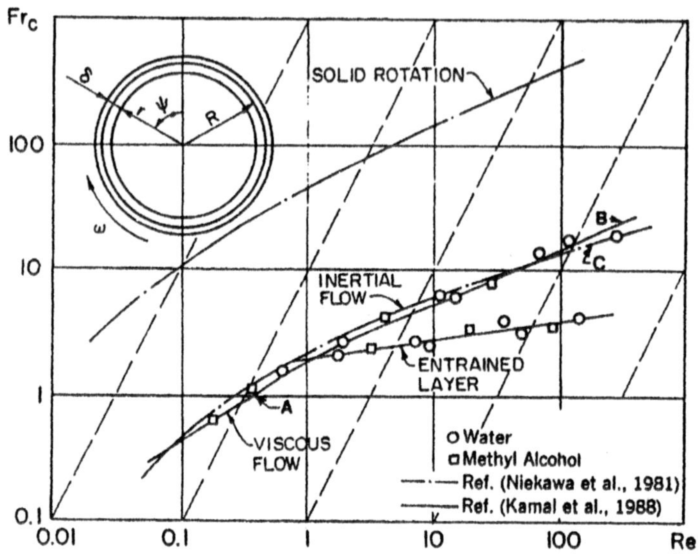

Figure 6.8 System of coordinates and monograph of flow regimes: (*A*) $Fr_c = 2.2\,Re^{0.77}$; (*B*) $Fr_c = 2.2\,Re^{0.44}$; (C) $Fr_c = 2.2\,Re^{1.13}$, $Fr_c = \omega^2 R/g$, $Re = \omega(\bar{\delta})^2/V$ (Wu and Peterson, 1991b).

velocity and film thickness. For the solid rotation region, the liquid film thickness was assumed constant; for the viscous flow region, the effects of inertia and surface tension were neglected; and for the inertial flow region, the liquid film was divided into the near-wall region and the free surface region. Using these assumptions, the momentum equation was then expressed as

$$\frac{\partial V_\phi}{\partial t} = V\,\frac{\partial^2 V_\phi}{\partial \epsilon^2} = -g \sin \omega t \tag{6.9}$$

In the entrained layer region, both viscous and inertial forces were considered and Equations 6.7 and 6.8 were solved. In addition, an empirical expression of the form

$$\frac{\delta_e}{r} = 2 \times 10^{-2}\,\frac{Ca^{0.68}}{\bar{\Delta}^{0.36}}\,e^{11.84} \tag{6.10}$$

was developed, where $Ca = \omega\mu_l r/\sigma$ and $\bar{\Delta} = \delta/r$. This expression and the numerical solution of the entrained layer region produced results that typically agreed to within 2.5%. The results of other investigations indicate that the numerical solution of Semena and Khmelev (1982, 1983) can be used to accurately predict the experimental data.

Vasiliev and Khrolenok (1990) solved the basic momentum equations for the liquid region in rotating heat pipes analytically to find the liquid film thickness for several different axial groove shapes. Using this approach, the temperature drop of the heat pipe was expressed as a function of power input, rotational speed, and the other physical parameters. To obtain experimental data, a copper heat pipe was constructed with 1-mm rectangular grooves. The heat pipe was tested over ranges of several parameters, including rotational speed, working temperature, and power input with several working fluids (i.e., water, ethanol, and acetone). The results indicated a fair agreement when compared with the theoretical prediction. In addition, an empirical formula was developed to calculate the resulting heat flux over the evaporator surface.

Krivosheev et al. (1980) attempted to develop an expression for the optimum rotational speed as a function of the maximum heat transfer coefficient by approximating the liquid film thickness using the Landau–Levich equation. The experimental results indicated that there was a large range of rotational speeds within which the heat transfer coefficient was at or near a maximum value. This observation supports the graphical representation illustrated previously in Fig. 6.7.

The results of these and other investigations for rotating heat pipes without an internal taper can be summarized as follows:

1. Four distinct flow patterns exist: pool flow, partial annular flow, annular flow, and hysteresis flow.
2. The critical speed occurs at a rotational velocity sufficient to result in a force of approximately 20*g* and the destruction speed occurs at a rotational velocity sufficient to result in a force of approximately 10*g*.
3. The inclination angle affects both the critical rotational speed and destruction rotational speed with increases in the inclination angle decreasing these values.
4. Variation of the inclination angle produces adverse effects on the heat transfer performance.

6.2.3 Revolving Heat Pipes

In contrast to rotating heat pipes, only a limited number of investigations have been conducted on revolving heat pipes, including investigations by Bontemps et al. (1984), Chen and Lou (1990), Chen and Tu (1986, 1987), Curtila and Chataing (1984), Keiyou and Maezawa (1990), Mochizuki and Shiratori (1980), and Niekawa et al. (1981). As was the case for rotating heat pipes, the speed of revolution was the most important parameter affecting the heat transfer performance.

When the revolving axis for these types of heat pipes is in a horizontal position, the gravitational force acts perpendicular to the heat pipe axis, causing an oscillatory gravitational field that alternatively either assists or retards the centrifugal force depending on the rotational position. Most of the test results indicate that both evaporator and condenser heat transfer coefficients are low when the heat pipe is at rest. The low evaporator heat transfer coefficient is the result of all the liquid remaining at the bottom of the heat pipe, limiting the area of evaporation to the pool surface. In the case of the condenser, without centrifugal acceleration, the condensate film is relatively thick, which causes the low condensation heat transfer coefficient.

As shown in Fig. 6.9, when the speed of revolution increases, both the evaporator and condenser heat transfer coefficients increase rapidly until the centrifugal acceleration reaches approximately 1*g*. At this point, the centrifugal acceleration resulting from the speed of revolution, acting together with the fixed gravitational acceleration, causes the liquid to splash, substantially enhancing the surface area of evaporation and condensation. When the centrifugal acceleration increases to above 1*g*, the heat transfer coefficient decreases and stabilizes at a level similar to that at which it started when the effect of gravity was negligible. In this region, no more splashing occurs and the evaporation surface is again limited to the pool surface. Further increasing the speed of revolution increases the condenser heat transfer coefficient because the greater centrifugal acceleration makes the condensate film thinner. The evaporator heat transfer coefficient, however, remains approximately the same.

Vertical revolving heat pipes were investigated for the case where the gravitational acceleration was always perpendicular to the centrifugal acceleration. Because of the low speed of revolution, no splashing occurred. However, as was the case for horizontal revolving heat pipes, at

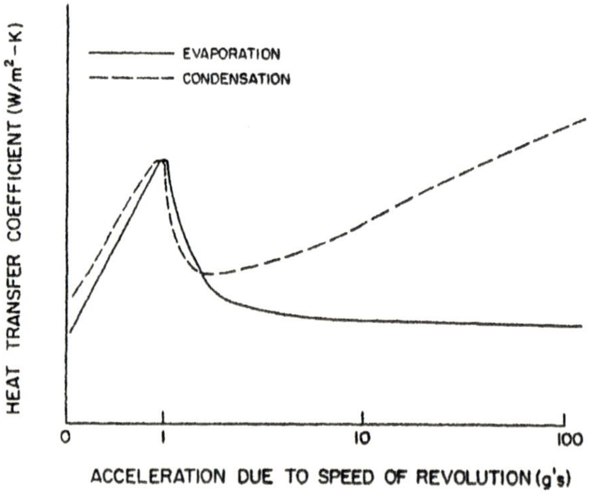

Figure 6.9 Effect of speed of revolution on evaporation and condensation coefficient (Wu and Peterson, 1991b).

speeds of revolution sufficient to cause centrifugal accelerations greater than 1g, the heat transfer coefficient increased gradually as the revolving speed increased. Although similar to the case of horizontal pipes, it is unlikely to expect splashing to occur because there is no driving mechanism. Instead, this increase in the heat transfer coefficient is probably due to a thinning of the liquid layer on the pipe wall caused by the increased speed of revolution.

Chen and Lou (1990) and Chen and Tu (1987) investigated the heat transfer performance of inclined revolving heat pipes. When the condenser of the inclined heat pipe was close to the revolving axis, the heat transfer performance improved because the centrifugal acceleration assisted the liquid flow back to the evaporator. In this investigation, the circumferential temperature distributions were reported, with lower temperatures found at the outside surfaces where the liquid pool was located.

The test results of Keiyou and Maezawa (1990) indicated that the optimum fill ratio for revolving heat pipes was approximately 10% by volume. However, Curtila and Chataing (1984) concluded that the fill ratio must be between 25 and 30%. This disparity may be the result of variations in the power input used in the two investigations. When the fill ratio is low, the heat transfer performance will be good provided the dryout limit has not been exceeded because the liquid pool is shallow. If, however, the power input exceeds the dryout limit, a higher fill ratio will be necessary to keep the evaporator wet.

A majority of the investigations concerned with revolving heat pipes attempted to predict the heat transfer coefficient theoretically. Nusselt types of analyses were used to compute the liquid film thickness, and several dimensionless semiempirical expressions were developed (i.e., the Nusselt number of the revolving heat pipe is a function of some other dimensionless numbers, such as the Galileo number, Prandtl number, acceleration ratio, and / or Reynolds number). Among these, the prediction of Keiyou and Maezawa (1990) best fits the accumulated experimental data. This empirical expression is in the form

$$\mathrm{Nu} = 0.35\left(\frac{\mathrm{Gr}\,\mathrm{Pr}\,G}{H}\right)^{0.25}\left(\frac{1}{V*}\right)^{0.4} \tag{6.11}$$

Figure 6.10 compares the experimentally measured condensation heat transfer coefficient with the value predicted by equation 6.11. As illustrated, the agreement is fairly good.

Bontemps et al. (1984) performed a kinematic analysis of revolving heat pipes and reached several qualitative conclusions. In this investigation, the momentum equations were solved based on Nusselt's theory, and an expression of power input under high rotational speeds was obtained as a function of the speed of revolution, the inclination angle, the temperature difference, the geometry, and the working fluid properties. The resulting

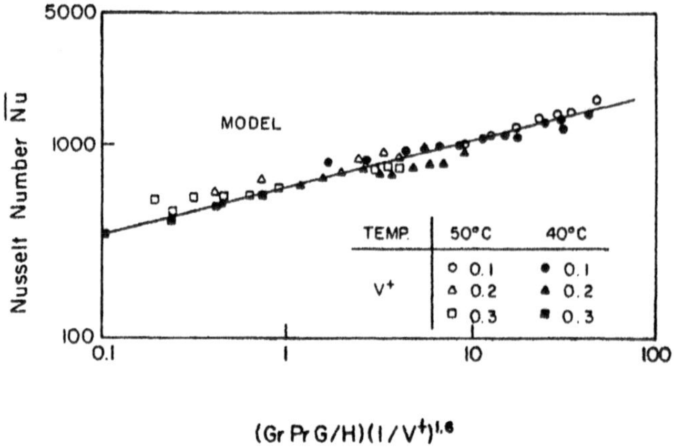

Figure 6.10 Prediction of condensation heat transfer coefficient (Keiyou and Maezawa, 1990).

equation indicates that the heat flux is proportional to the square root of the rotational speed, a result similar to that obtained for rotating heat pipes.

Several questions resulted from this investigation. First, although Bontemps et al. (1984) concluded that there was circulation inside the liquid pool when the centrifugal acceleration was very high compared to the gravitational acceleration, this may not be true because, if gravity were neglected, the centrifugal acceleration field would be a potential field and the liquid would stay in a position where the potential energy was minimized. This would require extra energy to maintain the liquid in circulation. Second, the power input expression mentioned above indicates that the heat flux is proportional to the one-fourth power of the inclined angle. When the heat pipe is parallel to the revolving axis, the heat flux becomes zero, and this is clearly not the case.

As before the results of these and other investigations conducted on revolving heat pipes can be summarized as follows:

1. When the revolving axis is in a horizontal position, an oscillatory gravitational field exists that can assist or retard the centrifugal force.
2. The evaporator and condenser heat transfer coefficients are lowest when the heat pipe is at rest. As the speed of revolution increases both the evaporator and condenser heat transfer coefficients increase rapidly until the centrifugal acceleration reaches approximately 1g.
3. At approximately 1g, the centrifugal acceleration acting together with the fixed gravitational acceleration causes the liquid to splash, substantially enhancing the surface area of evaporation and condensation.
4. Above 1g, the heat transfer coefficient decreases and stabilizes at a

level similar to that at which it started when the effect of gravity was negligible.

5. As was the case for horizontal heat pipes, increases in the speed of revolution results in increases in the heat transfer coefficient, due to a thinning of the liquid layer on the pipe wall.
6. The optimum fill ratio for revolving heat pipes is between 10 and 30% by volume, and depends on the power input used.

6.2.4 Summary of Rotating and Revolving Heat Pipes

A summary of the results of the investigations presented and discussed yields several general conclusions. First, the heat transfer characteristics of rotating and revolving heat pipes depend primarily on rotational speed (or the speed of revolution), geometry, properties of the liquid charge, and working fluid fill ratio, along with the position relative to gravity. Among these parameters, the rotational speed is most important. This is due to the dependence of the heat transfer coefficient on the liquid flow pattern of the working fluid inside. Most theoretical analyses indicate that the heat transfer rate is proportional to the square root of rotational speed (i.e., $q \propto \sqrt{\omega}$) and support the hypotheses presented here that the heat transfer coefficient behaves as described in Figs. 6.7 and 6.9 for rotating and revolving heat pipes, respectively.

The second general conclusion is that the heat transfer performance is enhanced for a rotating heat pipe with an internal taper because the centrifugal pumping force promotes flow from the condenser back toward the evaporator. In this case, increasing rotational speed will always increase the heat transfer coefficient. Rotating heat pipes without internal tapers perform well when the centrifugal acceleration is less than 20g. Above this level, however, the heat pipes perform poorly due to an annular flow pattern set up within the heat pipe. A hysteresis phenomenon is unique for heat pipes of this category.

Although revolving heat pipes perform well at low speeds of revolution, increasing speeds improve the condenser heat transfer performance by decreasing condensate film thickness. The evaporator heat transfer coefficients, however, are lower at higher speeds of revolution because the evaporation area is limited to the pool surface. As a result, there is no critical rotational speed for revolving heat pipes. Finally, as is the case for stationary heat pipes, the presence of noncondensible gases can seriously impair the heat transfer performance of both rotating and revolving heat pipes.

6.3 MICRO–HEAT PIPES

In 1984, Cotter first introduced the concept of very small "micro" heat pipes incorporated into semiconductor devices to promote more uniform temperature distributions and improve thermal control (Cotter, 1984). Since that initial conceptual description, which defined a micro–heat pipe as one "so small that the mean curvature of the liquid–vapor interface is necessarily comparable in magnitude to the reciprocal of the hydraulic radius of the total flow channel," significant additional work has been performed (Peterson, 1992).

To date, investigations have been conducted on many different types of relatively small heat pipes (Dean, 1976; Basiulis et al., 1987; Murase et al., 1987; Babin and Peterson, 1990; Sotani et al., 1990). Most of these devices, however, are in reality only miniaturized versions of larger more conventional heat pipes. To better understand what the term "*micro–heat pipe*" implies, Babin et al. (1990) expressed Cotter's initial definition of a micro–heat pipe mathematically as

$$K \propto \frac{1}{r_h} \tag{6.12}$$

where K is the mean curvature of the liquid–vapor interface and r_h is the hydraulic radius of the flow channel. Then by assuming a constant of proportionality of 1 and multiplying both the mean curvature of the liquid–vapor interface and the hydraulic radius by the capillary radius r_c, a dimensionless expression was developed (Peterson, 1988a):

$$\frac{r_c}{r_h} \geq 1 \tag{6.13}$$

This expression better defines a micro–heat pipe and helps to differentiate between small versions of conventional heat pipes and the more recently developed micro–heat pipes.

In practical terms, a micro–heat pipe consists of a small noncircular channel that utilizes the sharp angled corner regions as liquid arteries. While the initial application proposed by Cotter (1984) involved the thermal control of semiconductor devices, a wide variety of other uses have been investigated or proposed. These include the removal of heat from laser diodes (Mrácek, 1988) and other small localized heat-generating devices (Peterson, 1988b; Stulc and Horváth, 1988), the thermal control of photovoltaic cells (Peterson 1987a,b), the removal or dissipation of heat from the leading edge of hypersonic aircraft (Camarda et al., 1994), and applications involving the nonsurgical treatment of cancerous tissue through either hyper- or hypothermia (Anonymous, 1989; Fletcher and Peterson, 1990).

The fundamental operating principles of micro–heat pipes are essentially the same as those occurring in larger, more conventional heat pipes. Heat applied to one end of the heat pipe vaporizes the liquid in that region and forces it to move to the cooler end where it condenses and gives up the latent heat of vaporization. This vaporization and condensation process causes the liquid–vapor interface in the liquid arteries to change continually along the pipe, as illustrated in Fig. 6.11, and results in a capillary pressure difference between the evaporator and condenser regions. This capillary pressure difference promotes the flow of the working fluid from the condenser back to the evaporator through the triangular shaped corner regions. These corner regions serve as liquid arteries; thus no wicking structure is required.

Since the initial conceptualization in 1984, numerous analytical and experimental investigations of micro–heat pipes have been conducted. The following review presents a summary of these investigations in an attempt to assess the current state-of-the-art for investigators new to the field and to focus future investigations for those already working in this area. For clarity, this review is divided into two major sections: one that focuses on

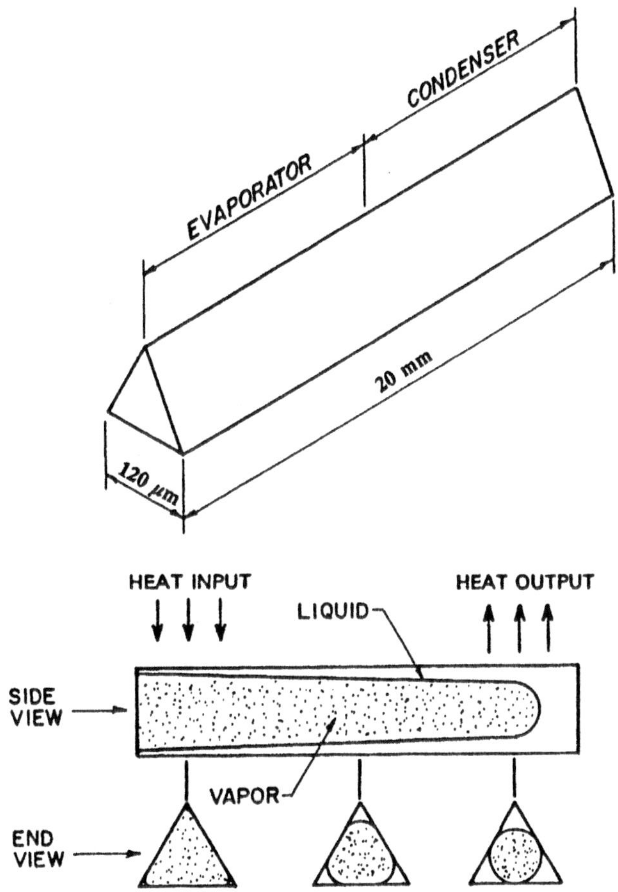

Figure 6.11 Micro–heat pipe operation.

analytical and numerical modeling and another that focuses on the experimental investigations conducted on micro–heat pipes.

6.3.1 Modeling Micro–Heat Pipe Performance

While many of the analytical models developed for use in predicting the steady-state and transient performance characteristics of micro–heat pipes are similar to those developed for larger, more conventional heat pipes, the relative importance of many of the parameters is quite different. For example, because of their size, micro–heat pipes are extremely sensitive to the amount of working fluid present, and care must be taken to ensure proper wetted conditions without flooding. Early attempts to model micro–heat pipes began with simple steady-state pressure balance models and proceeded to the development of both transient numerical models and three-dimensional models of micro–heat pipe arrays.

Steady-State Modeling

The first steady-state model was presented by Cotter (1984). Starting with the momentum equation, which for steady-state parallel flow simplifies to

$$\mu\left(\frac{\partial^2 w}{\partial x^2} + \frac{\partial^2 w}{\partial y^2}\right) = \frac{dP}{dz} \tag{6.14}$$

and assuming uniform cross-sectional area and no-slip conditions at the boundaries, this expression is transformed to

$$\frac{\delta w'}{\delta x'^2} + \frac{\delta w'}{\delta y'^2} = -1 \tag{6.15}$$

where x', y' and w' are given by

$$x = A^{0.5}x', \qquad y = A^{0.5}y', \qquad w = \frac{A}{\mu}\frac{dP}{dz}w' \tag{6.16}$$

Substitution of these terms into the local mass flow equation,

$$\dot{m} = \rho \int\int w\, dx\, dy \tag{6.17}$$

results in an expression for the mass flow rate as

$$\dot{m} = -\frac{K'\rho A^2}{8\pi\mu}\frac{dP}{dz} \tag{6.18}$$

where K' is a shape factor given by

$$K' = 8\pi \int \int w' \, dx' \, dy' \tag{6.19}$$

Solving for both the liquid and vapor pressure differential in the vapor and liquid cases and substituting yield

$$\frac{8\pi \nu_v \dot{m}_v}{K_v' A_v^2} = \frac{8\pi \nu_l \dot{m}_l}{K_l' A_l^2} - \frac{d}{dz}\frac{\sigma}{r_m} \tag{6.20}$$

Combining this relationship with the continuity expression results in a first-order ordinary differential equation, which relates the radius of curvature of the liquid–vapor interface to the axial position along the pipe, or

$$\frac{dr}{dz} = \frac{8\pi \dot{q} r h(z/L)}{\sigma\lambda}\left(\frac{\nu_v}{K_v'(A - \beta^2 r)^2} + \frac{\nu_l}{K_l' \beta^4 r^4}\right) \tag{6.21}$$

Assuming a constant cross-sectional area, this equation can be solved to yield

$$\dot{q}_m = \frac{0.16\beta(K_l' K_v')^{0.5}}{8\pi H(L)} \frac{\sigma\lambda}{\nu_l} \frac{\nu_l}{\nu_v}^{0.5} \frac{A^{0.75}}{L} \tag{6.22}$$

where $H(L)$ is the integral of $h(z/L)$ over the length of the pipe.

Somewhat later, Peterson (1988a) and Babin and Peterson (1990) developed a steady – state model for a trapezoidal micro–heat pipe with a cross-sectional area similar to the one illustrated in Fig. 6.12. Using the conventional steady-state modeling techniques outlined by Chi (1976), the maximum heat transport capacity of this heat pipe was found to be primarily governed by the capillary pumping pressure, or

$$\Delta P_c \geq \Delta P_+ + \Delta P_{\|} + \Delta P_l + \Delta P_v \tag{6.23}$$

where

$$\Delta P_c = \frac{2\sigma}{r_{ce}} \cos\theta \tag{6.24}$$

$$\Delta P_+ = \rho_l g d_v \cos\psi \tag{6.25}$$

$$\Delta P_{\|} = \rho_l g L \sin\psi \tag{6.26}$$

$$\Delta P_l = \left(\frac{\mu_l}{K A_l \lambda \rho_l}\right) L_{\text{eff}} q \tag{6.27}$$

and

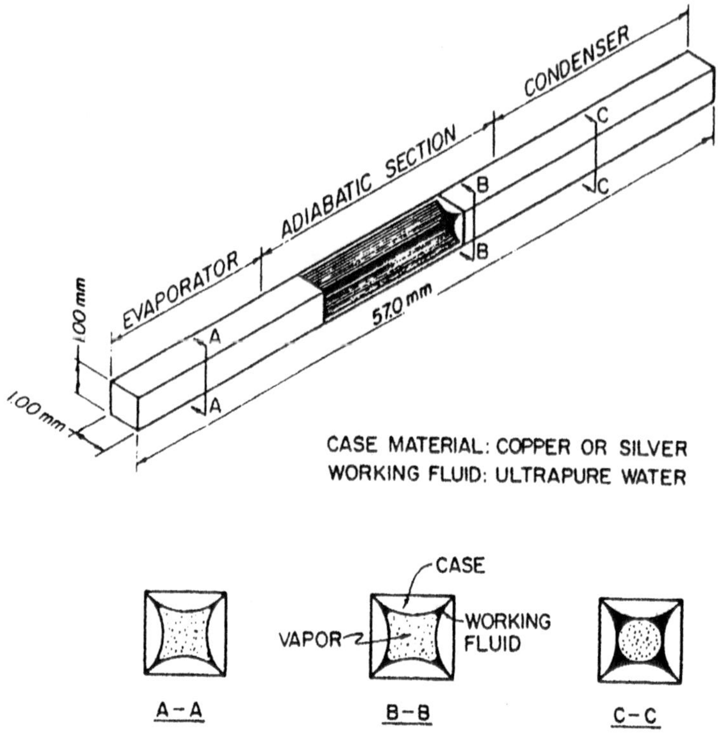

Figure 6.12 Trapezoidal micro–heat pipe modeled by Babin et al. (1990).

$$\Delta P_v = \left(\frac{C(f_v \, \mathrm{Re}_v)\mu_v}{2(r_{h,v})^2 A_v \rho_v \lambda} \right) L_{\mathrm{eff}} q \tag{6.28}$$

The performance limitations resulting from the model presented by Cotter (1984) and by Babin and Peterson (1990) are compared in Fig. 6.13. As illustrated, this comparison indicates significant differences in the capillary limit predicted by the two models. These differences have been analyzed and found to be the result of specific assumptions made in the initial model formulations.

Gerner (1989, 1990) provided a critique of these two steady-state models and indicated that the major contributions of the model developed by Babin et al. (1990) were the inclusion of the effect of gravity and the recognition of the significance of the vapor pressure losses. However, the assumption that the pressure gradient in the liquid flow passages was similar to that occurring in Hagen–Poiseuille flow was questioned. As a result, Gerner (1990) presented a force balance and scaling argument for the liquid pressure drop as

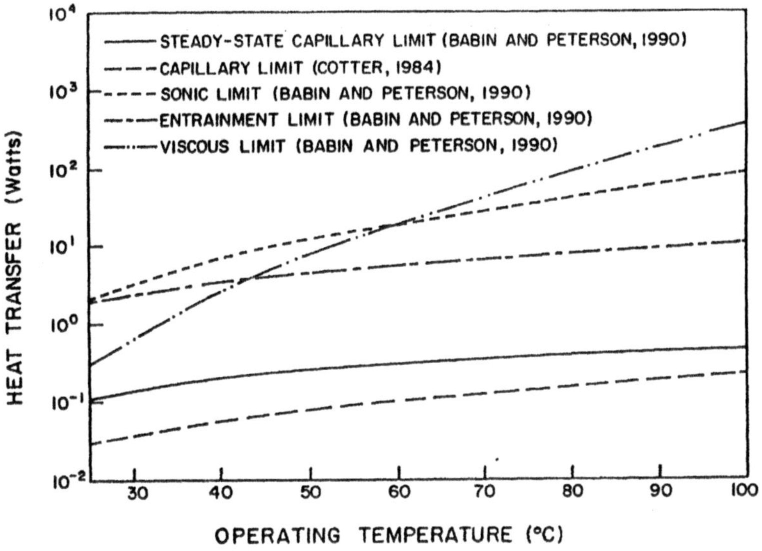

Figure 6.13 Limitations of trapezoidal micro–heat pipe with water at a horizontal orientation as predicted by Cotter (1984) and Babin et al. (1990).

$$\Delta P_l \frac{\pi}{4}[d^2 - (d - 2\delta)^2] = \tau_i \pi (d - 2\delta)L + \rho_l g \frac{\pi}{4}[d^2 - (d - 2\delta)^2]L \sin \psi \tag{6.29}$$

and estimated the average film thickness to be approximately one-fourth the hydraulic radius. This resulted in a capillary limitation of

$$\Delta P_c = \rho_l g L \sin \psi + \frac{8192}{3} \frac{V_v \dot{m}_v L}{\pi d^4} \tag{6.30}$$

In addition to this modification to the capillary limit, Gerner (1990) hypothesized that the capillary limit may in fact never be reached due to Kelvin–Helmholtz instabilities occurring at the liquid–vapor interface. This hypothesis was supported by an analysis that used the stability analysis for countercurrent flow in tubes developed by Tien et al. (1979). By minimizing the resulting equations, the wavelength was found to be

$$\lambda_{cr} = 2\pi \left(\frac{\sigma}{\rho_v g} \right)^{1/2} < 1 \tag{6.31}$$

or approximately 1 cm for atmospheric water and methanol. Because this length is long with respect to the characteristic wavelength

$$\frac{2\pi}{L}(\sigma\rho_v g)^{1/2} < 1 \tag{6.32}$$

gravity is assumed to be the predominant stabilization mechanism. The results of this technique indicated that if the inertial forces are of the same order of magnitude as the restoring forces, the instabilities will not be damped and flooding in the heat pipe will occur. This relationship is expressed in terms of the heat flux as

$$\dot{q}_{\max} = \rho_v \lambda \left(\pi \frac{d^2}{4} \right) \left(\frac{\rho_l}{\rho_v} \frac{gd}{16} \right)^{0.5} \tag{6.33}$$

The question of which limit is most appropriate, the traditional capillary limit proposed by Babin et al. (1990) or the interfacial instability limit proposed by Gerner (1990), can be resolved by evaluating the shape and physical dimensions of the specific micro–heat pipe being considered.

Transient Modeling

In addition to the steady-state models described above, a transient model has been developed by Wu and Peterson (1991a). This model utilizes the relationship presented by Collier (1981) and later by Colwell and Chang (1984) to determine the free molecular flow mass flux of evaporation, that is,

$$j = \left(\frac{M}{2\pi RT} \right)^{0.5} (P_s - P_v) \tag{6.34}$$

In the development of the numerical model, the evaporation–condensation rate is assumed to be proportional to the liquid–vapor interface area in each section of the heat pipe,

$$\Delta \dot{m}_{ec} = j\, \Delta A_{ec} = j\omega_{ec}\, \Delta x \tag{6.35}$$

and the amount of heat released or absorbed in any section is determined from an energy balance

$$\Delta Q_{ec} = \lambda\, \Delta \dot{m}_{ec} \tag{6.36}$$

Based on the momentum equation, the viscous pressure drop due to friction is expressed as

$$\Delta P_f = (2\mu\, \Delta x V)/r_h^2 \tag{6.37}$$

and the total pressure drop due to the momentum change is

$$\Delta P_m = (V\, \Delta \dot{m} + \dot{m}\, \Delta V + \Delta \dot{m}\, \Delta V)/A \tag{6.38}$$

The single boundary condition utilized in the evaporator section is the time dependent heat flux. For a specific input heat flux, the saturation pressure at a given location is obtained by using a combination of the mass flux expression and the energy conservation equation. Assuming that a temperature change at the boundary results in a corresponding temperature change of the liquid and neglecting the temperature gradient in the liquid, these equations take the following form:

$$P_s = \frac{\dot{q}_{ec}}{(M/2\pi RT_l)^{1/2}\,\Delta x\, W_{ec}\lambda} + P_v \tag{6.39}$$

$$\dot{q}_l = \frac{c_{p,l}\rho_l A_l\,\Delta x\,\Delta T_l}{\Delta t} \tag{6.40}$$

$$\dot{q}_s = \frac{c_{p,s}\rho_s A_s\,\Delta x\,\Delta T_s}{\Delta t} \tag{6.41}$$

$$\dot{q}_{in} = \dot{q}_{ec} + \dot{q}_l + \dot{q}_s \tag{6.42}$$

Because the liquid temperature T_l is a function of the saturation pressure P_s, these equations are coupled and can be solved using an iterative method with relaxation to obtain values for T_l, $\dot{q}_{ec}$, $\dot{q}_l$, and $\dot{q}_s$. Because the difference between the boundary and liquid temperatures, ΔT_{bl}, is proportional to the input heat flux, the boundary temperature can be obtained by adding the temperature drop across the boundary to the liquid temperature, or ΔT_{bl} to T_l.

In the adiabatic section, the governing equations for each element are expressed as

$$\dot{q}_{ec} = \lambda W_{ec}\,\Delta x\left(\frac{M}{2\pi RT_l}\right)^{1/2}(P_s - P_v) \tag{6.43}$$

$$\dot{q}_l = \frac{c_{p,l}\rho_l A_l\,\Delta T_l\,\Delta x}{\Delta t} \tag{6.44}$$

$$\dot{q}_s = \frac{c_{p,s}\rho_s A_s\,\Delta T_b\,\Delta x}{\Delta t} \tag{6.45}$$

$$\dot{q}_{ec} + \dot{q}_l + \dot{q}_s = 0 \tag{6.46}$$

By assuming no temperature drop at the liquid/case boundary (i.e., $T_l = T_b$), these equations can also be solved using an iterative technique to obtain $\dot{q}_{ec}$, $\dot{q}_l$, $\dot{q}_s$, and ΔT_l.

In the condenser region, the boundary temperature of the heat pipe is assumed to be constant. This results in governing equations similar to those used in the adiabatic section,

$$\dot{q}_{ec} = \lambda w_{ec}\, \Delta x \left(\frac{M}{2\pi R T_l} \right)^{1/2} (P_s - P_v) \tag{6.47}$$

$$\dot{q}_l = \frac{c_{p,l} \rho_l A_l\, \Delta T_l\, \Delta x}{\Delta t} \tag{6.48}$$

$$\dot{q}_s = \frac{c_{p,s} \rho_s A_s\, \Delta T_b\, \Delta x}{\Delta t} \tag{6.49}$$

$$\dot{q}_{ec} + \dot{q}_l + \dot{q}_s = \dot{q}_{out} \tag{6.50}$$

Because the evaporation/condensation rate is extremely sensitive to changes in the vapor pressure, a predictor–corrector treatment was applied to obtain a stable solution. Initially, the evaporation rate was assumed to be equal to the condensation rate and the vapor pressure was calculated. This vapor pressure was then used to compute the change in the mass of vapor and, hence, the difference between evaporation and condensation. The original assumption of equal evaporation and condensation rates was then checked and the values updated. The solution procedure consisted of initializing the variables and arrays and, by iteration, computing the boundary, liquid, and vapor temperatures along with the evaporation and condensation rates, the pressure drops, and the flow rates for each section. The resulting radii of curvature for the liquid–vapor interface were then computed and the liquid–vapor mass flow rate corrected using the resulting vapor pressure.

In a first attempt to verify the numerical model, the heat pipe originally modeled by Babin et al. (1990) and shown in Fig. 6.12 was analyzed using this transient model. The liquid and vapor cross-sectional areas were defined in terms of the geometry, as shown in Fig. 6.14. The results indicated several interesting phenomena. First, the vapor temperature variations were proportional to the heat input and without dryout or flooding, the effective conductivity of the heat pipe was independent of time shortly after startup. In contrast, a fairly long period of time was required for the other thermal hydraulic parameters to reach their steady-state values. For example, the time required for the outer case to reach steady state ranged from 10 to 30 s after full heat input. This time was found to be strongly dependent on the wetting angle with smaller wetting angles and higher temperatures resulting in shorter time intervals required (Wu and Peterson, 1991b).

Second and perhaps more interesting was the behavior of the liquid mass flow rate as a function of time. Figures 6.15 and 6.16 illustrate the changes in the liquid mass flow rate and liquid cross-sectional area, respectively, as a function of time. As illustrated, during the time immediately after the application of the heat to the evaporator, the liquid and vapor flow in the evaporator are both in the same direction. This reverse

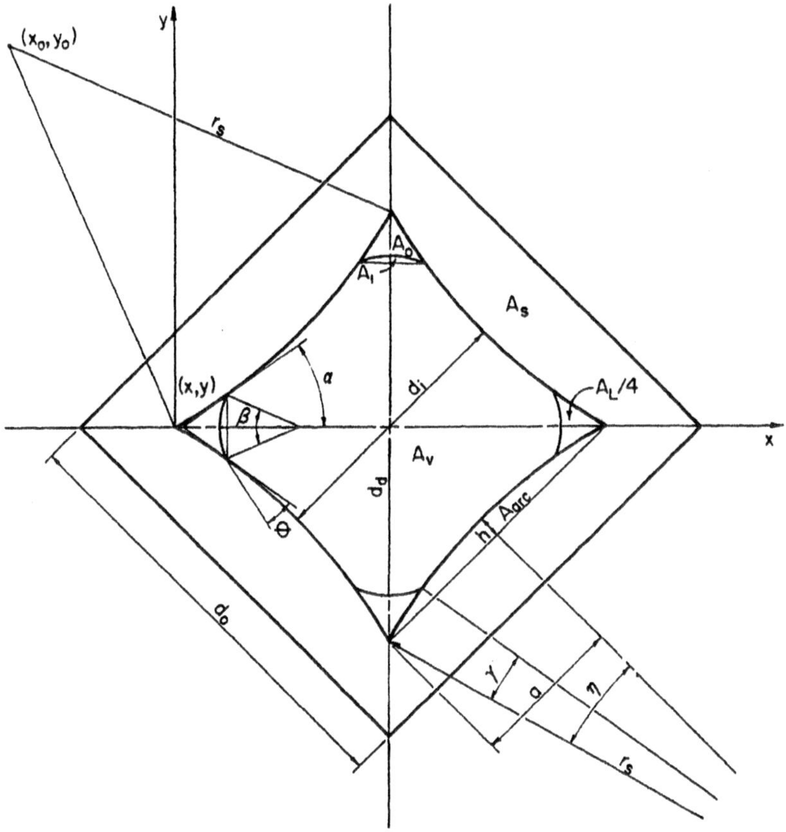

Figure 6.14 Cross-sectional dimensions of micro–heat pipe investigated by Babin et al. (1990).

liquid flow is the result of an imbalance in the total pressure drop and occurs because the evaporation rate does not provide an adequate change in the liquid–vapor interfacial curvature to compensate for the pressure drop. As a result, the increased pressure in the evaporator causes the meniscus to recede into the corner regions, forcing liquid out of the evaporator and into the condenser. Once the heat input reaches full load, the reverse liquid flow disappears and the liquid mass flow rate into the evaporator gradually increases until a steady-state condition is reached. At this time the change in the liquid mass flow rate is equal to the change in the vapor mass flow rate for any given section (Wu and Peterson, 1991a). This is more apparent in Fig. 6.16, which compares the pressure as a function of axial position 10 s after startup and for steady-state. As illustrated, 10 s after startup the pressures of both the liquid and vapor are higher in the evaporator and gradually decrease with position, promoting flow away from the evaporator. At steady state, the liquid pressure is

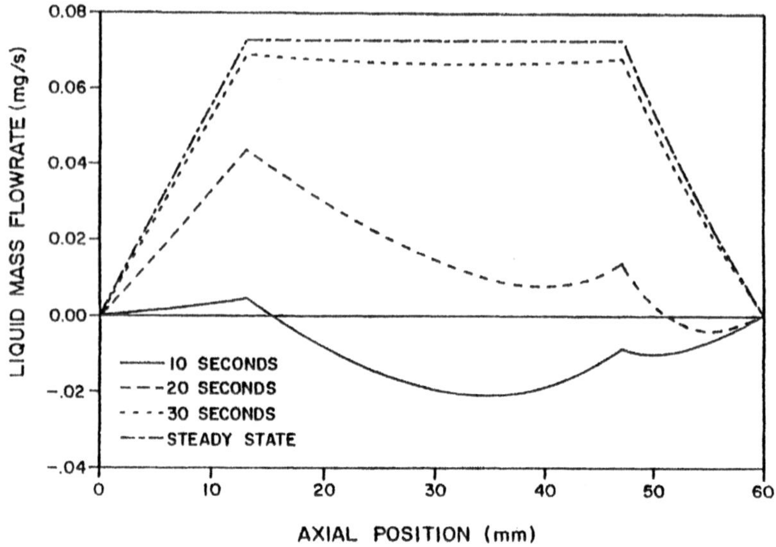

Figure 6.15 Liquid mass flow rate as a function of time and axial position (Wu and Peterson, 1991a).

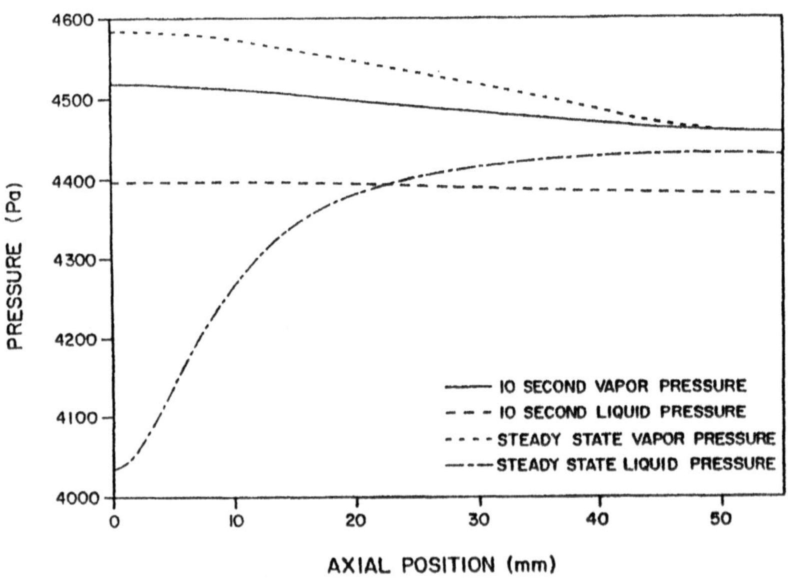

Figure 6.16 Liquid cross-sectional area as a function of time and axial position (Wu and Peterson, 1991a).

much lower in the evaporator and increases dramatically with position, promoting flow back to the evaporator.

Modeling of Heat Pipe Arrays

As noted previously, the original concept proposed by Cotter (1984) was to incorporate heat pipes as part of semiconductor devices. Although the previously discussed models can be used to predict the performance limitations and operational characteristics of individual micro–heat pipes, they do not address how the incorporation of an array of these devices in semiconductor devices might affect the overall temperature distribution and thermal performance. In an effort to determine the potential advantages of this concept Mallik et al. (1991) developed a three-dimensional numerical model capable of predicting the thermal performance of an array of parallel micro–heat pipes constructed as an integral part of semiconductor chips, as shown in Fig. 6.17. For comparative purposes, several different thermal loading configurations were modeled both with and without heat pipes in order to determine the reduction in the maximum surface temperature, the mean chip temperature, and the maximum temperature gradient across the chip. Using the steady-state and transient analyses of Babin et al. (1990), Wu and Peterson (1991a) and Wu et al. (1991), the effective thermal conductivity of the region occupied by the micro–heat pipes was found to be approximately ten times that of the silicon substrate if methanol was used as the working fluid.

Figure 6.18*a* illustrates the predicted temperature distribution for a 20×20-mm silicon wafer with 22.5 W applied to an 8×8-mm region located in the center of the wafer. For this test case, heat rejection from the wafer occurred over a 10-mm-wide region on each end of the wafer and perpendicular to the heat pipe array. As shown, the maximum and mean

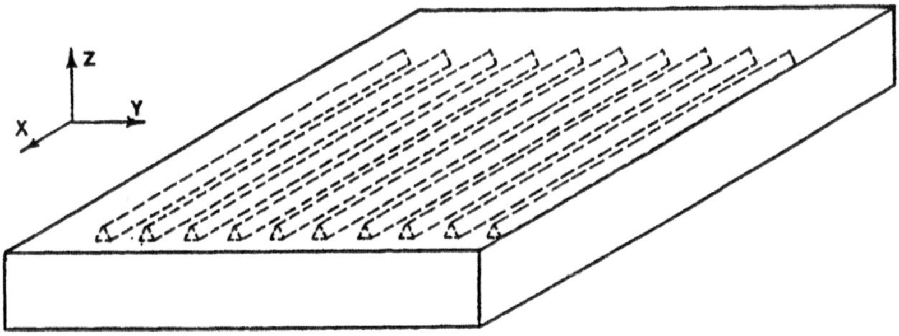

Figure 6.17 Micro–heat pipe array.

surface temperatures without any heat pipes were predicted to be 48.63 and 13.14°C, respectively. The addition of 19 uniformly spaced micro–heat pipes with an effective thermal conductivity of ten times that of the silicon substrate reduced the maximum and mean surface temperatures to 41.81 and 12.60°C, or reductions of 14.0 and 6.4%, respectively. More importantly, as illustrated in Fig. 6.18*b*, the maximum temperature gradient across the wafer was significantly reduced.

To determine how variations in the cross-sectional array density of the heat pipes (i.e., the percentage of cross-sectional area occupied by the heat pipes) affected the thermal characteristics, heat pipe arrays of 10 and 35 triangular micro–heat pipes, each 100 μm on a side, were also evaluated using this model. As shown in Figs. 6.19*a*,*b*, the maximum wafer surface temperature is significantly lower for the wafer with the array consisting of 35 micro–heat pipes than for the one with 10. The effect of variations in the number of heat pipes is more clearly illustrated in Fig. 6.20, where the maximum chip surface temperature is shown as a function of the number of heat pipes. The results indicate that increasing the number of heat pipes can significantly decrease the maximum chip temperature. Although as expected, the maximum chip temperature continues to decrease with an increasing number of chips, the rate of decrease drops off significantly in the region of 19 heat pipes. For this configuration, an array density of 12.8% was used and resulted in reductions in the maximum chip temperature of up to 40%. In addition, significant reductions in the transient response time were observed, further illustrating the effectiveness of an array of these micro–heat pipes in dissipating the heat over the entire chip surface and improving the heat removal capability.

Although the previous investigations of Babin et al. (1990), Wu and Peterson (1991a) and Wu et al. (1991) indicated that an effective thermal conductivity 10 times that of silicon was reasonable, additional analyses were conducted to determine the effect of variations in this value. Steady-state analyses were performed using a heat pipe array comprised of 19 parallel heat pipes. Using an effective thermal conductivity ratio of five, the maximum and mean surface temperatures were 37.69 and 4.91°C, respectively. With an effective thermal conductivity ratio of 10, the maximum and mean surface temperatures were 35.20 and 4.21°C, respectively. Using an effective thermal conductivity ratio of 15, the maximum and mean surface temperatures were 32.67 and 3.64°C, respectively. These results illustrate how the incorporation of an array of micro–heat pipes can reduce the maximum wafer temperature, reduce the temperature gradient across the wafers, and eliminate localized hot spots. In addition, this work highlights the significance the incorporation of these devices could have on semiconductor chips, particularly those constructed in materials with thermal conductivities significantly less than that of silicon, such as gallium arsenide.

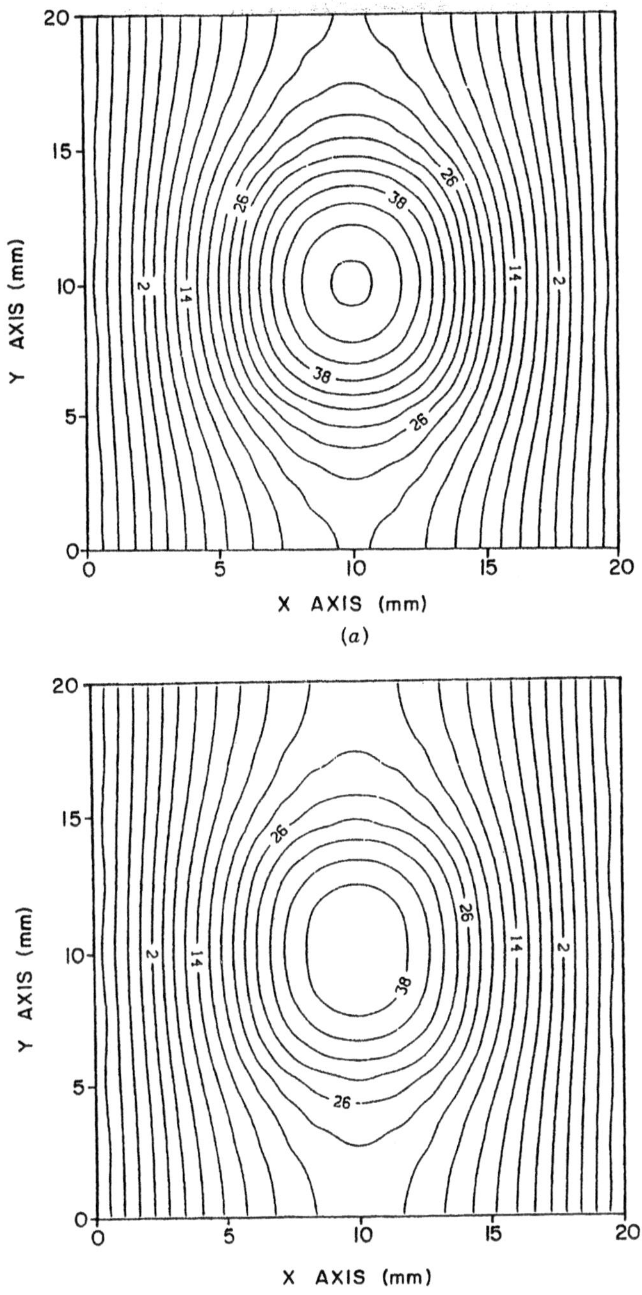

Figure 6.18 Steady-state temperature distribution for thermal loading configuration HC2 (*a*) without heat pipes and (*b*) with 19 parallel heat pipes (Mallik et al., 1992).

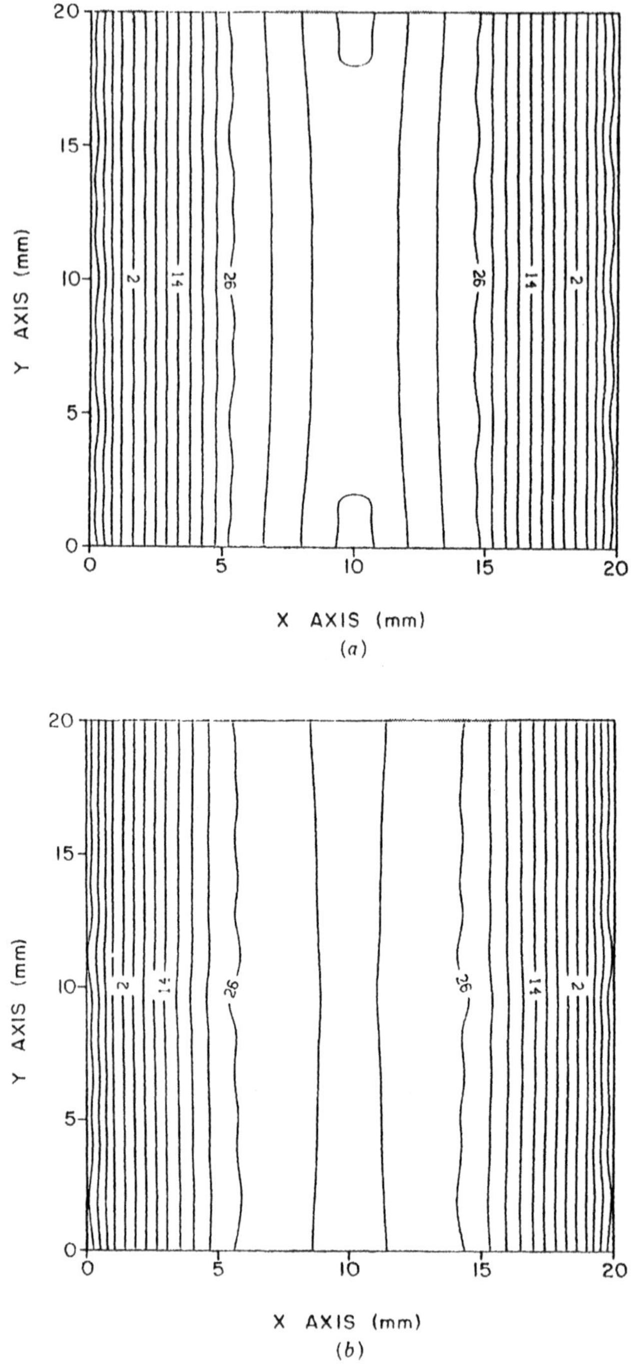

Figure 6.19 Steady-state temperature distribution for thermal loading configuration HC4 with (*a*) 19 heat pipes and (*b*) heat pipes (Mallik et al., 1991).

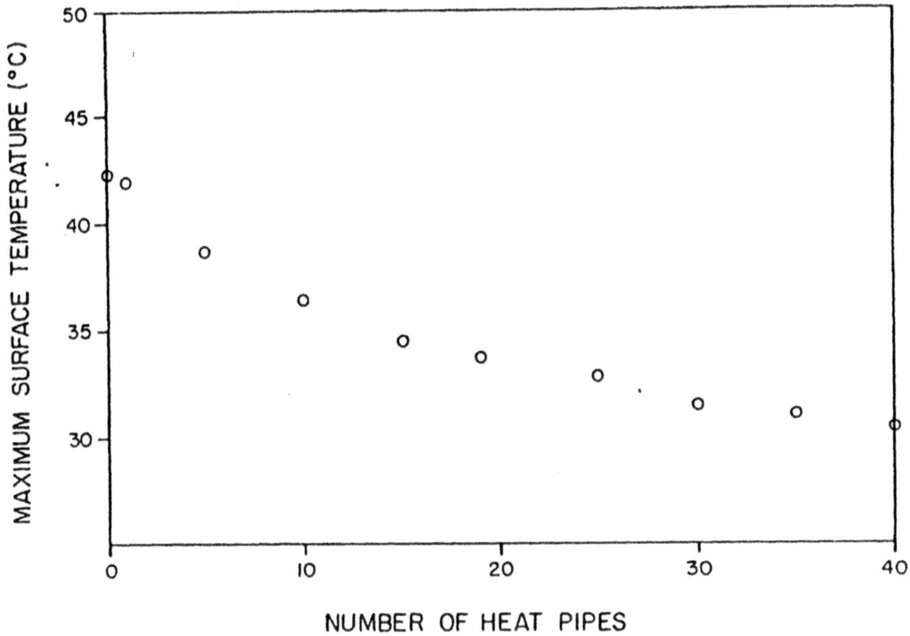

Figure 6.20 Effect of variations in number of heat pipes.

Microscopic Models

As discussed previously, the model developed by Mallik et al. (1991) utilized the macroscopic models of Babin et al. (1990), Wu and Peterson (1991a), and Wu et al. (1991), which are primarily concerned with bulk properties and behavior, to estimate an effective thermal conductivity of each individual heat pipe. Although these models have been shown to be reasonably accurate for heat pipes on the order of 1 mm in diameter, continued reductions in size will require the development of a microscopic model that focuses on the behavior of thin films and particularly the evaporation and condensation occurring in the thin film region. The precise location of this transition is not yet clear. However, for micro–heat pipes with diameters greater than 100 μm the percentage of the liquid–vapor interface in the thin-film region is still relatively small and the liquid flow retains the characteristics of the bulk fluid. For micro–heat pipes with diameters less than 100 μm, a majority of the liquid–vapor interface occurs over a region where the liquid would be considered in the thin-film regime and the fluid behavior may begin to deviate from the bulk liquid behavior. This indicates that the transition occurs somewhere near the point where the mean passage diameter is approximately 100 μm. As a result, accurate prediction of the thermal behavior of micro–heat pipes smaller than

100 μm in diameter will require development of models that utilize knowledge of the thin-film behavior and include the effects of the wetting angle, the disjoining pressure, and van der Waals forces.

Numerous investigators have studied evaporation and condensation in small triangular channels (Mirzamoghadam and Catton, 1988; Xu and Carey, 1990). More recently, Swanson and Peterson (1992) investigated the behavior of the intrinsic meniscus in triangular channels as directly applied to micro–heat pipes. In this investigation a mathematical model of the evaporating intrinsic meniscus in a V-shaped channel was developed to determine the effect of wedge half-angle and vapor mass transfer on the meniscus morphology, fluid flow, and heat transfer behavior. The Navier–Stokes and energy equations as well as the interfacial conditions were scaled using the lubrication approximation. The one-sided formulation applied to the gas phase was constant. This reduced the mathematical complexity of the problem and allowed the use of a simple mass transfer coefficient at the vapor-liquid interface. The scaled normal stress interfacial condition was examined in detail to compare the magnitudes of the surface tension forces and van der Waals forces both near the interline and at the edge of the intrinsic meniscus.

The scaled equation set showed that the hydrostatic pressure acting in both the radial and azimuthal directions can be neglected. Also, flow in the axial direction was found to be predominant in the intrinsic meniscus. The analysis of the normal stress interfacial condition, which was based on a copper–methanol–air system and the nonrelated form of the disjoining pressure, indicated that the disjoining pressure term could be neglected for thin films greater than approximately 500 Å. This film thickness roughly defined the edge of the intrinsic meniscus. Furthermore, near the interline ($\leq$ 250 Å) the magnitude of the surface tension term was found to be much less than that for the disjoining pressure. Formal matching of the intrinsic meniscus to the interline region was not attempted because a practical functional form for the disjoining pressure incorporating retardation effects was not available in the literature. Thus, the nonretarded disjoining pressure, evaluated at 250 Å, was used as an asymptotic boundary condition.

The scaled equation set was solved numerically to determine the Nusselt number, interface morphology, axial pressure gradient profile, and locus of axial dryout points as functions of the wedge half-angle and the surface mass transfer number. The endpoints defining the range of the nonretarded disjoining pressure for a copper–methanol–air system were used as parameters. The results showed that the Nusselt number was independent of the surface mass transfer number and passed through minimum points near α values of 18° and 21° for the limiting values of the nondimensional disjoining pressure. The magnitude of the Nusselt number at any given wedge half-angle was always largest for the larger disjoining pressure. This behavior was attributed to a stronger van der Waals attraction near the interline that forced the overall thickness of the intrinsic meniscus to decrease. The decrease in layer thickness caused a mild increase in the interface temperature and produced a larger evaporation surface area; both of these effects caused an enhancement in heat transfer. The axial pressure gradient decreased linearly downstream, with a slope that increased with increasing surface mass flux. The locus of liquid dryout locations were estimated based on points where the piezometric pressure was equal to zero: this represented a point of zero flow in the channel. As expected, the points where dryout occurred moved toward the inlet (or evaporator) as the surface mass flux (or heat input) was increased.

6.3.2 Experimental Investigations

Experimental investigations of micro–heat pipes have been somewhat limited and, unlike the analytical or numerical investigations discussed previously, have progressed fairly systematically from relatively large micro–heat pipes approximately 3 mm in diameter to micro–heat pipes in the 30-μm-diameter range. These investigations have included steady-state and transient investigations along with investigations of arrays of micro–heat pipes included as an integral part of silicon wafers.

Steady-State Experimental Investigations

One of the earliest experimental investigations was conducted by Murakami et al. (1987) in conjunction with the study of a flat-plate heat pipe several millimeters thick. As shown in Fig. 6.21, two types of heat pipe augmentation were evaluated: a conventional flat-plate heat pipe that used a series of rectangular grooves machined into the upper and lower plate as the wicking structure and one comprised of a series of triangular grooves machined into the plate to form an array of micro–heat pipes 3.2 mm on a side. The evaporator and condenser sections for both test articles were 50 mm long and 150 mm wide. As illustrated in Fig. 6.22, only a very limited amount of data was obtained for the flat-plate heat pipe that used triangular micro–heat pipes. Although the results of these tests demonstrated successful heat pipe operation using the sharp angled corner regions of a noncircular channel, the heat transport capability was quite small when compared to the values obtained for the more conventional flat-plate heat pipe. The principal cause for this performance degradation was attributed to slugging occurring in the condenser region of the triangular heat pipes, and as a result, further investigations were not conducted. In addition to the experimental tests, an analytical model similar to those described previously was developed. For comparison, the maximum heat transfer rate as predicted by this model is also shown in Fig. 6.22. As illustrated, the predicted values compare quite favorably with experimental results.

At about the same point in time, Babin et al. (1990) conducted an

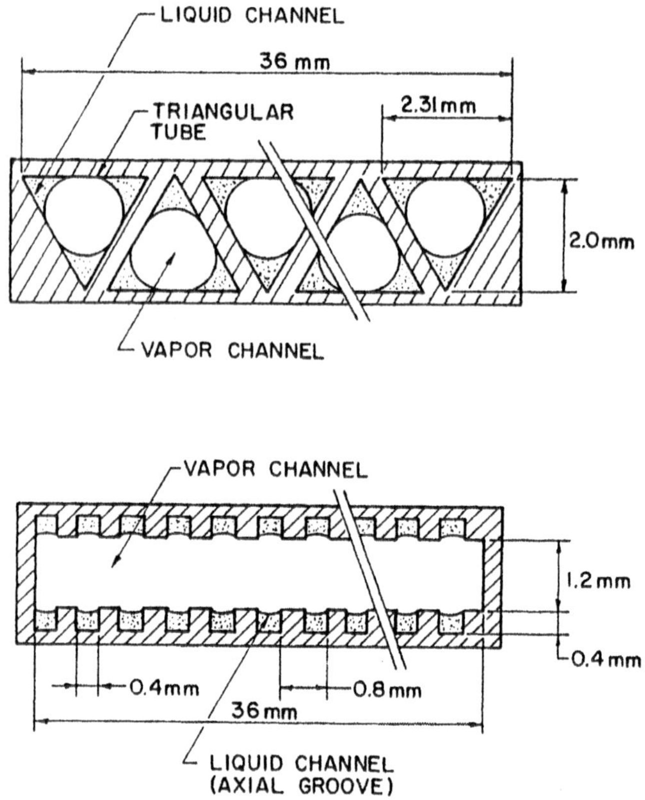

Figure 6.21 Flat-plate heat pipe configurations of Murakami et al. (1987): (*a*) flat-plate heat pipe; (*b*) micro–heat pipe array.

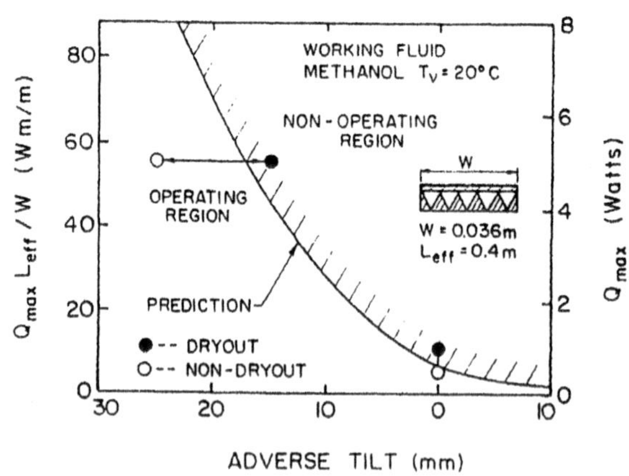

Figure 6.22 Comparison of experimental and analytical results of flat-plate heat pipe of Murakami et al. (1987) with triangular micro–heat pipes.

experimental investigation of several individual micro–heat pipes each approximately 1 mm in diameter. The primary purpose of this investigation was to determine the accuracy of the previously described steady-state modeling techniques, verify the micro–heat pipe concept, and determine the maximum heat transport capacity. The test articles used in this investigation were constructed by Itoh Research and Development Company, Osaka, Japan (Itoh, 1988). The general size and shape of these test articles are illustrated in Fig. 6.12, and the critical cross-sectional dimensions are shown in Fig. 6.23. Four test articles were evaluated, two constructed from silver and two from copper. Two of the test pipes, one silver and one copper, were charged with 0.0032 g of distilled, deionized water. The other two were evaluated without a fluid charge to determine

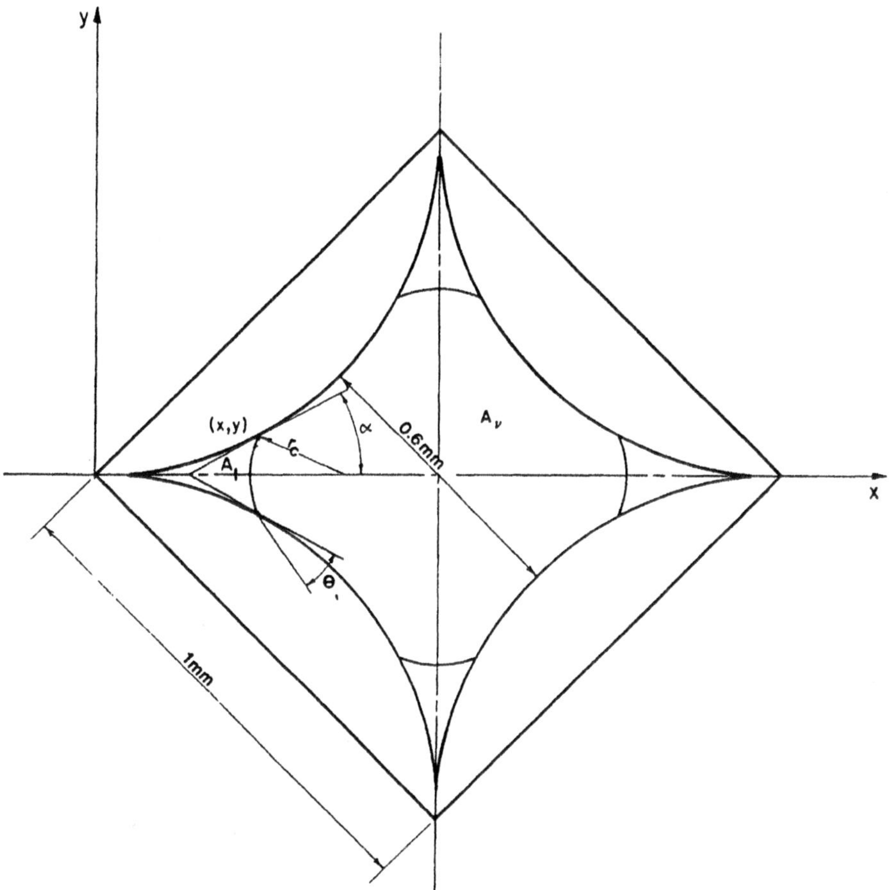

Figure 6.23 Cross-sectional shape of trapezoidal micro–heat pipe tested by (Peterson, 1988a).

the amount of heat conducted through the pipe wall and to establish a baseline value for determining the improvement in performance due to the vaporization and condensation of the working fluid.

The steady-state tests were conducted at tilt angles that both assisted and hindered the return of liquid to the evaporator. Electrical resistance heating was used in the evaporator and a constant temperature ethylglycol solution was circulated over the condenser to provide a heat sink. The temperature distribution of the heat pipe was monitored by a series of five thermocouples located on the outer surface of the heat pipe: three at equidistant positions on the evaporator; one on the condenser, and one on the outer surface of the adiabatic section. During the tests, the heat flux was incremented and the temperature of the coolant bath adjusted to maintain a constant adiabatic wall temperature. This procedure was followed until the conductance of the heat pipe, defined as the measured input power divided by the measured temperature difference between the evaporator and the condenser sections, approached a constant value.

All four test pipes were evaluated at a total of six adiabatic wall temperatures. As illustrated in Fig. 6.24, as dryout occurred, the thermal conductance of the charged copper heat pipe decreased rapidly with respect to the evaporator temperature and asymptotically approached a constant value slightly above the values obtained for the uncharged copper pipe. This difference was attributed to the expansion of the evaporator portion as dryout progressed (Babin et al., 1990). Using a mathematical technique, the maximum curvature for each operating temperature was determined. The region on both sides of the point of maximum curvature

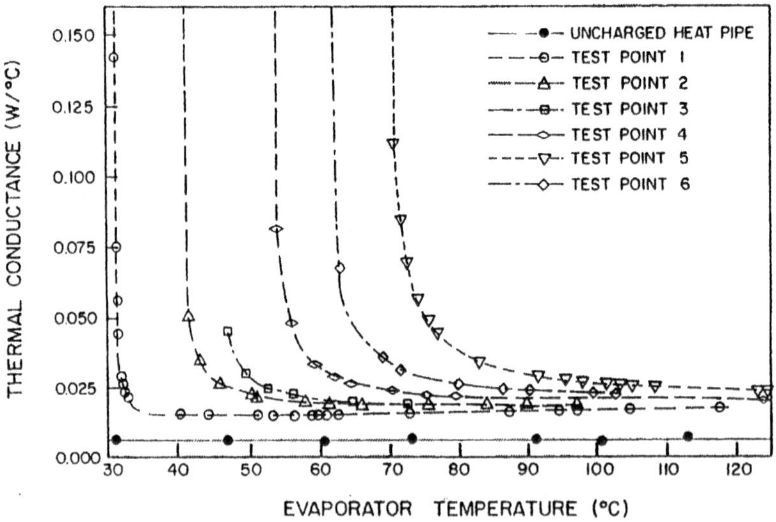

Figure 6.24 Measured thermal conductance of a trapezoidal micro–heat pipe as a function of evaporator temperature (copper, 0.0032 g H_2O) (Babin et al., 1990).

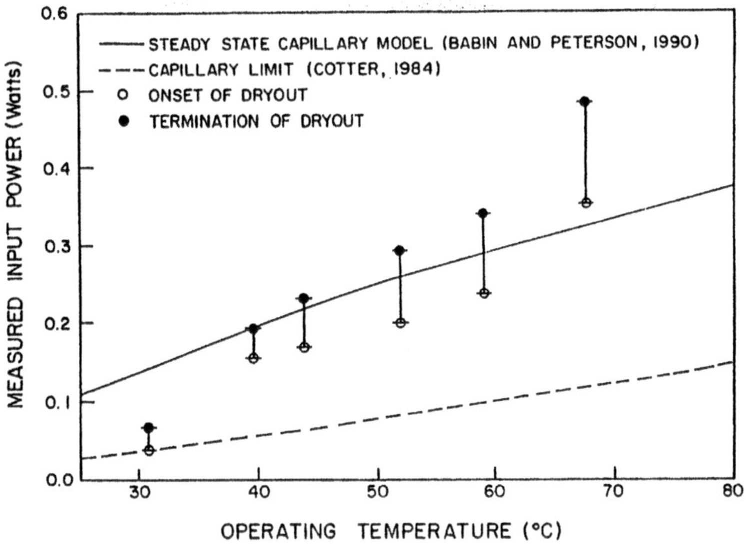

Figure 6.25 Comparison of maximum heat transport capacity of a trapezoidal micro–heat pipe as a function of operating temperature (copper, 0.0032 g H_2O) (Babin et al., 1990).

was assumed to be the operating temperature range over which dryout occurred, with the initial point being the onset of dryout and the final point, the termination of dryout.

As shown in Fig. 6.25, the steady-state model overpredicted the experimentally determined heat transport capacity at operating temperatures below 40°C and underpredicted it at operating temperatures above 60°C. It should be noted that although the experimental tests were conducted in a vacuum, when the condenser temperature was greater than the ambient temperature, a small amount of power was required to maintain a uniform temperature along the pipe. Therefore, to determine the heat transferred through the test pipe, this initial power was subtracted from the measured power input.

Transient Experimental Investigations

Although the analytical model developed by Babin et al. (1990) was shown to predict the steady-state performance limitations and operational characteristics of the trapezoidal heat pipe reasonably well for operating temperatures between 40 and 60°C, no transient information was obtained. For this reason, further investigations were conducted to evaluate the transient model developed by Wu and Peterson (1991a). These tests again utilized a micro–heat pipe test article developed by Itoh (1988). However, this test pipe was specifically designed for use in the thermal control of ceramic chip carriers (Wu et al., 1991) and was slightly different in shape. As

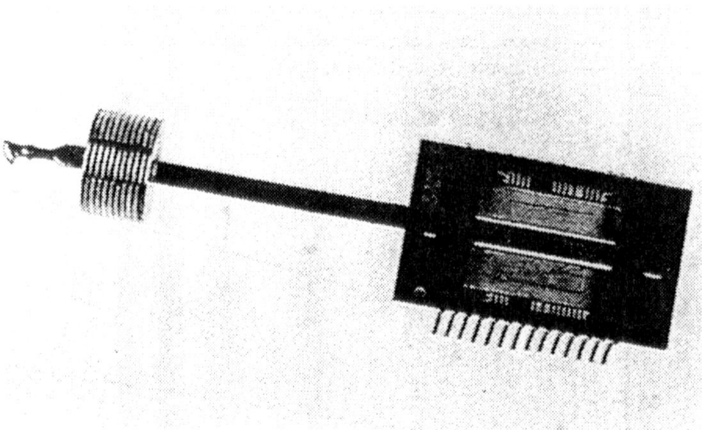

Figure 6.26 Tapered micro–heat pipe (Wu et al., 1991).

shown in Fig. 6.26, the heat pipe was designed to fit securely under the chip and was attached to the chip carrier. Small fins located at the condenser end of the heat pipe promote the cooling of the condenser by free or forced convection.

The experimental test facility used in this investigation was similar to the one used for the trapezoidal heat pipe. However, because the transient response was of interest, a thin nichrome strip heater located on the top portion of the heat pipe was used for heat input. Further, in addition to a series of nine thermocouples attached to the underside of each test pipe, an infrared thermal measurement system was used to monitor the temperature distribution in the heat pipe during both startup and transient operation. To determine the accuracy of the numerical model at steady-state conditions, the experimentally measured steady-state temperature distribution was compared with the distribution predicted by the numerical model for a power level of 0.12 W. As illustrated in Fig. 6.27, the temperature data closely follow the predicted trend, and although some heat rejection occurs in the adiabatic section, the condenser temperature remains relatively constant. Throughout the entire length the predicted and measured temperature distributions are extremely close to the predicted value, with the largest deviation being less than 0.3°C.

Once this had been accomplished, startup and transient tests were conducted. In these tests, the time dependent temperature distributions were measured to determine the transient response characteristics of the heat pipe as a function of incremental power increases, tilt angle, and mean operating temperature. Figure 6.28 compares the predicted and measured temperature difference for the tapered micro–heat pipe shown in Fig. 6.26 as a function of time. As illustrated, the transient model predicted a much more rapid response than was actually measured. However, as time

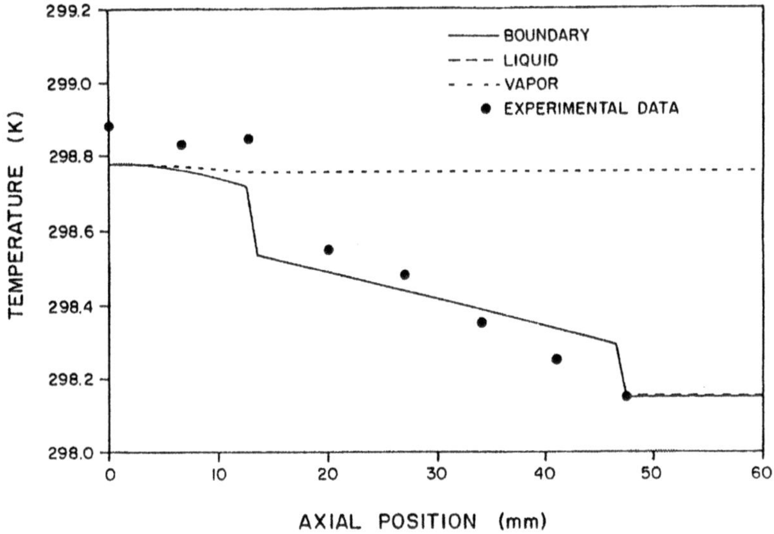

Figure 6.27 Steady-state comparison of predicted and measured temperature distribution, (power = 0.12 W, T_c = 24.5°C, no tilt) (Peterson, 1988a).

increased, the measured temperature drop between the evaporator and the condenser approached and exceeded the predicted value by approximately 15%. As discussed by Wu and Peterson (1991a), the initial deviation of the actual response from that predicted may in part be due to the time

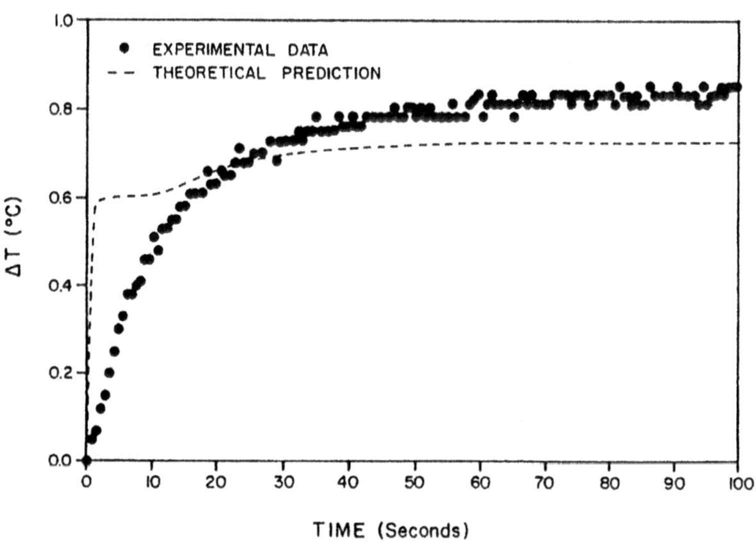

Figure 6.28 Comparison of predicted and measured transient response for tapered micro–heat pipe (Peterson, 1988a).

constant associated with the nichrome heater. It is interesting to note that the reverse liquid flow apparent in Figs. 6.15 and 6.16 is also apparent in Fig. 6.28 (the slight inflection at 12 s) but is not apparent in the experimental results.

The results of an extensive experimental investigation on a series of micro–heat pipes ranging in size and shape from 1 to 3 mm in diameter and 30 to 150 mm in length have been presented by Itoh and Polásek (1990a,b). While a majority of these heat pipes use cross-sectional configurations similar to those presented previously, several additional pipes with conventional internal wicking structures were evaluated (Polásek, 1990; Fejfar et al., 1990). Some of these pipes were as small as 1.2 mm in diameter with lengths up to 30 mm. All of the test pipes investigated utilized ultrapure water as the working fluid and either silver, copper, or a copper–beryllium alloy as the case material. The tests were conducted to determine the maximum transport capacity of the various types of pipes under adverse tilt conditions.

In addition to measurement of the temperature distribution and maximum transport capacity, extensive neutron radiography tests were conducted to determine the distribution of the working fluid within the heat pipes (Itoh and Polásek, 1990a,b; Ikeda, 1990). Two types of neutron radiography testing were conducted; one involved a static film method while the other utilized real time neutron television (Ikeda, 1990). Using these techniques, the amount and distribution of the working fluid within the heat pipes could be evaluated during real time operation. In addition to the location of the working fluid, the location and behavior of noncondensible gases could be observed, along with boiling and / or reflux flow. The results of these tests indicated several important results: (1) as is the case for conventional heat pipes, the maximum heat transport capacity is principally dependent upon the mean adiabatic vapor temperature and the inclination angle of the pipe; (2) micro–heat pipes with smooth inner surfaces were found to be more sensitive to overheating than those with grooved capillary systems; (3) the wall thickness of the individual micro–heat pipes had greater effect on the thermal performance than did the casing material; and (4) the maximum transport capacity of heat pipes utilizing axial channels for return of the liquid to the evaporator was found to be superior to that utilizing a formal wicking structure.

6.3.3 Integrated Micro–Heat Pipe Arrays

As mentioned previously, Mallik et al. (1991) developed a numerical model to determine the potential advantages of constructing an array of micro–heat pipes as an integral part of semiconductor chips. The principal function of this array is to act as a highly efficient heat spreader. Heat generated by a single active device on the chip surface can be dissipated over the entire chip area or transported to a condenser region where it can be rejected to an external heat sink. In this type of application, micro–heat pipes serve to eliminate localized hot spots, reduce temperature gradients, and improve the reliability of the chip. The results of the modeling effort indicated that significant reductions in the maximum chip temperature, thermal gradients, and localized heat fluxes could be obtained through the incorporation of these heat pipe arrays.

Fabrication Methods

Several techniques have been proposed by which micro–heat pipes on the order of 30–100 μm could be incorporated as an integral part of silicon or gallium arsenide wafers, and a majority of the most recent investigations have focused on the fabrication, development, and testing of these devices (Weichold et al., 1993; Peterson et al., 1991; Gerner, 1990). Several fabrication techniques have been proposed including the formation of a series of parallel triangular grooves using directionally dependent etching (Peterson, 1988a; Gerner, 1990) or a fabrication technique that utilizes a multisource vapor deposition process (Mallik et al., 1991).

In the first method, anisotropic or orientation dependent etching is used to produce a series of parallel microchannels in a silicon wafer. While wet chemical etching processes have been used extensively in semiconductor processing (Bean, 1978; Kendall, 1979), Mallik et al. (1991) utilized an orientation dependent etching process that allowed etching of silicon wafers in one particular direction at a higher rate as compared to the other directions. This orientation dependent etching occurs because the crystal structure of silicon is in the diamond cubic group, which causes the crystallographic directions, indicated by the notation $\langle hkl \rangle$, to be perpendicular to the crystal planes, indicated by the notation (hkl). As a result, the packing density is considerably higher in the $\langle 111 \rangle$ direction, which is perpendicular to the (111) plane, as compared with the $\langle 100 \rangle$ direction. This results in an etching rate that is faster in the $\langle 100 \rangle$ direction than in the $\langle 111 \rangle$ direction. If an orientation dependent etchant is used on a (100)-oriented slice covered by an oxide mask, it will create V-shaped grooves limited by the (111) planes. The etching process will stop at the point where the (111) planes intersect and the (100) surface reduces to a line.

In the investigation of Mallik et al. (1991), a series of V-shaped microchannels were etched in (100) silicon using a solution of KOH-normal propanol–water. A scanning electron micrograph of one of the V-shaped cross-sections is shown in Fig. 6.29. As shown, the V-shaped channels have a depth of 80 μm and a width of approximately 84 μm. Once the microchannels were etched into the silicon wafer, a 200-μm-thick clear glass cover plate was bonded to the surface to form the closed triangular channels using either an ionic bonding process (for bonding silicon to silicon) or an ultraviolet bonding process (for bonding silicon and glass).

The second fabrication technique proposed (Weichold et al., 1993) utilizes a dual source vapor deposition process. The process begins with

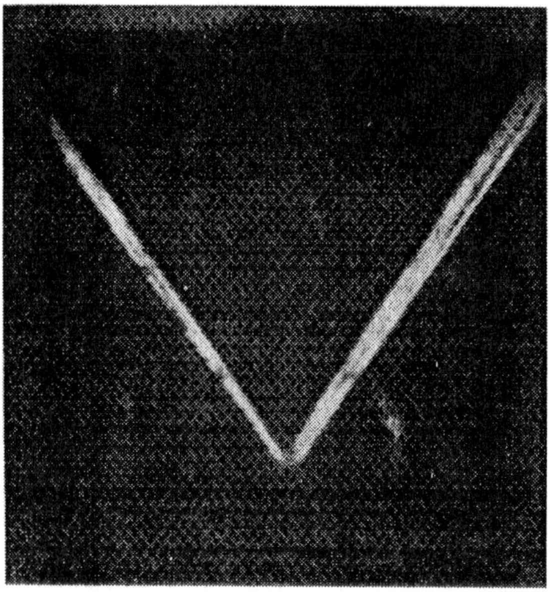

50 μm

Figure 6.29 Scanning electron micrograph of V-grooves in (100) silicon.

the fabrication of a series of square or rectangular grooves either machined or etched in a silicon wafer, as shown in Fig. 6.30*a,b*, respectively. Then using a dual E-beam vapor deposition process, the grooves are closed, creating an array of long narrow channels of triangular cross-section and open on both ends. Figure 6.31 shows a photomicrograph of the end view of a vapor-deposited micro–heat pipe that has not quite been completely closed at the top. As shown, the micro–heat pipes are lined with a thi: layer of copper, significantly reducing problems associated with the migration of the working fluid (methanol) throughout the semiconductor material. Using copper as the material for the vapor deposited heat pipes on silicon substrate ensured less than 2.0% differential expansion during the deposition process. The computer code Simulation and Modelling of Evaporated Deposition Profiles (SAMPLE) was used to simulate the metal step coverage and predict the final heat pipe cross-sectional profile (Sung, 1979). Using this model, the vapor deposition process was optimized and construction of an array of vapor-deposited micro–heat pipes was successfully completed in grooves with an aspect ratio of 1.0 (Mallik et al., 1991).

Charging Procedures

Regardless of the fabrication technique used, it was necessary to charge the

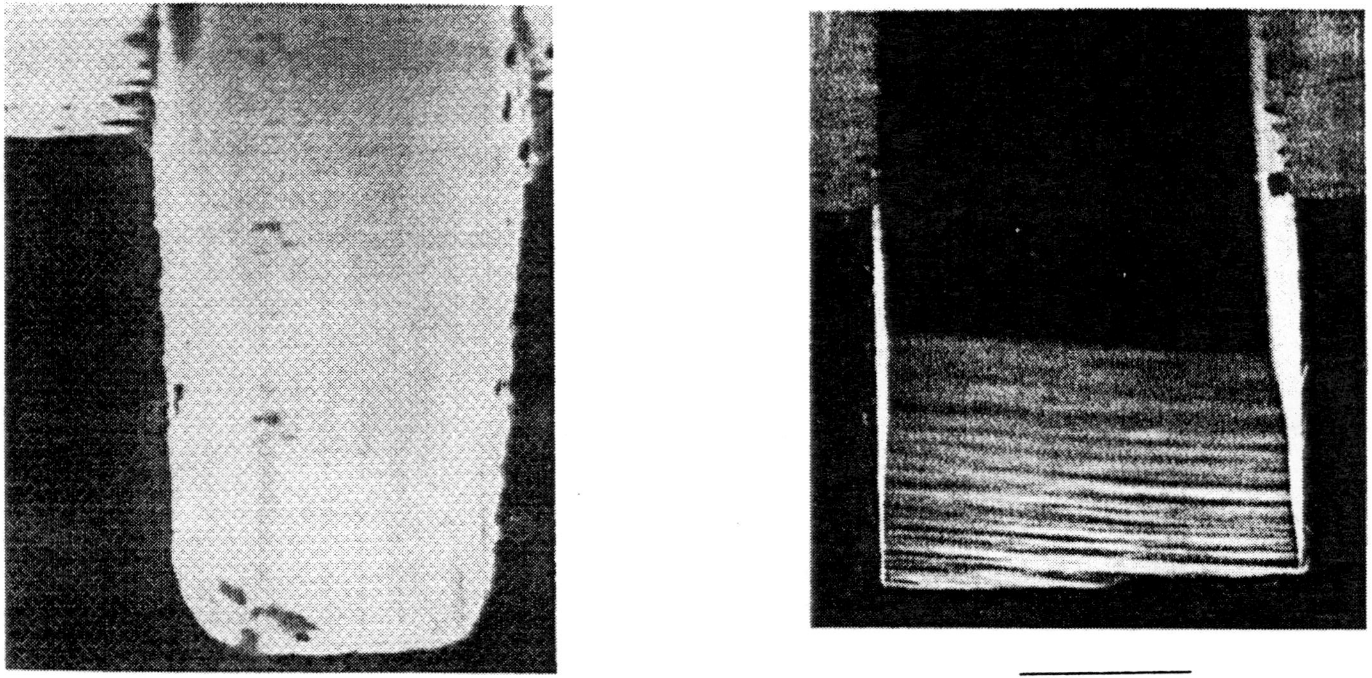

Figure 6.30 Scanning electron micrographs of (*a*) etched rectangular grooves in (110) silicon. (*b*) machined rectangular grooves in silicon.

Figure 6.31 Scanning electron micrographs of vapor deposited micro–heat pipe fabricated using a dual vapor deposition process and an etched rectangular groove.

heat pipes with the correct amount of working fluid. Several charging methods have been proposed (Peterson and Duncan, 1993). The most reliable of these is to place the micro–heat pipe array in a high pressure chamber. After evacuating the chamber, a predetermined amount of working fluid is added and the chamber is heated to a point above the critical temperature of the working fluid. In this state all of the working fluid exists as a vapor and is uniformly distributed throughout the chamber and micro–heat pipe array. The heat pipes are then sealed using an ionic or ultraviolet bonding process. When the chamber is then opened and the temperature reduced, a portion of the working fluid condenses. By controlling the volume of the pressure vessel and the quantity of the working fluid added after initial evacuation, the final amount of fluid in each heat pipe can be precisely controlled.

An alternative method, that is simpler but somewhat less accurate has been developed by Peterson et al. (1993) and consists of sealing one end of the micro–heat pipe array and positioning the wafer in a vacuum chamber vertically with the open end of the micro–heat pipes approximately 1 mm above a small charging trough. After evacuating the vacuum chamber, a small amount of working fluid is slowly injected into the charging trough using a microsyringe. As the chamber pressure increases and approaches the saturation pressure corresponding to the temperature of the chamber, the working fluid forms a pool in the charging trough and wicks up into the heat pipe channels, effectively sealing off the channel and trapping vapor in the upper portion of the heat pipe. The micro–heat pipe array can then be removed from the vacuum chamber and the open ends sealed. The final amount of liquid and vapor present in the microgrooves after sealing can be controlled by varying the temperature of the wafer, the height of the wafer above the bottom of the charging trough, and the rate at which the working fluid is injected into the vacuum chamber.

A third charging method has been proposed by Gerner (1990). This method is similar to the methods utilized on larger, more conventional heat pipes and consists of filling a small reservoir with working fluid and then evacuating the micro–heat pipe array and backfilling from the reservoir. Although this procedure has proved to be quite satisfactory on larger heat pipes, some problems with the control of the working fluid inventory and non-condensible gasses may occur in smaller heat pipes such as those under consideration.

Experimental Results

An experimental investigation was conducted by Peterson et al. (1991) to verify the micro–heat pipe concept and determine the potential advantages of constructing an array of micro–heat pipes as an integral part of semiconductor devices. Several silicon wafers were fabricated with distributed heat sources on one side and an array of micro–heat pipes on the other. These micro–heat pipe arrays were evacuated and charged with a predetermined amount of methanol. Fabrication of the micro–heat pipe array consisted of three separate operations. First, a series of 39 parallel rectangular channels 30 μm wide, 80 μm deep, and 19.75 mm long were machined into a silicon wafer 2 cm square and 0.378 mm thick with an interchannel spacing of 500 μm; second, a clear glass cover plate was bonded to the surface to close off the micro–heat pipe array; and third, the heat pipe array was charged using the procedure developed by Peterson et al. (1991). This resulted in a micro–heat pipe array that occupied approximately 2% of the cross – sectional area, perpendicular to the longitudinal axis of the wafer.

Once fabricated and charged, two silicon wafers were located adjacent to a common heat sink, as shown in Fig. 6.32. One of the wafers had the heat pipe array, and the other was a plain ungrooved wafer with a glass cover bonded to the top surface. Both wafers had a small 15.2×8-mm thermofoil heater located at one end of the micro–heat pipe array. These heaters were located on the surface of the wafer opposite the heat pipe array, much like the active devices would be in an actual application. The

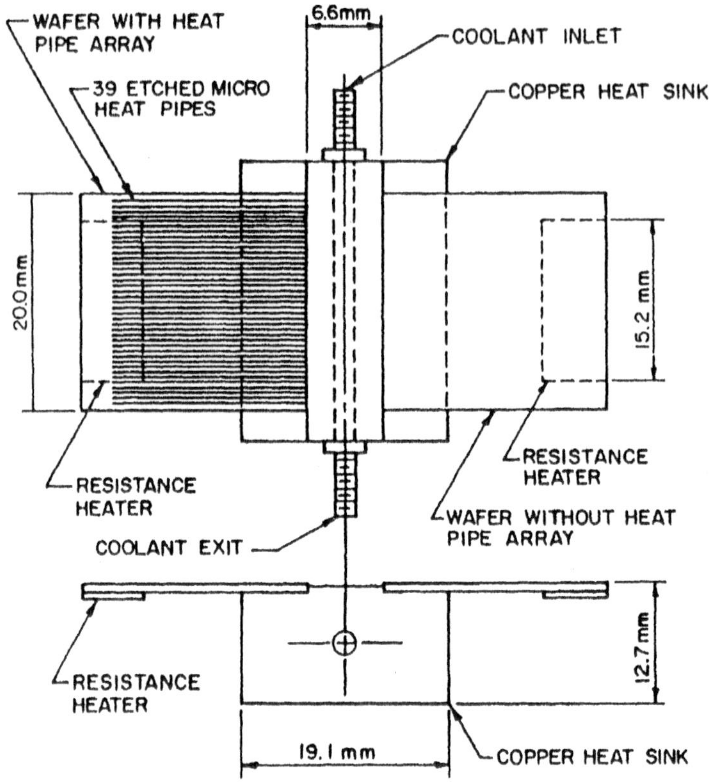

Figure 6.32 Experimental test facility used by Peterson et al. (1991).

two heaters, one on each chip, were wired in parallel to ensure that a constant power was supplied to each wafer and the voltage and current were measured by two digital multimeters.

As mentioned previously, heat was removed from the wafers by a common copper heat sink. This was accomplished by attaching the end of the heat pipe array opposite the heaters to the copper heat sink. Coolant from a constant temperature circulating bath was pumped through the copper heat sink to maintain it at a temperature of approximately 15°C.

The temperature distribution across the surface of the wafer was obtained using a Hughes Probeye TVS model 3000 infrared thermal imaging system. The resolution of this system as specified by the manufacturer was ±0.05°C, allowing precise resolution of the temperature gradients across the chip surface. The uncertainty of the actual measured temperatures was a function of the emissivity. However, using a previously developed calibration technique, the uncertainty of these values was estimated to be within ±0.10°C. Throughout all of the tests, the infrared camera was situated so that both the wafer with the micro–heat pipe array and the plain ungrooved wafer were in the field of view. In this manner, the temperature gradient and maximum wafer temperature for both wafers could be observed simultaneously for different power levels. This allowed the values obtained for the wafer with the micro–heat pipe array to be compared with the values obtained for the plain ungrooved silicon wafer in order to determine the effectiveness of the micro–heat pipe array. In addition, reductions in the maximum chip temperature and changes in the effective thermal conductivity were readily apparent.

Figure 6.33 illustrates a typical steady-state temperature profile for the two wafers and compares the temperature distribution at steady state for both the wafer with the micro–heat pipe array and the plain ungrooved wafer at an input power of 4.0 W per wafer and a bath temperature of 15°C. As illustrated, the maximum temperature for both wafers occurs at a point directly under the heaters. For these conditions, the maximum temperature for the wafer with the micro–heat pipe array is approximately 11°C lower than for the plain ungrooved wafer. In addition, it is apparent that the temperature gradient that exists across the surface of the wafer is significantly smaller for the wafer with the micro–heat pipe array than for the plain ungrooved wafer.

To determine the effect of variations in the input power on the maximum chip temperature, a series of tests were conducted in which the cooling bath temperature was held constant at 15°C and the input power was varied in approximately 0.5-W increments. Using the thermal imaging system, the magnitude and location of the maximum chip surface temperature were identified and recorded. Figure 6.34 illustrates the results of this series of tests. As shown, the maximum surface temperature for the wafer

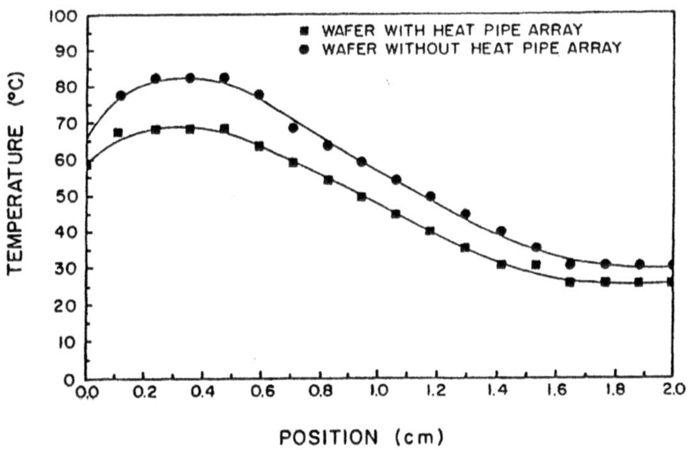

Figure 6.33 Measured temperature profile for wafer with and without micro–heat pipe array at an input power of 4.0 W and a sink temperature of 15°C (Peterson et al., 1991).

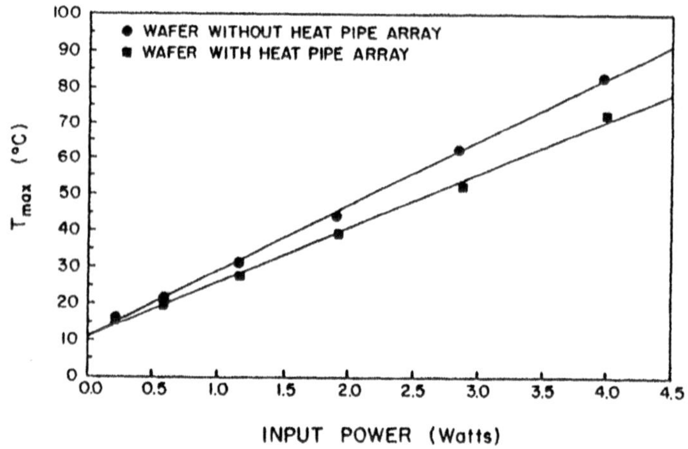

Figure 6.34 Comparison of maximum chip temperature for wafer with and without the micro–heat pipe array as a function of input power (Peterson et al., 1991).

with the micro–heat pipe array was significantly lower than for the plain ungrooved wafer. In addition, the size of this temperature difference increased in direct proportion to the input heat flux and varied from almost zero at low power levels (0.25 W per wafer) to a difference of approximately 11°C at 4.0 W per wafer. This figure illustrates the effectiveness of the micro–heat pipe array and clearly indicates the types of reductions that can be expected at higher power levels.

In an attempt to better quantify the experimental results, an effective wafer conductivity was computed from the experimental results. This value was defined as the difference between the maximum and minimum wafer temperature at a given power input divided by the distance between the locations at which these two temperatures were measured. As illustrated in Fig. 6.35, the effective thermal conductivity of the plain ungrooved silicon wafer decreases from approximately 200 W/m · °C at an input power of 0.5 W per wafer to a value of approximately 175 W/m · °C at an input power of 4.0 W per wafer. The measured values correspond quite well with the available thermal conductivity data for silicon available in the open literature. Although the effective thermal conductivity in Fig. 6.35 is shown as a function of input power, it is clearly apparent that increases in wafer temperature result in decreases in the effective thermal conductivity. This trend is also consistent with the published thermophysical property data for silicon available in the literature.

In contrast, measurement of the effective thermal conductivity of the wafer with the micro–heat pipe array resulted in a nearly constant value with almost no variation occurring with variations in the input heat flux and/or the mean wafer temperature. The combination of decreasing

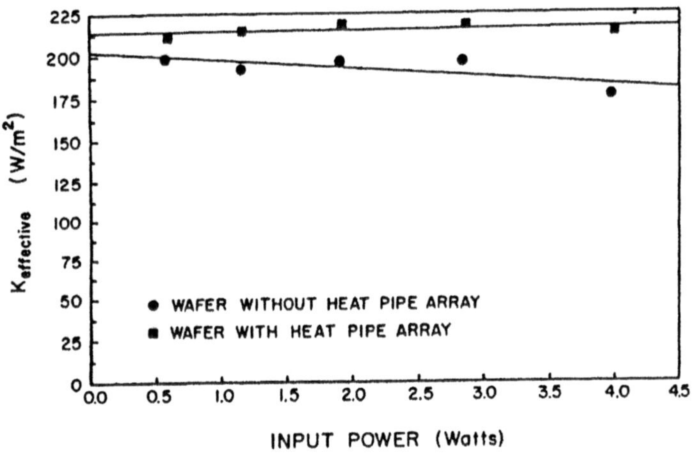

Figure 6.35 Comparison of effective thermal conductivity of wafer with and without micro–heat pipe array as a function of input power (Peterson et al., 1991).

effective thermal conductivity in the plain ungrooved wafer and constant effective thermal conductivity of the wafer with the micro–heat pipe array results in a linearly increasing difference between the effective thermal conductivity of the two wafers, which further increases the rate at which heat can be removed from localized hot spots by the micro–heat pipe array.

Although not quantified in the investigation of Peterson et al. (1991), the transient thermal response of the two wafers was significantly faster for the wafer with the micro–heat pipe array than for the plain ungrooved wafer. This decrease in the thermal time constant is a direct result of the increase in the effective thermal conductivity caused by the vaporization and condensation occurring in the micro–heat pipe array.

6.3.4 Closure on Micro–Heat Pipes

As evidenced by the preceding discussion, the micro–heat pipe concept has been demonstrated to be feasible from both an experimental and theoretical perspective. Steady-state and transient models have been developed and verified experimentally that are capable of predicting the operational limits and performance characteristics of micro–heat pipes with diameters less than 100 μm with a high degree of reliability. These models are currently being expanded for use in both individual heat pipes and also with arrays of heat pipes constructed as an integral part of semiconductor devices.

Experimentally, heat pipes on the order of 1 mm in diameter have been tested and shown to be effective in dissipating and transporting heat from localized heat sources. The results of tests conducted on these heat pipes have provided the background for the adoption of these heat pipes in commercial applications. In addition, arrays of micro–heat pipes on the order of 35 μm have been successfully fabricated, charged, and tested. When incorporated as an integral part of semiconductor devices, these heat pipe arrays have been shown to be an effective method for dissipating localized heat fluxes, eliminating localized hot spots, reducing the maximum wafer temperatures, and thereby improving the wafer reliability.

While many of the obstacles preventing the widespread use of micro–heat pipes have been overcome, several issues have not been addressed. These include determination of the effect of the precise size and shape of the micro–heat pipe, the effect of variations in the heat pipe array density, and performance degradation with respect to time. It is clear from these results that incorporating relatively large micro–heat pipes (1 mm diameter) or arrays of micro–heat pipes as an integral part of silicon chips is a feasible alternative cooling scheme for numerous heat transfer applications.

6.4 CRYOGENIC HEAT PIPES

Within three years after the first technical article on heat pipes (Grover et al., 1964), the results of an experimental investigation that used nitrogen heat pipes to control the temperature of an infrared detector were published (Haskins, 1966). Since that time, the use of cryogenic heat pipes has increased steadily. Most of the investigations involving low temperature heat pipes have been focused either on space applications involving the cooling and thermal control of optical surfaces (Sherman, 1982) or infrared scanning systems (Schoch, 1985) or on the thermal control of large superconducting magnets.

Small thermal gradients in optical surfaces can result in dramatic variations in the wavefront quality and pointing accuracy of the optical element. These thermal gradients can and have been successfully controlled through the use of heat pipes, such as those used on the Orbiting Astronomical Observatory (Harwell et al., 1974) launched in the early 1970s. In this type of application, heat pipes are utilized to maintain the optical surface at a constant temperature and hence improve the optical quality. In satellite infrared scanning systems, heat pipes are used not only to eliminate temperature variation but also to remove heat from both the infrared detectors and the aperture shroud. These devices must be cooled to a temperature ranging from 5 to 120 K. In the detector, this low temperature provides greater resolution, while in the aperture shroud it helps eliminate background radiation or thermal "noise." These two applications along with others such as electronic cooling have resulted in a renewed interest in the operation and use of cryogenic heat pipes for thermal control (Joy, 1970), particular in zero- or reduced gravity environments.

6.4.1 Working Fluids

Cryogenic heat pipes use either a chemically pure material such as helium, argon, krypton, nitrogen, or oxygen or a chemical compound such as methane, ethane, or freon as the working fluid. The operating temperature range and liquid figure of merit for these fluids are shown in Figs. 6.36 and 6.37, respectively. New working fluids such as superfluid helium (He II) have been used in cooling systems for superconducting magnets in accelerators and fusion reactors (Khalil, 1983). These new fluids are being developed to fill the gaps in the operating temperature ranges that exist between the critical point of helium 4 (5.2 K) and the freezing point of hydrogen (13.8 K), the critical point of neon (44.4 K) and the freezing point of fluorine (53.5 K) (Dunn and Reay, 1978; Prager and Basiulis, 1980).

The use of cryogenic fluids in heat pipes presents several problems that do not normally occur either at ambient temperature or in liquid metal heat pipes. Most of these problems can be directly attributed to the relatively low surface tension, thermal conductivity, and latent heat of vaporization or to the high liquid viscosity present in cryogenic heat pipes. The most significant of these problems involves the ground testing of cryogenic heat pipes that have been designed for spacecraft applications. In the development of a cryogenic optimization model (Joy, 1970), it was discovered that heat pipes optimized for zero-gravity operation would not properly prime in a 1g environment due to the low surface tension of the working fluids.

A second problem specific to cryogenic working fluids is that of the high vapor pressures encountered during storage. Increased wall thicknesses result in an increase in the thermal resistance of the wall / wick

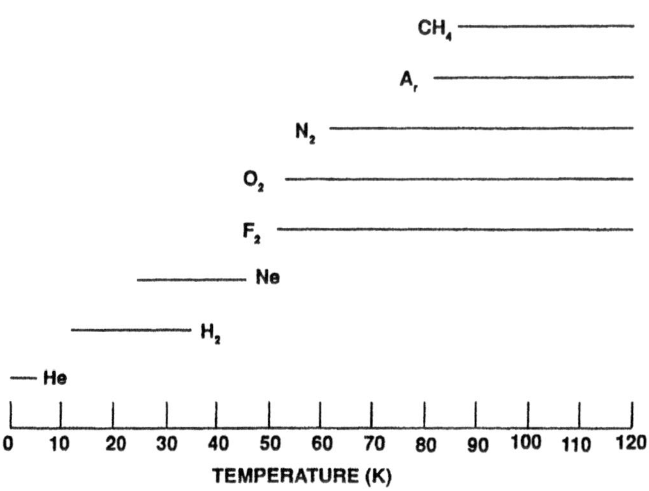

Figure 6.36 Operating temperature range for various cryogenic fluids.

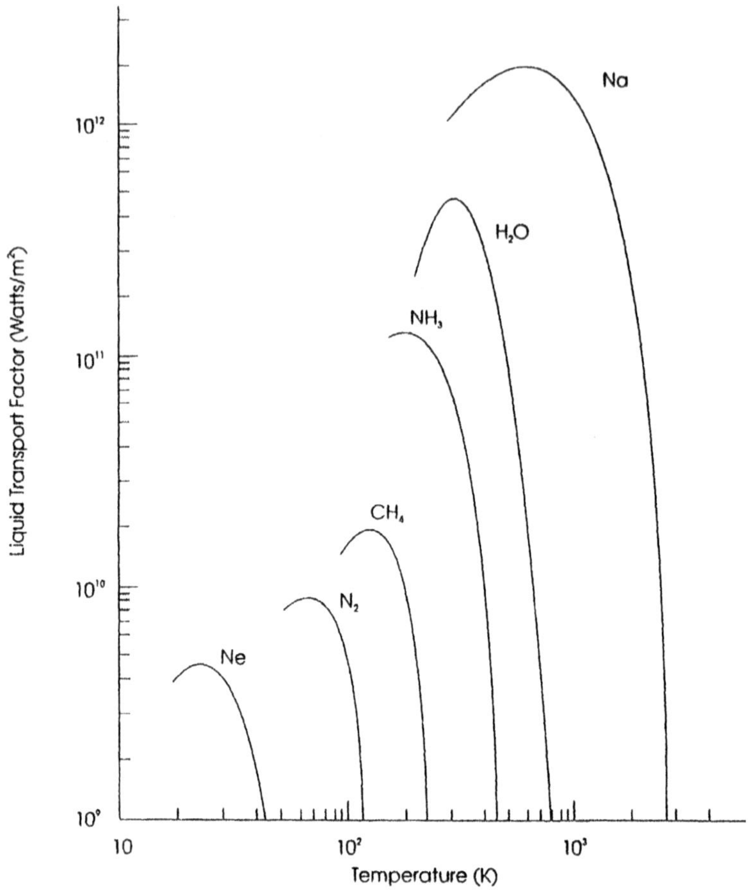

Figure 6.37 Liquid figure of merit for cryogenic working fluids.

combination and hence a reduction in the overall performance. One final problem is that due to the extremely low operating temperature range of cryogenic fluids, the methods available for cooling the condenser are quite limited. As a result of these problems and the difficulty of working with cryogenic fluids, nitrogen has been used in a majority of the heat pipes operating in the cryogenic region.

Several studies of heat pipe working fluids in the cryogenic range have been conducted (Joy, 1970; Prager and Basiulis, 1980; Chi and Cygnarowicz, 1970; Kline et al., 1978). The results of these can be summarized as follows:

1. The operating temperature range of a typical cryogenic working fluid is quite narrow and lies between the triple point and the critical temperature.
2. Cryogenic fluids undergo significant property change due to small

temperature variations. These property changes may result in large fluctuations in the fluid's transport capacity over a fairly narrow temperature range.

3. The reactive nature of such fluids as oxygen, hydrogen, and helium requires that care be taken in order to select a compatible container and working fluid.
4. The heat transport capability of cryogenic heat pipes is very limited when compared to ambient or high temperature heat pipes.
5. The radial evaporator heat flux is essentially limited to less than $1 \times 10^4\ m^{-2}$ for currently available cryogenic heat pipes.
6. Boiling may occur in cryogenic heat pipes with superheat in the range of 2–10 K.

Heat pipes can be limited in one of four ways: the capillary limitation, in which there is insufficient capillary pressure to pump the working fluid back to the condenser; the sonic limitation, in which the vapor velocity inside of the heat pipe reaches the sonic limitation causing choked flow; the entrainment limitation, in which liquid droplets are entrained in the vapor flow and carried back to the condenser; and the boiling limitation, in which the evaporator heat flux is so high that boiling occurs in the evaporator wick. Of these four limitations, cryogenic heat pipes typically encounter either a boiling or capillary limitation.

Capillary limitations result from the relatively low surface tension present in cryogenic fluids. For this reason, considerable effort has been expended to develop wicking structures that increase the capillary pressure while at the same time minimize the frictional pressure drop.

The boiling limitation in cryogenic heat pipes differs from the capillary limitation in that it is a function of the radial heat flux, while the capillary limitation is a function of the axial heat transport. Recent experimental investigations (Semena and Levterov, 1978; Anonymous 1983) and analytical studies (Peterson, 1985) have been performed; however, considerable work is necessary to understand the phenomenon of boiling in porous media for both normal and reduced gravity environments. As was the case with the capillary limitation, the boiling limitation is directly affected by the type of wicking structure utilized.

Traditional wick structures can be divided into two major categories: homogeneous wicks such as screens, fibers, or powders and arterial or axially grooved wicks. In 1974, a new type of homogeneous wick designed specifically for use in cryogenic heat pipes was reported (Sherman and Brennan, 1974). This wick structure was termed a slab wick and consisted of several layers of screen or fibrous material located in the center of the heat pipe. Several analytical and experimental studies have been conducted using this type of wick configuration (Hare, 1975; Colwell, 1977).

A method frequently used to evaluate the different types of wicking structures is to compare their predicted transport length values (the product of the maximum axial transport capacity and the transport length) to see which type of wick structure provides the greatest heat transport capacity. One comparison of this type was made between an axially grooved heat pipe and one with a homogeneous wick (Joy, 1970). Using oxygen as the working fluid, a greater than twofold increase in this value was predicted for the grooved heat pipe due to the reduction in the viscous pressure drop of the liquid in the wick.

More recently, modifications to the axial grooves have been investigated. In one investigation, experimental tests on a series of seven cryogenic heat pipes was performed (Molt, 1981). Each heat pipe had a different wick structure and was tested using two different working fluids, CH_4 and CF_4. The results indicated that a wick comprised of screen-covered axial grooves, a structure not previously used in cryogenic heat pipes, resulted in the highest heat transport value prior to reaching the capillary limitation. In another investigation (Alario et al., 1978), the traditional trapezoidal groove was replaced with a groove of circular cross section, referred to as a "*reentrant groove.*" This modification improved the capillary pumping while at the same time reducing the viscous pressure losses. Using this configuration and hydrogen as the working fluid, the overall conductance was increased almost 400% over a conventional trapezoidal grooved heat pipe.

The results of these investigations seem to indicate that the most efficient type of wick design for use in reduced gravity environments is that of an open arterial channel. However, conclusions based upon experimental investigations (Ivonovskii et al., 1980) involving this type of wick structure indicate that bubbles trapped in the arteries could result in a partial blockage of the liquid flow, thereby reducing the heat pipe efficiency. In an attempt to determine the effect of these bubbles on long term reduced gravity operation, an analytical model capable of predicting the condensation rate of these vapor bubbles was also presented.

6.4.2 Problems Specific to Cryogenic Heat Pipes

The startup of cryogenic heat pipes presents some unique problems, since it normally occurs from the supercritical state as opposed to the frozen state, as is the case with ambient or high temperature heat pipes. Modeling techniques that describe the transient behavior of cryogenic heat pipes during startup and recovery from partial dryout have been developed (Chang and Colwell, 1985; Colwell and Chang, 1984; William, 1979) and although these efforts have been going on for some time, the startup phenomenon is still not clearly understood.

In 1982, the effect of vibration (both sine wave and random) on cryogenic heat pipes was investigated by a series of tests on 3 and 6-mm-diameter heat pipes using methane, ethane, nitrogen, and oxygen as the working fluids (Mathieu, 1982). In these tests, the priming behavior, burst pressure, operation, and maximum heat transport capacities were evaluated. Both types of vibrations tended to break up the liquid / vapor interface, resulting in a reduction in the maximum heat transport capacity.

6.5 VARIABLE CONDUCTANCE HEAT PIPES

As is the case with most two-phase cycles, the presence of noncondensible gases creates a problem due to the partial blockage of the condensing area. Heat pipes are no exception. During normal operation, any noncondensible gases present are carried to the condenser and remain there, reducing the effective condenser area. This characteristic, although normally undesirable, can be used to control both the direction and amount of heat transferred (Tien and Chen, 1984).

6.5.1 Gas-loaded Heat Pipes

Figure 6.38 illustrates one such application of this concept, referred to as a gas-loaded, variable conductance heat pipe. In this type of device the thermal conductance of the heat pipe varies as a function of the gas front position. As the heat available at the evaporator varies, the vapor temperature varies and the gas contained within the gas reservoir expands or contracts, moving the gas front. This in turn results in a variation in the thermal conductance, that is, as the heat flux increases, the gas front recedes and the thermal conductance increases due to the larger condenser surface area. In this way, the temperature drop across the evaporator and condenser can be maintained fairly constant even though the evaporator heat flux may fluctuate.

Figure 6.39 illustrates a variable conductance heat pipe (VCHP) constructed from stainless steel with ammonia as the working fluid. This application was designed to control the temperature and dissipate heat in a

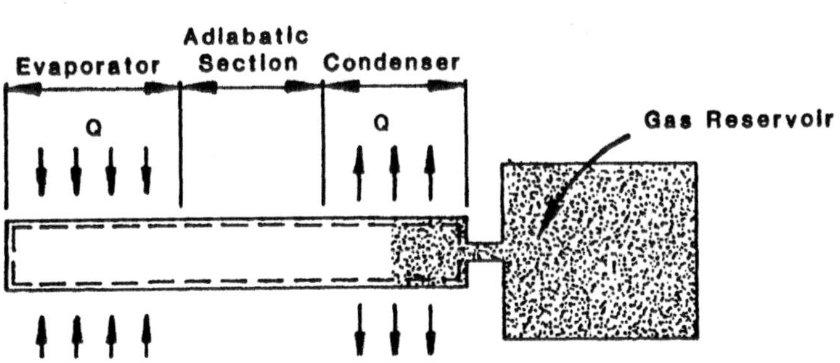

Figure 6.38 Passively controlled gas-loaded heat pipe (Chi, 1976).

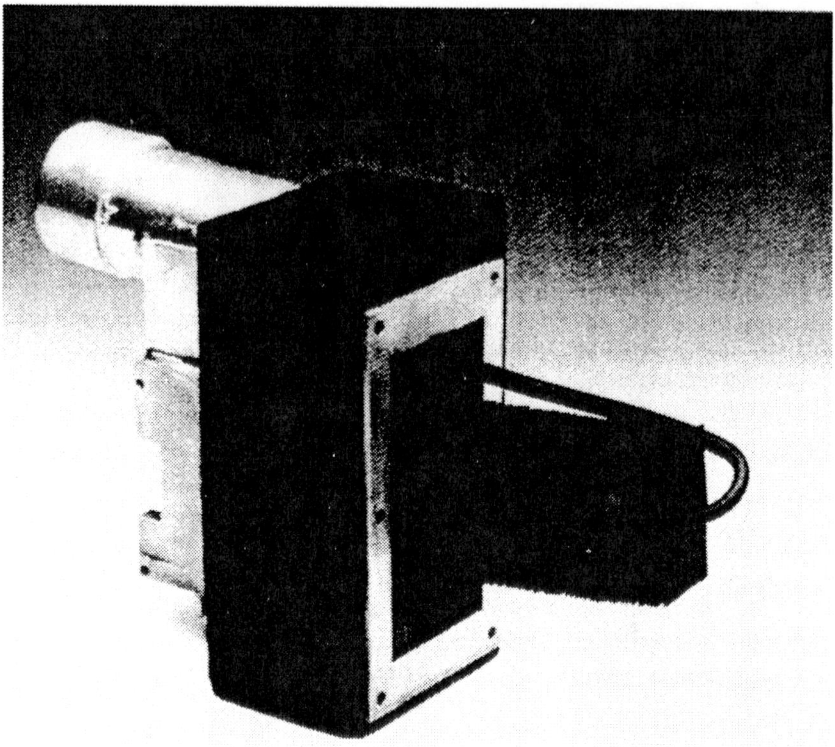

Figure 6.39 Variable conductance heat pipe (courtesy of Thermacore, Inc.)

cesium clock utilized at a ground station of the Global Positioning System. The performance specifications allowed the removal of approximately 5 W while maintaining a constant temperature bandwidth of 57 ± 13°C while the ambient conditions ranged from −25 to +55°C.

While in most VCHP applications heat pipes operate in a passive manner, adjusting the heat flow rate to compensate for the temperature difference between the evaporator and condenser (Van Buggenum and Daniels, 1987), several active control schemes have been developed (Sakuri et al., 1984). Most notable among these are (1) gas-loaded heat pipes with some type of feedback system, (2) excess-liquid heat pipes, (3) vapor flow-modulated heat pipes, and (4) liquid flow–modulated heat pipes (Chi, 1976). Figure 6.40 illustrates an example of an actively controlled, gas-loaded heat pipe in which the gas volume at the reservoir end can be controlled externally. In the example shown, a temperature-sensing device at the evaporator provides a signal to the reservoir heater. This heater

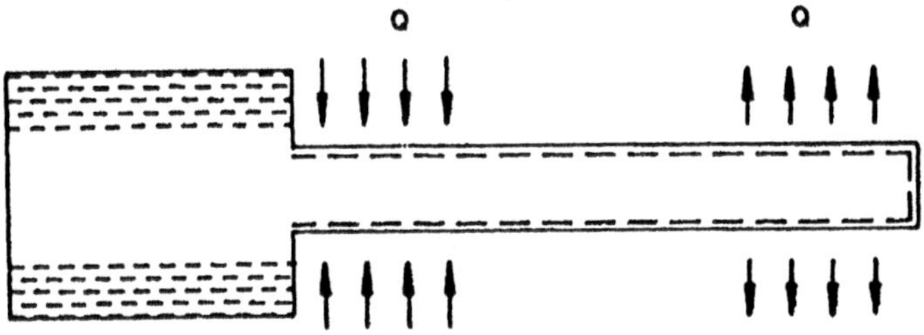

Figure 6.40 Excess liquid heat pipe (Chi, 1976).

when activated, can heat the gas contained in the reservoir, causing it to expand and thereby reducing the condenser area.

6.5.2 Excess-Liquid Heat Pipes

Excess-liquid heat pipes operate in much the same manner as gas-loaded heat pipes, but utilize excess working fluid to block portions of the pipe and control the condenser size or prevent reversal of heat transfer. Figure 6.41 illustrates the principle used in excess liquid or liquid flow-modulated heat pipes. This type of heat pipe has two separate wicking structures, one to transport liquid from the evaporator to the condenser and another that serves as a liquid trap. As the temperature gradient is reversed, the liquid moves into the trap and starves the evaporator of fluid (Groll et al., 1982; Brost and Mack, 1987).

6.5.3 Vapor Flow–modulated Heat Pipes

Vapor flow–modulated heat pipes utilize some mechanism to control the vapor flow from the evaporator to the condenser. This mechanism can be in the form of an actively controlled valve, as shown in Fig. 6.42, where the liquid and vapor flow through separate tubes and the vapor passage has a

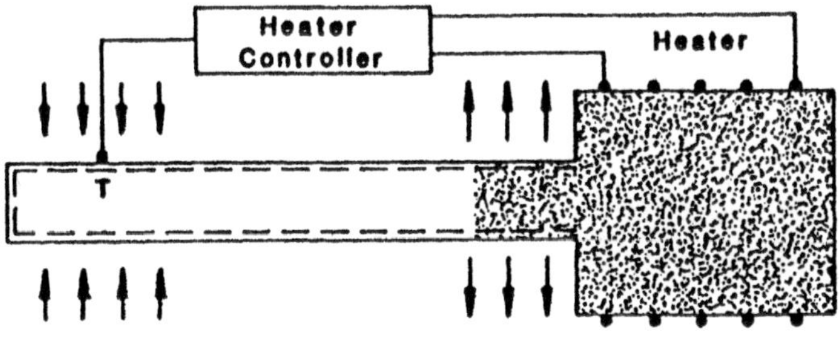

Figure 6.41 Actively controlled vapor flow–modulated heat pipe (Chisholm 1971).

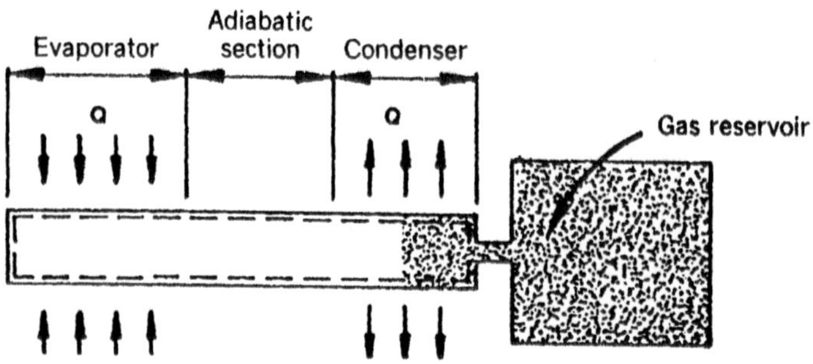

Figure 6.42 Passively controlled vapor flow–modulated heat pipe. (Chi, 1976)

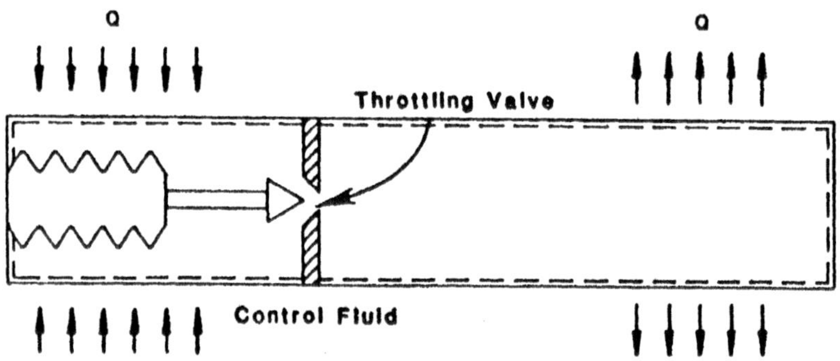

Figure 6.43 Throttled variable conductance heat pipe (Chi, 1976).

control valve that adjusts the flow to maintain the required temperature (Chisholm, 1971), or in the form of a passive throttling valve, as illustrated in Fig. 6.43. In this type of application, the throttling valve controls the amount of vapor leaving the evaporator. Increased evaporator temperatures result in an expansion of the bellows chamber containing the control fluid. This in turn closes down the throttling valve and reduces the flow of vapor to the condenser. This type of device is typically applied in situations where the evaporator temperature varies and a constant condenser temperature is desired and was originally developed for use in spacesuit temperature control. In addition to these liquid–vapor control schemes, the quantity and direction of heat transfer can also be controlled through internal or external pumps (Furukawa et al., 1987) or through actual physical contact with the heat sink (Ollendorf, 1983).

6.6 THERMAL SWITCHES AND DIODES

Two final types of heat pipes fairly unique to applications in the electronics industry are thermal switches and thermal diodes. In the former, the heat pipe is designed in such a manner as to make only intermittent contact with the heat source or heat sink. In this manner, the direction can be controlled by eliminating the thermal path between the source and sink. In thermal diodes, which are similar and sometimes indistinguishable, the heat pipe allows heat to be transferred in only one direction. Conceptually, the devices illustrated in Figs. 6.38–43 can all be classified as thermal switches or thermal diodes. Chapter 9 will discuss these devices in more detail and identify several specific applications where they are or have been utilized.

NOMENCLATURE

A	Cross-sectional area (m^3)
a	Constant
Ca	Capillary number
c_p	Specific heat $\left(\frac{J}{kg \cdot K}\right)$
d	Pipe diameter (m)
Dr	Friction number
Fr	Froude number
f	Friction factor, constant related to Reynolds number, $=$ cons./ Re^a
G	Acceleration ratio
Gr	Galileo number, $= \rho g d_v^3 (\rho_v - \rho) / \mu^2$
g	Gravitational acceleration $\left(\frac{m}{s^2}\right)$
H	$c_p(T_s\text{-}T_c)/L$ $\left(\frac{m}{s^2}\right)$
$H(L)$	Integrated value of $h(z/L)$ (W · m)
h	External heat transfer coefficient $\left(\frac{W}{m^2 \cdot K}\right)$
$h(z/L)$	Fraction of axial heat transport (W)
j	Evaporation rate $\left(\frac{kg}{m^2 \cdot s}\right)$
K	Permeability, curvature (m^2), (m^{-1})
K'	Shape factor
k	Thermal conductivity $\left(\frac{W}{m \cdot K}\right)$
L	Length (m)

M	Molecular weight $\left(\frac{\text{kg}}{\text{k mol}}\right)$
$\dot{m}$	Mass flow rate $\left(\frac{\text{kg}}{\text{s}}\right)$
Nu	Nusselt number
P	Pressure Pa
Pr	Prandtl number
q	heat transfer or heat flow (W)
$\dot{q}$	Heat transfer rate $\left(\frac{\text{W}}{\text{s}}\right)$
R	Universal gas constant $\left(\frac{\text{J}}{\text{mol} \cdot \text{K}}\right)$
R_o	Radius of evaporator end of heat pipe (m)
Re	Reynolds number
Re_v	Two-phase Reynolds number, $= \rho(U_v x \cos \alpha) / \mu$
r	Radius (m)
Sh	Sherwood number
T	Temperature (K)
T_s	Saturation temperature of vapor (K)
$T\infty$	Ambient coolant temperature (K)
ΔT	Temperature difference (K)
t	Time (s)
$\bar{v}_v$	Mean vapor velocity (m / s)
V	Velocity (m / s)
V^*	Fill charge ratio (%)
w	Axial velocity component (m / s)
x	Length or distance (m)
y	Length or distance (m)
z	Length or distance (m)

Greek Symbols

α	Half-cone angle, Half included angle (deg)
β	Dimensionless parameter (Fig. 6.4)
δ	Average film thickness (m)
δ_o	Thickness of entrained thin layer (m)
ϵ	Variable in equation 6.7
λ	Wavelength, latent heat of vaporization (m), $\left(\frac{1}{\text{kg} \cdot \text{K}}\right)$
μ	Viscosity $\left(\frac{\text{N} \cdot \text{s}}{\text{m}^2}\right)$

ν	Kinematic viscosity (m^2/S)
ρ	Density (kg/m^3)
σ	Surface tension (N/m)
τ	Shear stress (N/m^2)
ϕ	Azimuthal angle, tilt angle (deg)
θ	Liquid wetting angle (deg)
ω	Groove width, rotational speed (m), (s^{-1})

Subscripts

A	Adiabatic
c	Capillary, curvature
e	Evaporator, entrainment
ec	Evaporation–condensation
eff	Effective
f	Friction
h	Hydraulic
i	Interfacial
l	Liquid
m	Momentum, mean, maximum
n	nucleation
p	Pipe
s	Sonic, solid
v	Vapor, viscous
w	Wall
x	Axial direction
$\parallel$	Parallel
$+$	Radial
ϕ	Rotational
bl	Boundary layer
in	Into
out	Out of

REFERENCES

Alario, J., Kossan, R., and McCreight, C., 1978, "A re-entrant Groove Hydrogen Heat Pipe," Proc. 3rd Int. Heat Pipe Conf. Bologna, Italy, pp. 194–202.

Al-Jumaily, F. K., 1973, "Investigation of the Factors Affecting the Performance of a Rotating Heat Pipe," Ph.D. Dissertation, University College of Swansea, Wales.

Anonymous, 1989, "Application of Micro Heat Pipes in Hyperthermia," Annual Report of the Itoh R and D Laboratory, Osaka, Japan, 1989.

Anonymous, 1983, *Proceedings of the Environmental and Thermal Control Systems for*

Space Vehicles International Symposium, Vol. 1, European Space Agency, Special Report 200, Ispra, Italy.

Babin, B. R., and Peterson, G. P., 1990 "Experimental Investigation of a Flexible Bellows Heat Pipe for Cooling Discrete Heat Sources," *ASME Journal of Heat Transfer*, Vol. 112, No. 3, August, pp. 602–607.

Babin, B. R., Peterson, G. P., and Wu, D., 1990, "Steady-State Modeling and Testing of a Micro Heat Pipe," *ASME Journal of Heat Transfer*, Vol. 112, No. 3, August, pp. 595–601.

Ballback, L. J., 1969, "The Operation of a Rotating, Wickless Heat Pipe," M.S. Thesis, Naval Postgraduate School, Monterey, CA.

Basiulis, A., Tanzer, H., and McCabe, S., 1987, "Heat Pipes for Cooling of High Density Printed Wiring Boards," Proc. 6th Int. Heat Pipe Conf., Grenoble, France, pp. 531–536.

Bean, K. E., 1978, "Anisotropic Etching of Silicon," *IEEE Transactions on Electron Devices*, Vol. ED-25, No. 10, pp. 1185–1193.

Bontemps, A., Goubier, C., Marquet, C. and Solecki, J. C., 1984, "Theoretical Analysis of a Revolving Heat-Pipe," Proc. 5th Int. Heat Pipe Conf., Tsukuba Science City, Japan, May 14–18, pp. 274–279.

Brost, O., and Mack, H., 1987, "A Liquid Controlled Variable Conductance Heat Pipe," Proc. 6th Int. Heat Pipe Conf., pp. 549–555.

Camarda, C. J., Rummler, D. R., and Peterson, G. P., in press, "Multi Heat Pipe Panels," *NASA Tech Briefs*, LAR-14150, 1994.

Chang, W. S., and Colwell, G. T., 1985, "Mathematical Modeling of the Transient Operating Characteristics of a Low-Temperature Heat Pipe," *Numerical Heat Transfer*, Vol. 8, pp. 169–186.

Chen, J., and Lou, Y. S., 1990, "Investigation of the Evaporation Heat Transfer in the Rotating Heat Pipe," Proc. 7th Int. Heat Pipe Conf., Minsk, USSR, May 21–25.

Chen, J., and Tu, C., 1986, "Theoretical and Experimental Research of Condensation Heat Transfer in Parallel Rotating Heat Pipe," Int. Heat Pipe Symposium, Osaka, Japan, pp. 155–165.

Chen, J., and Tu, C., 1987, "Condenser Heat Transfer in Inclined Rotating Heat Pipe," Proc. 6th Int. Heat Pipe Conf., Grenoble, France, May 25–29.

Chi, S. W., 1976, *Heat Pipe Theory and Practice*, McGraw-Hill, New York.

Chi, S. W., and Cygnarowicz, T. A., 1970, "Theoretical Analysis of Cryogenic Heat Pipes," Paper No. 70-HT / SpT-6, American Society of Mechanical Engineers, New York.

Chisholm, D., 1971, *The Heat Pipe*, Mills and Boon, London, England.

Collier, J. G., 1981, *Convective Boiling and Condensation*, McGraw-Hill, New York.

Colwell, G. T., 1977, "Prediction of Cryogenic Heat Pipe Performance," Final Report No. NSG-2054, National Aeronautics and Space Administration. Washington, D.C.

Colwell, G. T., and Chang, W. S., 1984, "Measurements of the Transient Behavior of a Capillary Structure under Heavy Thermal Loading," *Internatioal Journal of Heat and Mass Transfer*, Vol. 27, No. 4, pp. 541–551.

Cotter, T. P., 1965, "Theory of Heat Pipes", Report No. LA-3246-MS, Los Alamos National Laboratory, Los Alamos, NM.

Cotter, T. P., 1984, "Principles and Prospects of Micro Heat Pipes," Proc. 5th Int'l. Heat Pipe Conf., Tsukuba, Japan, pp. 328–335.

Curtila, R., and Chataing, T., 1984, "Experimental Study of a Revolving Heat Pipe," Proc. 5th Int. Heat Pipe Conf., Tsukuba Science City, Japan, May 14–18, pp. 268–273.

Daley, T. J., 1970, "The Experimental Design and Operation of a Rotating, Wickless Heat Pipe," M.S. Thesis, Naval Postgraduate School, Monterey, CA.

Daniels, T. C., and Al-Baharnah, N. S., 1979, "Temperature and Heat Load Distribution in Rotating Heat Pipes," *AIAA Journal*, Vol. 18, No. 2, pp. 202–207.

Daniels, T. C., and Al-Jumaily, F. K., 1972, "Investigations of the Factors Affecting the Performance of a Rotating Heat Pipe," *ASME Journal of Heat Transfer*, Vol. 94, May, pp. 155–162.

Daniels, T. C., and Williams, R. J., 1976, "Theoretical and Experimental Analysis of Non-Condensible Gas Effects on a Rotating Heat Pipe," Proc. 2nd Int. Heat Pipe Conf., Bologna, ESA Scientific and Technical Information Branch, Netherlands.

Daniels, T. C., and Williams, R. J., 1977, "Condensation Heat Transfer in Rotating Heat Pipes," *Mec. Appl.*, Vol. 22, pp. 497–521.

Daniels, T. C., and Williams, R. J., 1978, "Experimental Temperature Distribution and Heat Load Characteristics of Rotating Heat Pipe," *International Journal of Heat and Mass Transfer*, Vol. 21, pp. 193–201.

Daniels, T. C., and Williams, R. J., 1979, "The Effect of External Boundary Conditions on Condensation Heat Transfer in Rotating Heat Pipes," *International Journal of Heat Mass Transfer*, Vol. 22, pp. 1237–1241.

Dean, D. S., 1976, "An integral heat pipe package for microelectronic circuits," Proc. 2nd Int. Heat Pipe Conf., ESTEC, Noordwijk, The Netherlands, pp. 481–502.

Dunn, P. D., and Reay, D. A., 1978, *Heat Pipes*, Pergamon, New York.

Fejfar, K., Polásek, F., and Stulc, P., 1990, "Tests of Micro Heat Pipes," Annual Report of the SVÚSS, Prague.

Fletcher, L. S., and Peterson, G. P., 1990, "A Micro Heat Pipe Catheter for Local Tumor Hyperthermia," Patent Pending, approved by U.S. Patent and Trademark Office on May 2.

Gerner, F. M., 1989, "Flow Limitations in Micro Heat Pipes," AFSOR Final Report No. F49620-88-6-0053, Wright-Patterson Air Force Base, Dayton, OH.

Gerner, F. M., 1990, "Micro Heat Pipes," AFSOR Final Report No. S–210–10MG-066, Wright-Patterson Air Force Base, Dayton, OH.

Giessler, F., Sattler, P. K., and Thoren, F., 1987, "Heat Pipe Cooling of Electrical Machines," Proc. 6th Int. Heat Pipe Conf., Grenoble, France, Paper No. 2.3.40.

Groll M., Supper, W., and Savage, C. J., 1982, "Shutdown Characteristics of an Axial-grooved Liquid-Trap Heat Pipe Thermal Diode, *Journal of Spacecraft*, Vol. 19, No. 2, pp. 173–178.

Grover, G. M., Cotter, T. P., and Erikson, F. F., 1964, "Structures of Very High Thermal Conductance," *Journal of Applied Physics*, Vol. 35, pp. 1990–1997.

Hare, J. D., 1975, "Performance of a Nitrogen Heat Pipe with Various Capillary Structures", M.S. Thesis, Georgia Institute of Technology.

Harwell, W., Edelstein, F., McIntosh, R., and Ollendorf, S., 1974, "Orbiting Astronomical Observatory Heat Pipe Flight Performance Data," *Progress in Astronautics and Aeronautics*, Vol. 35, pp. 445–467.

Haskins, W. L., 1966, "Cryogenic Heat Pipe," Report No. AFFDL-TR-228 Dayton, OH.

Ikeda, Y., 1990, "Neutron Radiography Tests of Itoh's Micro Heat Pipes," private communications of the Nagoya University to F. Polásek.

Itoh, A., 1988, "Micro Heat Pipes," Prospectus of the Itoh R and D Laboratory, Osaka, Japan.

Itoh, A., and Polásek, F., 1990a, "Development and Application of Micro Heat Pipes," Proc. 7th Int. Heat Pipe Conf., Paper No. 123, Minsk, USSR, May 21–25.

Itoh, A., and Polásek, F., 1990b, "Micro Heat Pipes and Their Application in Industry," Proc. of the Czechoslovak-Japanese Symposium on Heat Pipes, Rícany, Czechoslovakia.

Ivonovskii, M. N., Privezentsev, V. V., Sorokin, V. P., and Chulkov, B., 1980, "Continuity of the Flow of Liquids in Arterial heat Pipes," *Teplofiz (USSR)*, Vol. 17, pp. 406–416.

Jakes, J., Oslejsek, O., and Polasek, F., 1984, "Intensification of Cooling of Electrical Rotating Machines by Heat Pipes," Proc. 5th Int. Heat Pipe Conf., Tsukuba Science City, Japan, May 14–18.

Joy, P., 1970, "Optimum Cryogenic Heat Pipe Design," Paper No. 70-HT / SpT-7, American Society of Mechanical Engineers, New York.

Kamal, A. R., Ismail, A., and Miranda, R. F., 1988, "Theoretical Analysis of Rotating Heat Pipe," Proc. 3rd Int. Heat Pipe Symposium, University of Tsukuba, Tsukuba Science City, Japan, September 12–14, 157–162.

Katsuta, M., Kigami, N., Nagata, K., Sotani, J., and Koizumi, T., 1984, "Performance and Characteristics of Rotating Heat Pipes," Proc. 5th Int. Heat Pipe Conf., Tsukuba, Tokyo, Japan, pp. 106–112.

Katsuta, M., Wanniarachchi, A. S., and Marto, P. J., 1987, "Condensation Heat-Transfer Measurements in Co-Axial Rotating Heat Pipes," Proc. 6th Int. Heat Pipe Conf., Grenoble, France, Paper No. 2.3.37.

Keiyou, G., and Maezawa, S., 1990, "Heat Transfer Characteristics of Parallel Rotating Heat Pipe," Proc. 7th Int. Heat Pipe Conf., Minsk, USSR, May 21–25.

Kendall, D. L., 1979, "Vertical Etching of Silicon at Very High Aspect Ratios," *Annual Review of Material Science*, pp. 373–403.

Khalil, A., 1983, "Effect of Channel Restrictions on the Axial Transport in Subcooled Superfluid Helium," *Cryogenics*, Vol. 18, pp. 67–71.

Kline, M., Battrick, B., and Rowely, C., 1978, "Spacecraft and Environmental Control System," *Proceedings of the Spacecraft and Environmental Control Systems Conference*, Paris, France, Vol. 10, pp. 630–638.

Krivosheev, B. N., 1983, "Development of the Method of Calculating Heat Transfer Characteristics of Cylindrical Centrifugal Heat Pipes," Ph.D. Thesis, Technical Science Institute, Moscow.

Krivosheev, B. N., Kukharskii, M. P., and Portnov, V. D., 1979, "Heat Transfer in the Evaporator Section of a Rotating Heat Pipe at Low Rotational Speeds," translated from *Inzhenerno-Fizicheskii Zhurnal*, Vol. 37, No. 1, July, pp. 773–778.

Krivosheev, B. N., Kukharskii, M.P., and Portnov, V. D., 1980, "Investigation of the Heat Transfer in a Centrifugal Heat Pipe with Optimized Thickness of the Layer of Liquid," *Transactions of the Mechanical Engineers Institute*, No. 448, pp. 32–35.

Kukharskii, M. P., Krivosheev, B. N., and Kosheleva, G.V., 1977, "Heat Transfer in a Centrifugal Heat Pipe," translated from *Inzhenerno-Fizicheskii Zhurnal*, Vol. 33, No. 3, September, pp. 1004–1007.

Leppert, G., and Nimmo, B., 1968, "Laminar Film Condensation on a Finite Horizontal Surface Normal to Body of Inertial Forces," *Journal of Heat Transfer*, Vol. 90, pp. 178–179.

Lou, Y. S., and Chen, J., 1990, "Condensation Heat Transfer in a Rotating Heat Pipe," Proc. 5th AIAA / ASME Joint Meeting, Seattle, WA.

Loynes, J. L., 1976, "Design Improvements on a Rotating Heat Pipe Apparatus," M.S. Thesis, Naval Post Graduate School, Monterey, CA.

Mallik, A. K., and Peterson, G. P., 1991, "On the Use of Micro Heat Pipes as an Integral Part of Semiconductors," *Third ASME–JSME Thermal Engineering Joint Conference Proceedings*, Reno, Nevada, March 17–22, Vol. 2, pp. 394–401.

Mallik, A. K., Peterson, G. P., and Weichold, M. H., 1992, "On the Use of Micro Heat Pipes as an Integral Part of Semiconductor Devices," *ASME J. of Electronic Packaging*, Vol. 114, No. 4, 1992, pp. 436–442, (also in *3rd ASME–JSME Thermal Engineering Joint Conf. Proc.*, Vol. 2, pp. 394–401, March 1991).

Marto, P. J., 1973a, "An Analytical and Experimental Investigation of Rotating, Non-Capillary Heat Pipes," Report No. CR–130373, September, National Aeronautics and Space Administration. Washington, D.C.

Marto, P. J., 1973, "Laminar Film Condensation on the Inside of Slender, Rotating Truncated Cones," *Journal of Heat Transfer*, Vol. 95, pp. 270–272.

Marto, P. J., 1976, "Performance Characteristics of a Wickless Rotating Heat Pipe," Proc. 2nd Int. Heat Pipe Conf., Bologna, Italy.

Marto, P. J., 1984, "Rotating Heat Pipes," *Heat and Mass Transfer in Rotating Machinery*, D. E. Metzger and N. H. Afgan, (Eds.), Hemisphere, Washington, DC. pp. 609–632.

Marto, P. J., and Wagenseil, L.L., 1979a, "Increasing the Heat-Transfer Coefficient in the Condensation Zone of Rotating Heat Pipes," *Raket. Tekh. Kosmon.*, Vol. 17, No. 6, pp. 117–124.

Marto, P. J., and Wagenseil, L. L., 1979b, "Augmenting the Condenser Heat-Transfer Performance of Rotating Heat Pipes," *AIAA Journal*, Vol. 17, No. 6, pp. 647–652.

Marto, P. J., and Wanniarachchi, A. S., 1987, "The Influence of Internal Axial Fins on Condensation Heat Transfer in Co-Axial Rotating Heat Pipes," *Heat Transfer and Fluid Flow in Rotating Machinery*, Wen-Jei Yang (Ed.), Hemisphere, Washington, DC, pp. 235–244.

Marto, P. J., Daley, T. J., and Ballback, L. J., 1970, "An Analytical and Experimental Investigation of Rotating, Non-Capillary Heat Pipes—Annual Report," Report No. NPS–59MX70061A, June.

Mathieu, J. P., 1982, "Development of Small Cryogenic Heat pipes—Phase 3 Report," European Space Agency, Contract No. ESA 3737 / 78, Final Report.

Mirzamoghadam, A., and Catton, I., 1988, "A Physical Model of the Evaporating Meniscus," *ASME Journal of Heat Transfer*, Vol. 110, No. 1, pp. 201–208.

Mochizuki, S., and Shiratori, T., 1980, "Condensation Heat Transfer within a Circular Tube under Centrifugal Acceleration Field," *Transactions of the ASME*, Vol. 102, pp. 158–162.

Molt, W., 1981, "Operating Behavior and the Limits of Performance of Cryogenic Heat Pipes," Report No. NTIS PC A07 / Mf A01.

Morris, W. D., 1964, "Heat Transfer Characteristics of a Rotating Thermosyphon," Ph.D. Dissertation, University College of Swansea, Wales.

Mrácek, P., 1988, "Application of Micro Heat Pipes to Laser Diode Cooling," Annual Report of the SVÚSS, Prague, Czechoslovakia.

Murakami, M., Ogushi, T., Sakurai, Y., Masumoto, H., Furukawa, M., and Imai, R., 1987, "Heat Pipe Heat Sink," *Proc. 6th Int. Heat Pipe Conf.*, Grenoble, France, May 25–28, Vol. 2, pp. 537–542.

Murase, T., Tanaka, S., and Ishida, S., 1987, "Natural Convection Type Long Heat Pipe Heat Sink 'POWERKICKER-N' for the Cooling of GTO Thyristor," *Proc. 6th Int. Heat Pipe Conf.*, Grenoble, France, May 25–28, Vol. 2, pp. 537–542.

Nakayama, W., Ohtsuka, Y., Itoh, H., and Yoshida, T., 1984a, "Optimum Charge of Working Fluids in Horizontal Rotating Heat Pipes," *Heat and Mass Transfer in Rotating Machinery*, D. E. Metzger and N. H. Afgan, (Eds.), Hemisphere, Washington, DC, pp. 633–644.

Nakayama, W., Ohtsuka, Y., and Yoshikawa, T., 1984b, "The Effects of Fine Surface Structures on the Performance of Horizontal Rotating Heat Pipes," Proc. 5th Int. Heat Pipe Conf., Tsukuba Science City, Japan, May 14–18, pp. 121–125.

Newton, W. H., 1971, "Performance Characteristics of Rotating, Non-Capillary Heat Pipes," M.S. Thesis, Naval Postgraduate School, Monterey, CA.

Niekawa, J., Matsumoto, K., Koizumi, T., Hasegawa, K., and Kaneko, H., 1981, "Performance of Revolving Heat Pipes and Application to a Rotary Heat Exchanger," *Advances in Heat Pipe Technology*, D. A. Reay (Ed.), pp. 225–234. Pergamon Press, London, England

Ollendorf, S., 1983, "Heat Pipe Thermal Switch," *NASA Technical Briefs*, Vol. 8, No. 1, p. 84.

Patankar, S. V., 1980, *Numerical Heat Transfer and Fluid Flow*, Hemisphere, Washington, DC.

Peterson, G. P., 1985, "Heat Pipe Modeling and Simulation," American Institute of Aeronautics and Astronautics, Paper No. 85-0152, Washington, D.C.

Peterson, G. P., 1987a, "Analysis of a Heat Pipe Thermal Switch," *Proc. 6th Int. Heat Pipe Conf.*, Grenoble, France, May 25–28, Vol. 1, pp. 177–183.

Peterson, G. P., 1987b, "Heat Removal Key to Shrinking Avionics," *Aerospace America*, No. 8, October, pp. 20–22.

Peterson, G. P., 1988a, "Investigation of Miniature Heat Pipes," Final Report, Wright Patterson Air Force Base, Contract No. F33615-86-C-2733, Task 9.

Peterson, G. P., 1988b, "Heat Pipes in the Thermal Control of Electronic Components," invited paper, Proc. 3rd Int. Heat Pipe Symp., Tsukuba, Japan, September 12–14, pp. 2–12.

Peterson, G. P., 1992, "An Overview of Micro Heat Pipe Research," *Applied Mechanics Review*, Vol. 45, No. 5, pp. 175–189.

Peterson, G. P., and Duncan, A. B., 1993, *Micro Heat Pipe Priming and Charging Processes,* Pending U.S. Patent, submitted August 1, 1993.

Peterson, G. P., Duncan, A. B., Ahmed, A. K., Mallik, A. K., and Weichold, M. H. 1991, "Experimental Investigation of Micro Heat Pipes in Silicon Devices," 1991 ASME Winter Annual Meeting, Vol. DSC–32, Atlanta, GA, December 1–6, American Society of Mechanical Engineers, New York, pp. 341–348.

Peterson, G. P., Duncan, A. B., and Weichold, M. H., 1993, "Experimental Investigation of Micro Heat Pipes Fabricated in Silicon Wafers," *ASME Journal of Heat Transfer.* Vol. 115, No. 3, pp. 751–756.

Polásek, F., 1990, "Testing and Application of Itoh's Micro Heat Pipes," Annual Report of the SVÚSS, Prague, Czechoslovakia.

Prager, R. C., and Basiulis, A., 1980, "State-of-the-Art Cryogenic Heat Pipes," American Institute of Aeronautics and Astronautics, Washington, D.C., Paper No. 80–0211, AIAA.

Quao, Z., 1990, "Design and Research of Heat-Pipe Type Drill," Proc. 7th Int. Heat Pipe Conf., Minsk, USSR, May 21–25.

Reddy, R. S., 1989, "Studies on Rotating Heat Pipe and its Application to Cooling of Electrical Machines," Ph.D. Dissertation Indian Institute of Technology, Madras.

Reddy, R. S., Venkateswarlu, P., and Sastri, V. M. K., 1987, "Experimental Heat Transfer Characteristics of a Rotating Heat Pipe with an Air-Cooled Condenser," Proc. 6th Int. Heat Pipe Conf., Grenoble, France, May 25–29, Paper No. 2.3.39.

Reddy, R. S., Venkateswarlu, P., and Sastri, V. M. K., 1990, "Effect of Inclination on the Performance of Rotating Heat Pipe," Proc. 7th Int. Heat Pipe Conf., Minsk, USSR, May 21–25, Paper No. B–2P.

Sakuri, Y., Masumoto, H., Kimura, H., Furukawa, M., and Edwards, D. K., 1984, "Flight Experiments for Gas-loaded Variable Conductance Heat Pipe on ETS-III Active Control Package," Proc. 5th Int. Heat Pipe Conf., Tsukuba, Japan, pp. 26–32.

Salinas, D., and Marto, P. J., 1991, "Analysis of an Internally Finned Rotating Heat Pipe," *Numerical Heat Transfer,* Part A, Vol. 19, pp. 255–275.

Schoch, K. F., 1985, "Bulk Storage and Transfer of Cryogenic Fluids—A Survey," Paper No. 85-WA / HT-64, American Society of Mechanical Engineers, New York.

Semena, M. G., and Khmelev, Y. A., 1982, "Hydrodynamic Regimes of a Liquid in a Smooth-walled Rotating Heat Pipe: Part I," *Journal of Engineering Physics,* Vol. 43, No. 5, November, pp. 1235–1242.

Semena, M. G., and Khmelev, Y. A., 1983, "Investigation of the Hydrodynamic Regimes of a Liquid in a Smooth-walled Rotating Heat Pipe: Part II," *Journal of Engineering Physics,* Vol. 44, No. 1, pp. 4–9.

Semena, M. G., and Levterov, A. I., 1978, "Investigation of the Thermophysical Characteristics of Cryogenic Heat Pipes with a Metal Fiber Wick," *Journal of Engineering Physics,* Vol. 35, pp. 48–53.

Sherman, A., 1982, "History, Status, and Future Applications of Spaceborne Cryogenic Systems," *Advances in Cryogenic Engineering,* Vol. 27, Plenum, New York.

Sherman, A., and Brennan, P. J., 1974, "Cryogenic and Low Temperature Heat Pipe Cooler Studies for Spacecraft Application," Paper No. 74–129, American Institute of Aeronautics and Astronautics, Washington, D.C.

Shimizu, A., and Yamazaki, S., 1987, "Helical Guide-Type Rotating Heat Pipe," Proc. 6th Int. Heat Pipe Conf., Grenoble, France, May 25–29, Paper No. 2.3.38.

Sotani, J., Susa, Y., Tanaka, S., Sato, K., and Kimura, Y., 1990, "Micro Heat Pipe," Proc. 7th Int. Heat Pipe Conf., Minsk, USSR, May 21–25, Paper No. A-41.

Stulc, P., and Horváth, L., 1988, "Heat Pipes for Cooling of Electronic Elements," Annual Report of the SVÚSS, Prague.

Sung, C., 1979, "Simulation and Modelling of Evaporated Deposition Profiles," Ph.D. Thesis, Department of Electrical Engineering and Computer Sciences, University of California, Berkeley, CA.

Swanson, L., and Peterson, G. P., 1992, "The Evaporating Extended Meniscus in a V-Shaped Channel," ASME Paper HTD–Vol. 200, American Society of Mechanical Engineers, New York, NY, pp. 9–22.

Tien, C. L., and Chen, S. J., 1984, "Noncondensible Gases in Heat Pipes," Proc. 5th Int. Heat Pipe Conf., Tsukuda, Japan, pp. 97–101.

Tien, C. L., Chung, K. S., and Lui, C. P., 1979, "Flooding in Two-phase Countercurrent Flows, Report No. EPRI NP–1283, Electric Power Research Institute, Palo Alto, CA.

Tucker, R. S., 1974, "Heat Transfer Characteristics of a Rotating Two-Phase Thermosyphon," M.S. Thesis, Naval Postgraduate School, Monterey, CA.

Van Buggenum, R. I. J., and Daniels, D. H. V., 1987, "Development, Manufacturing and Testing of a Gas Loaded Variable Conductance Heat Pipe," Proc. 5th Int. Heat Pipe Conf., pp. 242–249.

Vasiliev, L. L., and Khrolenok, V. V., 1983, "Study of a Heat Transfer Process in the Condensation Zone of Rotating Heat Pipes," *Heat Recovery Systems*, Vol. 3, No. 4, pp. 281–290.

Vasiliev, L. L., and Khrolenok, V. V., 1990, "Heat Transfer in Rotating Heat Pipes," Proc. 7th Int. Heat Pipe Conf., Minsk, USSR, May 21–25.

Vasiliev, L. L., Kalita, V. V., and Khrolenok, V. V., 1984, "Intensification of Heat Exchange in the Condensation Zone of Centrifugal Heat Pipes," *Journal of Engineering Physics*, Vol. 46, No. 4, pp. 382–387.

Wagenseil, L. L., 1976, "Heat Transfer Performance of Various Rotating Heat Pipes," M.S. Thesis, Naval Postgraduate School, Monterey, CA.

Weichold, M. H., Peterson, G. P., and Mallik, A., 1993, "Vapor Deposited Micro Heat Pipes", U.S. Patent 5,179,043, issued January 12, 1993.

Williams, R. J., 1976, "Influence of Inert Gases on the Performance of Rotating Heat Pipes," Ph.D. Dissertation, University College of Swansea, Wales, UK.

Williams, R. J., 1979, "Transient Shutdown Analysis of Low-Temperature Thermal Diodes," Technical Paper 1369, N79–18287, National Aeronautics and Space Administration.

Woodard, J. S., 1972, "The Operation of Rotating, Non-Capillary Heat Pipes," M.S. Thesis, Naval Postgraduate School, Monterey, CA.

Wu, D., and Peterson, G. P., 1991a, "Investigation of the Transient Characteristics of a Micro Heat Pipe," *AIAA Journal of Thermophysics and Heat Transfer*, Vol. 5, No. 2, April, pp. 129–134.

Wu, D., and Peterson, G. P., 1991b, "A Review of Rotating and Revolving Heat Pipes," 1991 National Heat Transfer Conference, Paper No. 91-HT-24, Minneapolis, MN, American Society of Mechanical Engineers, New York.

Wu, D., Peterson, G. P., and Chang, W. S., 1991, "Transient Experimental Investigation of Micro Heat Pipes," *AIAA Journal of Thermophysics and Heat Transfer*, Vol. 5, No. 4, October, pp. 539–545.

Xu, X., and Carey, V. P., 1990, "Film Evaporation from a Micro-Grooved Surface—An Approximate Heat Transfer Model and Its Comparison with Experimental Data," *AIAA Journal of Thermophysics and Heat Transfer*, Vol. 4, No. 4, pp. 512–520.

Chapter

7

DESIGN AND MANUFACTURING CONSIDERATIONS

7.1 INTRODUCTION

The design and manufacture of heat pipes is an extremely complex process involving many different physical variables such as size, shape, weight, and volume; thermophysical properties such as working fluid, wicking structure, and case material properties; and other design aspects, such as thermal load, transport length, evaporator / condenser length, acceptable temperature drop, operating temperature range, gravitational environment, source–sink interfaces, fluid inventory, life / reliability, and safety. Several early references published in the mid to late 1970s presented an excellent overview of many of the problems associated with these and other issues related to the design and manufacture of heat pipes (Feldman et al., 1973; Brennan and Kroliczek, 1979).

In addition to these specific areas, the design and manufacture of heat pipes is governed by three operational considerations: the effective operating temperature range, which is determined by the selection of the working fluid; the maximum power the heat pipe is capable of transporting, which is determined by the ultimate pumping capacity of the wick structure (for the capillary wicking limit); and the maximum evaporator heat flux, which is determined by the point at which nucleate boiling occurs.

Because, as illustrated in Fig. 7.1, all three of these operational considerations must be included, the design process first requires that the design specifications for the problem under consideration be clearly identified. Once this has been accomplished, preliminary selection of the working

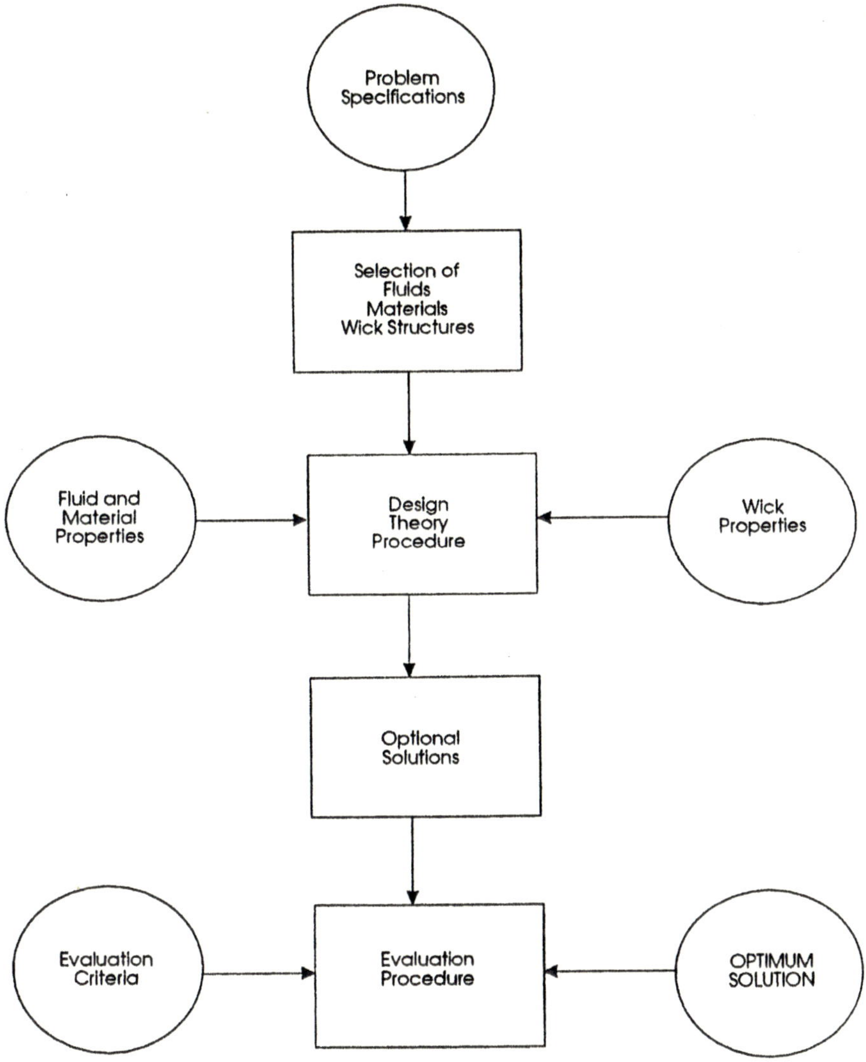

Figure 7.1 Heat pipe design flow chart.

fluid, wicking structure, and case materials can be performed. Then using the techniques outlined in Chapters 3 and 4, the various transport limitations can be estimated. Finally, using an iterative process, various combinations of working fluids, evaporator and condenser sizes, and case / wick material combinations can be evaluated. While experience is extremely helpful, the new designer can, using the guidelines outlined in this chapter, develop an improved, if not optimal, design.

Perhaps the most difficult part of the design process is determining how the various components utilized in heat pipe and thermosyphon construc-

TABLE 7.1 Heat Pipe Components and Their Influence on Design Requirements

Design Requirements	Working Fluid	Wick Material	Case Material
Thermal performance			
Transport capacity	S	S	W
Operating temperature range	S	W	W
Temperature drop	M	W	W
Mechanical			
Physical requirements (size, weight, etc.)	W	W	M
Wall thickness–internal pressure	W	N	S
Sink–source interface	N	N	S
Dynamic / static loads	W	S	M
Reliability and Safety			
Material compatibility	S	S	S
External corrosion	N	N	
Fabrication	M	M	M
Pressure containment / leakage	W	M	S
Toxicity	S	W	W
Gravitational environment	S	S	S
>1g	M	M	W
1g	W	M	W
<1g			

S, strong factor; M, moderate factor; W, weak factor; N, negligible.

tion affect the different design requirements. Table 7.1 presents a matrix which gives some indication of how each of the three primary components, the working fluid, wick material, and heat pipe case material, affect the various design requirements. As shown, no single component appears to be more important than any other, and very few design requirements are affected by only one of these three components.

7.2 WORKING FLUID SELECTION

Because the basis for operation of a heat pipe is the vaporization and condensation of the working fluid, selection of a suitable working fluid is perhaps the most important aspect of the design and manufacture process. Factors that affect the selection of an appropriate working fluid include the operating temperature range, the vapor pressure, the thermal conductivity, the compatibility with the wick and case materials, the stability, and the toxicity.

As discussed previously, the theoretical operating temperature range for a given heat pipe is typically between the critical temperature and triple state of the working fluid. Above the critical temperature, the working fluid exists in a vapor state and no increase in pressure will force it to return to a liquid state. As a result, when working fluids are above their critical temperature, the capillary pumping mechanism provided by the wicking structure ceases to function. Similarly, when the operating temperature is below the triple state, the working fluid exists in the solid and vapor states. While some heat transfer may occur due to sublimation (Ochterbeck and Peterson, 1992), operation in this temperature range should be avoided.

Although the critical temperature and triple state provide theoretical upper and lower limits, the actual optimum operating temperature range may be significantly less than the region defined by these temperatures. This fact is best demonstrated in Fig. 7.2, which shows a constant volume heat addition process. If the volume of the heat pipe container and the mass of the working fluid are such that the specific volume inside that heat pipe is less than the specific volume at the critical point, constant volume heat addition will result in a reduction in quality and hence a maximum operating temperature, which is below the critical point, as illustrated in Fig. 7.2 as path *A–B*. Likewise for situations where the specific volume is substantially greater than the specific volume at the critical point, which is more frequently the case, a similar phenomenon occurs except that here the working fluid moves from a relatively low quality to a somewhat higher quality, as illustrated by path *C–D*. While it is clear from this illustration that the maximum operating temperature is a strong function of the specific volume at the critical point, determination of the specific volume of the working fluid inside the heat pipe container is not typically a factor in the charging or priming process.

Perhaps more important than the determination of the maximum operating temperature of a given working fluid is the determination of an

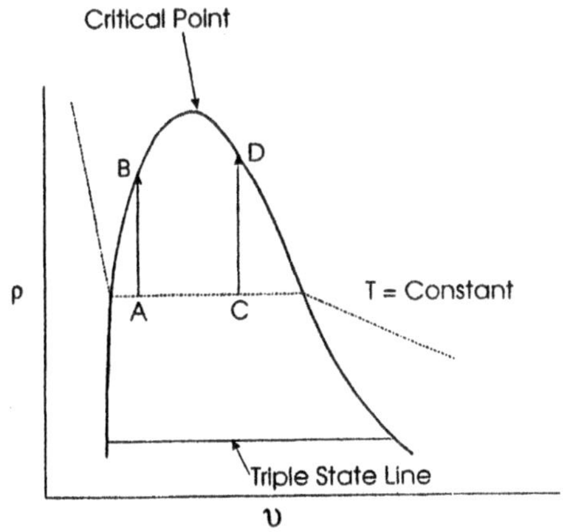

Figure 7.2 Effect of initial quality on heat pipe operation.

acceptable operating temperature range. Figure 7.3 illustrates the typical operating temperature range for various heat pipe working fluids. As shown, three fairly distinct regions or classifications of working fluids have been identified: cryogenic fluids such as helium, neon, oxygen, or nitrogen; low to moderate temperature working fluids, such as Freons, methanol, ammonia, or water; and high temperature or liquid metal working fluids, such as potassium, lithium, or sodium. As discussed by Marto and Peterson (1988), most heat pipe applications for electronic thermal control require the selection of a working fluid with boiling temperatures between 250 and 375 K. This would typically limit the selection to fluids such as ammonia, acetone, methanol, and water or dielectric fluids such as Freon-11 or -113.

For capillary wick limited heat pipes, which are the type most frequently encountered in electronic applications, the characteristics of a good working fluid are a high latent heat of vaporization, a high surface tension, a high liquid density and a low liquid viscosity. Chi (1976) combined these properties into a parameter referred to as the liquid transport factor or figure of merit,

$$N_l = \frac{\rho_l \sigma \lambda}{\mu_l} \tag{7.1}$$

which can be used to evaluate the effectiveness of various working fluids at specific operating temperatures. Figure 7.4 illustrates the figure of merit for working fluids in all three of the operating temperature regions described above. As shown, sodium provides the highest liquid figure of merit over a wide range of temperatures but is typically inappropriate at temperatures less than 500 K. In the low or moderate temperature range, similar to that anticipated for use in electronic thermal control systems, water appears to have the highest figure of merit and for many applications may be the most suitable working fluid. In many situations, however, the possibility of small amounts of water leaking onto electronic circuitry devices may be unacceptable. In these situations, other fluids such as ammonia, methanol, or dielectric fluids, such as Freons, may be more appropriate. In any case, the figure of merit defined in equation 7.1 can be utilized to evaluate the transport capacity of various fluid alternatives as a function of operating temperature. The concept of a parameter for evaluating working fluids has been extended to show that the thermophysical properties of the liquid–vapor equilibrium state can be reduced to three independent parameters.

As will be discussed in subsequent sections, the presence of impurities in the working fluid can have significant detrimental effects on the operation of heat pipes and thermosyphons. These effects can be subdivided into two general classifications: performance degradation caused

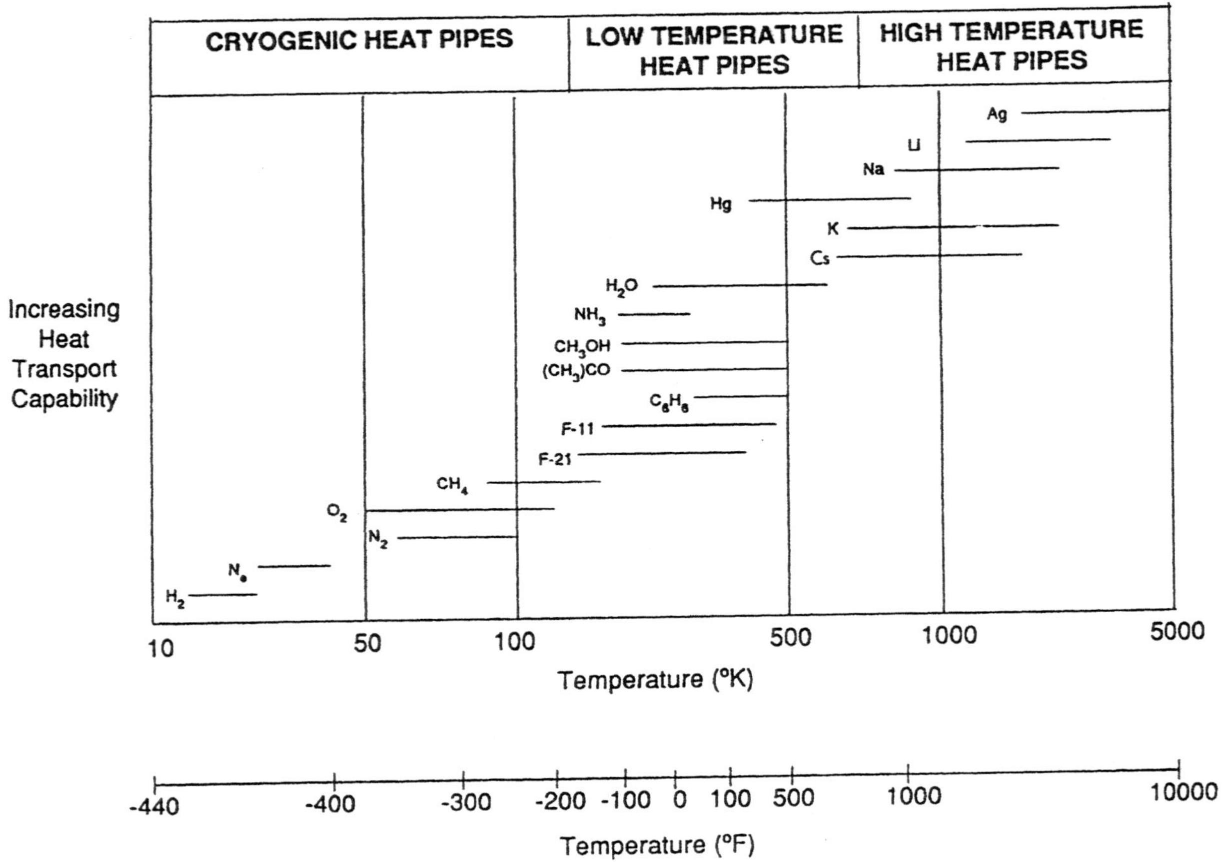

Figure 7.3 Typical operating range for heat pipe working fluids.

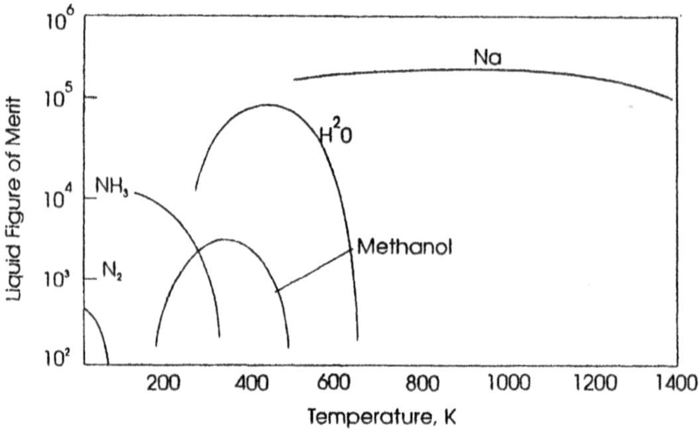

Figure 7.4 Figure of Merit for various working fluids (Dunn and Reay, 1982).

by the presence or formation of noncondensible gases and performance degradation caused by solid or liquid impurities.

In the first of these, the presence of noncondensible gases typically results in decreased performance due to partial or complete blockage of the condenser. Figure 7.5 illustrates the effects of either the presence of noncondensible gas (NCG) as a result of improper charging or NCG generation caused by impurities. Prior to operation, the noncondensible gas is uniformly distributed throughout the vapor space of the heat pipe. Once heat is added to the evaporator end, the vapor flow sweeps the NCG along with it toward the condenser, where it becomes trapped and builds up, blocking the condenser region. As shown in Fig. 7.5, the presence of NCGs is characterized by a sharp decrease in the temperature profile along the axial length of the heat pipe. The low temperature region at the condenser end will typically be quite close to the temperature of the

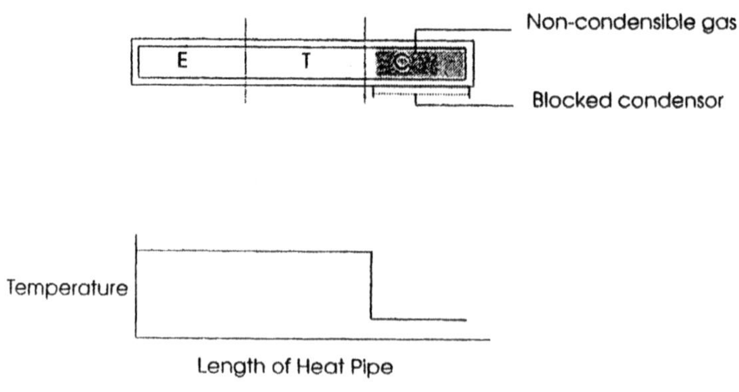

Figure 7.5 Effect of non-condensible gas on temperature distribution.

condenser coolant and will vary in length as a function of the adiabatic vapor temperature.

The second classification of impurities is solid or liquid impurities resulting from improper cleaning or charging procedures. These impurities can degrade performance due to changes in the wetting angle or viscosity of the working fluid, clogging of the fine pores in the capillary wicking structure, or internal corrosion of the heat pipe case material. On occasion, the latter of these, internal corrosion of the heat pipe case material, will also result in generation of NCGs, further degrading performance.

In addition to the thermophysical properties of the working fluid, consideration must be given to the ability of the working fluid to wet the wick and wall materials. Further criteria for the selection of the working fluids have been presented by Roesler et al. (1987), who consider a number of other factors, including the liquid and vapor pressure and the compatibility of the various materials.

7.2.1 Measurement of Surface Tension and Wetting Angle

As discussed in Chapter 2, the surface tension and wetting angle of a fluid play an extremely important role in determining the amount of heat that can be transferred by a given heat pipe. For this reason, it is helpful to discuss the ways in which these characteristics can be quantified and measured. The simplest and most straightforward method for measuring the surface tension of a working fluid is to measure the capillary rise height. Using the hydrostatic pressure difference and equating that to the capillary pressure across the interface as defined by the Young–Laplace equation presented in Chapter 2 (equation 2.10) yields

$$\rho g z = P_v - P_l = \frac{2\sigma \cos\theta}{r} \tag{7.2}$$

where z is the capillary rise height. While these measurements are simple and relatively accurate, it must be remembered that the capillary rise height is a strong function of the surface cleanliness and also is strongly dependent upon whether the meniscus is advancing or receding.

This technique is not only appropriate for open capillary channels such as axial grooves or liquid arteries but may also be applied to porous wicking structures consisting of sintered metal powders, metal fibers, screen mesh, or other types of porous materials. In addition, this measurement technique yields only a value for the capillary pressure rise and does not distinguish between the individual effects of the surface tension σ, wetting angle θ, and effective capillary radius r_c.

To distinguish between the various parameters that affect the capillary pressure difference and hence the capillary rise height, several different procedures can be utilized. The first of these, shown in Fig. 7.6, involves

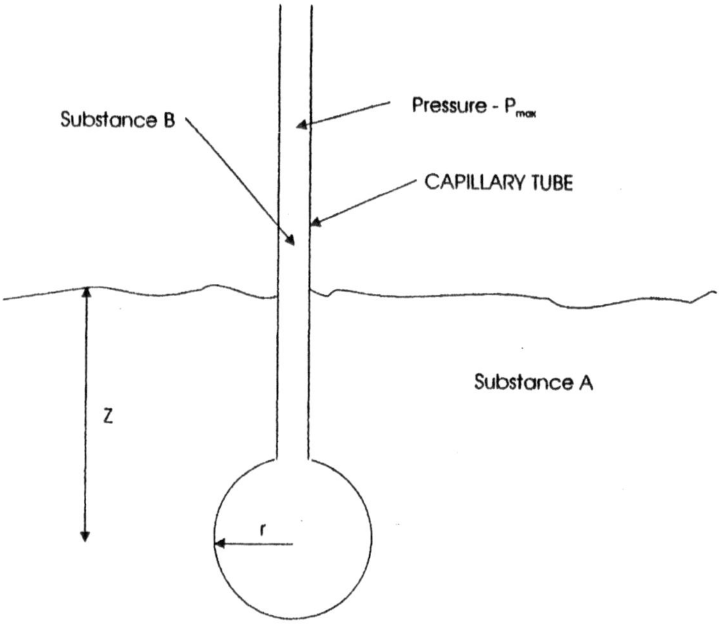

Figure 7.6 Maximum buble method for measuring surface tension (Bohdanski and Schins, 1967).

measurement of the surface tension and is referred to as the maximum bubble pressure method (Bohdanski and Schins, 1967). In this method a small capillary tube filled with liquid A is immersed in another liquid, gas, or vapor of the same substance, B. The pressure in the tube is slowly increased until a maximum pressure is achieved and the bubble or liquid droplet breaks away from the capillary tube. Utilizing a hydrostatic pressure balance, the pressure at which the bubble or droplet breaks away or the maximum pressure P_{max} can be defined as the summation of the capillary pressure caused by the bubble radius, r_B, plus the hydrostatic pressure caused by the pressure difference in the two phases. This relationship can be expressed mathematically as

$$P_{max} = \frac{2\sigma_l}{r_B} + \rho_A g z \tag{7.3}$$

In a similar manner, the surface tension of two liquids can be related by evaluating the droplet size at which the liquid breaks away from a small capillary tube into either an air or vapor environment. Using two vertical tubes of the same capillary radius, liquid droplets are allowed to build up on the end of the tube and eventually break free due to the gravitational force. By measuring the radius of the two droplets and then comparing the ratio of the two masses, the surface tension ratio can be approximated (Dunn and Reay, 1982):

$$\frac{m_1 \rho_{lA}}{m_2 \rho_{lB}} = \frac{\sigma_A}{\sigma_B} \tag{7.4}$$

This method is particularly useful when trying to measure the surface tension of liquids for which little or no data are available.

For applications involving very fine capillary structures or micro–heat pipes, it may be helpful to have a more complete understanding of the thin-film behavior and properties of a particular working fluid. Properties such as the disjoining pressure and dispersion forces) can be measured using the procedure illustrated in Fig. 7.7. In this method, a liquid droplet is located between two tubes or cylinders, as shown in Fig. 7.7. By observing and monitoring the axial force, separation distance, and film thickness between the two tubes as they are separated, measurements of the thin-film behavior and properties can be obtained.

As a result of the methods described above, considerable literature exists on the surface tension, characteristic radius, and contact angle for many types of wicking structures and materials. While, ideally, one would expect liquids with high surface tensions such as water to exhibit contact angles of nearly zero on clean metallic surfaces in which no chemical reaction occurs (Adamson, 1990), this is typically not the case. Careful examination of the experimental data indicates large inconsistencies. These inconsistencies may in part be due to oxide layers or impurities either in the liquid or on the solid surface. As discussed in Chapter 2, small quantities of impurities or even very thin oxide layers can significantly affect the way a particular liquid behaves when in contact with a particular solid surface. While, typically, these variations in the wetting properties and surface tension can lead to somewhat reduced performance, for some cases (i.e., micro–heat pipes) variations in these parameters, particularly the contact angle, may lead to improper priming or complete failure (Peterson and Babin, 1989).

One way in which the contact angle for a particular solid–liquid

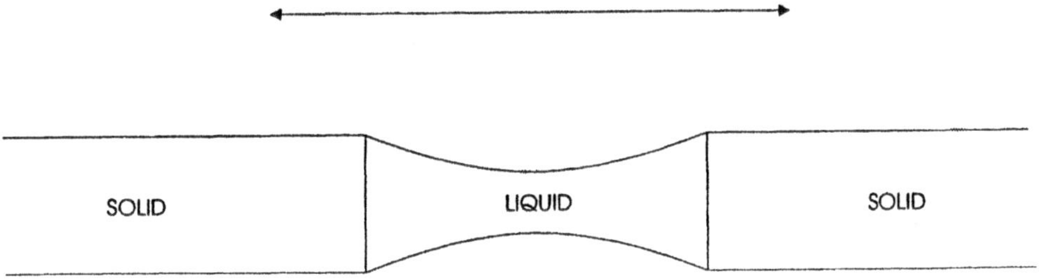

Figure 7.7 Measurement technique for determining thin film properties.

combination can be determined is to measure the difference in the contact angle for advancing and receding menisci. This can be done using a simple test fixture fabricated from Plexiglas, as shown in Fig. 7.8. The chamber shown has internal dimensions of approximately $2.54 \times 2.54 \times 8$ cm, but the size is not of any particular relevance. The front and back walls of the chamber are transparent while the left and right sides are lined with a thin layer of the solid materials under consideration. The chamber is filled through the fill port located in the top with approximately 10 cm^3 of the distilled working fluid and then evacuated for approximately 5 min to reduce the amount or NCGs present. The fill port is then closed, leaving

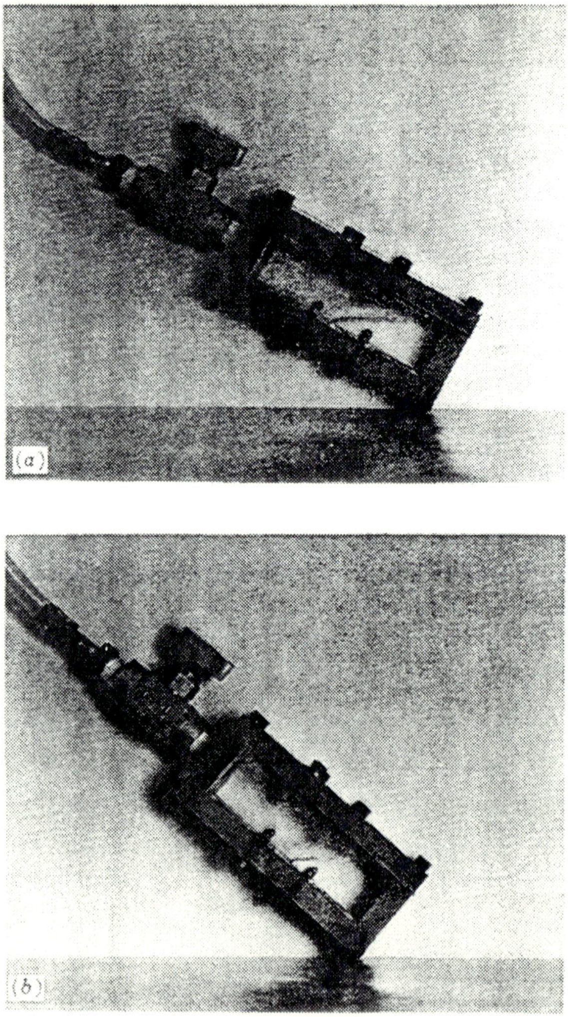

Figure 7.8 Surface tension measurement technique.

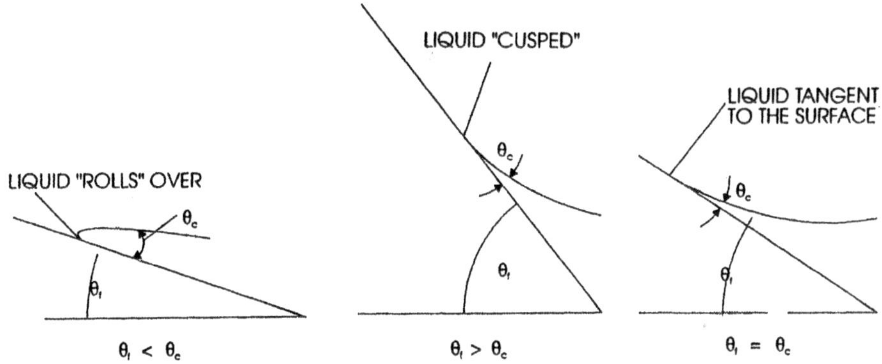

Figure 7.9 Determination of wetting angle.

only the pure working fluid in the liquid and vapor states. By tilting the chamber until the liquid at the interface forms a flat surface, as shown in Fig. 7.9, the contact angle can be measured. When the contact angle θ_c is less than the tilt angle of the fixture, θ_f, the meniscus rolls over at the interface, as shown in Fig. 7.9*a*. When the fixture angle is too large, the meniscus takes on a slightly cusped shape, as shown in Fig. 7.9*b*. Finally, the correct measurement of the surface tension can be determined when the meniscus appears to be tangent with the solid surface (i.e., $\theta_c = \theta_f$), as shown in Fig. 7.9*c*. In this manner, accurate determinations of the contact angle between the actual surface and liquid in question can be made. Experimental results using this procedure have indicated contact angles of 55° and 45° for water in contact with copper and silver, respectively (Peterson and Babin, 1989).

7.3 WICK STRUCTURES

The wicking structure has two functions in heat pipe operation: It is both the vehicle and the mechanism through which the working fluid returns from the condenser to the evaporator and it ensures that the working fluid is evenly distributed circumferentially over the entire evaporator surface. Figure 7.10 illustrates several common wicking structures presently in use, along with several composite and high capacity concepts under development. An excellent summary of conventional heat pipe wicking structures and their properties has been prepared by Swanson (1990) and is illustrated in Table 7.2.

As shown in Fig. 7.10, the various wicking structures can be divided into three broad categories: homogeneous, current composite, and advanced designs. Homogeneous wicks are typically comprised of a single material and are distributed uniformly along the axial length of the heat

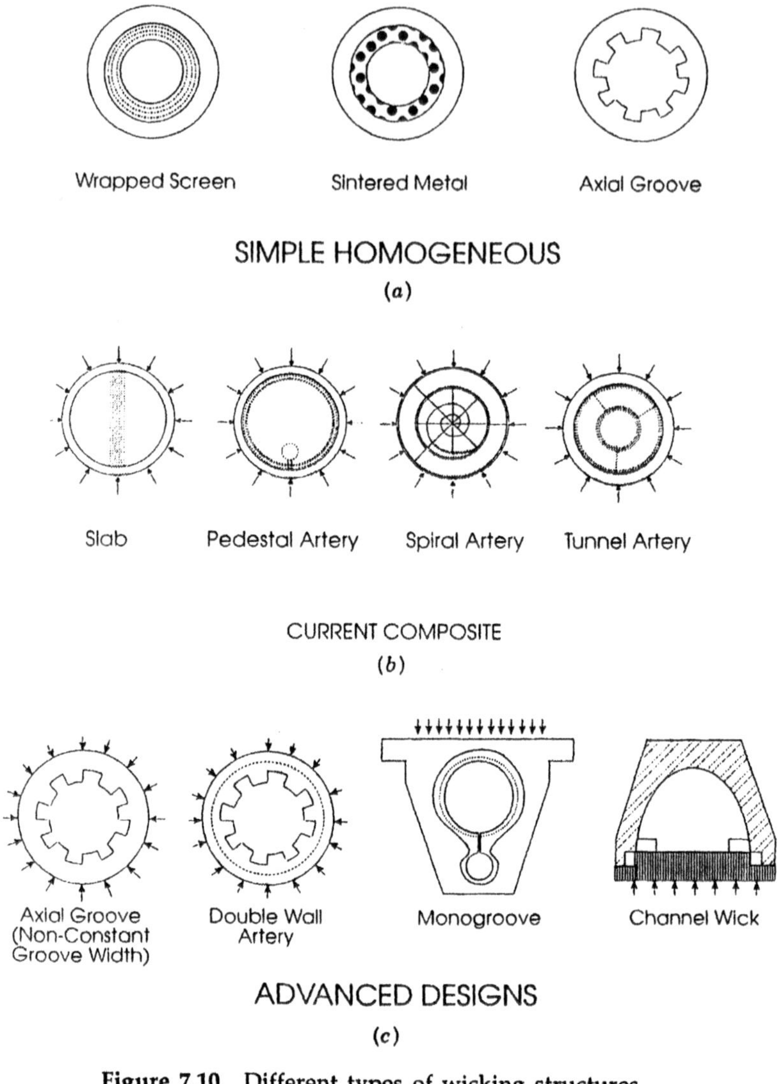

Figure 7.10 Different types of wicking structures.

pipe. The most common types of homogeneous wicks include wrapped screen, sintered metal, axially grooved, and crescent and are shown in Fig. 7.10*a*.

In order to provide a flow path with low flow resistance through which the liquid can be returned from the condenser to the evaporator, an open porous structure with a high permeability is desirable. However, to increase the capillary pumping pressure, a small pore size is necessary. Solutions to this apparent dichotomy can be achieved through the use of a nonhomogeneous wick made of several different materials or through a composite wicking structure. Composite wicks are typically comprised of a combination of several types or porosities of materials and / or configurations. Examples of these types of wick structures are illustrated in Fig. 7.10*b*.

Udell and Jennings (1984) proposed and formulated a model for a heat pipe with a wick consisting of porous media of two different permeabilities oriented parallel to the direction of the heat flux. This wick structure provided a large pore size in the center of the wick for liquid flow and a smaller size pore for capillary pressure. Composite wicking structures accomplish the same type of effect. As shown in Fig. 7.10, the capillary pumping and axial fluid transport are handled independently. In addition to fulfilling this dual purpose, several of the wicks illustrated in Fig. 7.10*b* separate the liquid and vapor flow. This results from an attempt to eliminate the viscous shear force that occurs during countercurrent, liquid–vapor flow. In the pedestal artery heat pipe illustrated in Fig. 7.10*b*, for example, the longitudinal liquid flow occurs primarily in the open liquid artery while the capillary pumping capability is developed in the circumferential wicking structure. In this situation the open artery yields a high permeability and low flow resistance axially while the circumferential wick provides a high capillary pumping pressure.

The final category of heat pipe wicking structures is of high capacity designs. Most of these are relatively new and consist of variations on the composite wicking structures shown in Fig. 7.10*b*. Again, the two functions of the wicking structure (i.e., circumferential distribution and axial fluid transport) are achieved by different segments of the capillary structure. For example, in the monogroove heat pipe presented and described by Alario et al. (1984) the circumferential distribution and axial flow of the working fluid are accomplished through two separate mechanisms. The basic design of this high capacity configuration consists of two large axial channels, one for vapor flow and the other for liquid flow. These two channels are separated by a small monogroove slot that creates a high capillary pressure difference and causes liquid to be "pumped" from the liquid channel to the fine circumferential grooves machined in the evaporator and condenser sections of the vapor channel. These circumferential grooves provide liquid distribution over the evaporating surface and also maximize the heat transfer area for the vaporization occurring in the evaporator and the condensation occurring in the condenser. Because vaporization occurs at the liquid–vapor interface and condensation occurs at the metal–vapor interface, the number of circumferential grooves can be varied in the evaporator and condenser to optimize the conditions for both situations.

In this type of heat pipe, several improvements result from the separation of the liquid and vapor channels. First, because the axial liquid transport can be handled independently from the circumferential distribution, a high heat transport capacity can be achieved. Second, by separating the two channels, the viscous pressure drop normally associated

TABLE 7.2 Physical Properties of Various Wick Structures (Revised from ESDU, Anon, 1979).

Wick type[a]		Thermal conductivity: Equation	Thermal conductivity: References	Porosity
Single-layer wire mesh screens (heat pipe axis in plane of paper in this sketch) 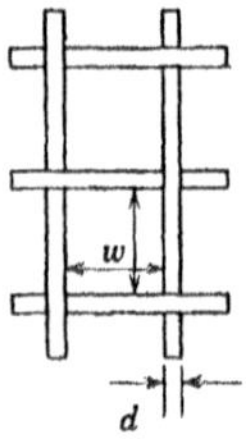	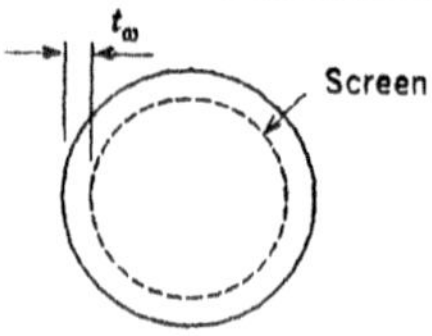	$k_{eff} = k_1$	—	$\epsilon = 1$
	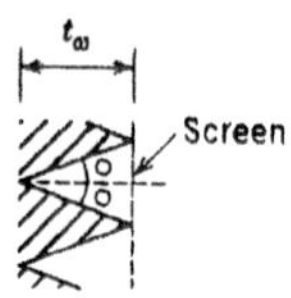	$k_{eff} = 0.85k_s(k_1/k_s \sin\alpha)^{0.63}$	Feldman and Berger, 1973; Chiakovsky et al., 1978	$\epsilon = 0.5$
$1/N = d + w$ N = number of apertures per unit length	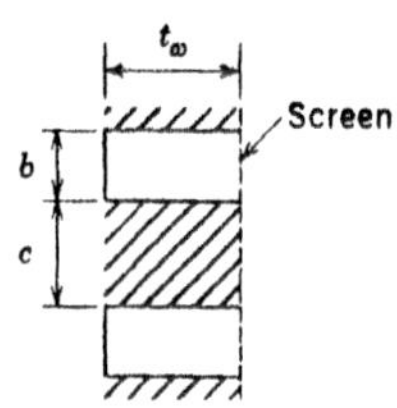	$k_{eff} = \epsilon k_1 + (1 - \epsilon)k_s$	Chi, 1976; Dunn and Reay, 1982	$\epsilon = b/b + c$
 Multiple wire mesh screens,[b] plain orsintered (screen dimensions as for single) layers illustrated above		$k_{eff} = \frac{k_1[k_1 + k_s - (1 - \epsilon)(k_1 - k_s)]}{k_1 + k_s + (1 - \epsilon)(k_1 - k_s)}$	Chi 1976; Ferrell et al., 1973; Keser, 1973; Vansant and Malet, 1975	Best obtained experimentally but can be estimated from the equation for single screens: $\epsilon = 1 - [(\pi N d)/4]$
 Unconsolidated packed spherical particles (d = average particle diameter)	Plain	$k_{eff} = \frac{k_1[2k_1 + k_s + 2(1 - \epsilon)(k_1 - k_s)]}{2k_1 + k_s + (1 - \epsilon)(k_1 - k_s)}$	Chi, 1976; Goring and Churchill, 1961	Depends on packing mode and should be measured; very rough estimate can be obtained assuming cubic packaging: $\epsilon = 0.48$
	Sintered	$k_{eff} = \frac{k_s[2k_s + k_1 - 2\epsilon(k_s - k_1)]}{2k_s + k_1 + \epsilon(k_s - k_1)}$	Dunn and Reay, 1982; Ferrell et al., 1973; Kesser, 1973	
 Sintered felt, metal fibers[b] (d = fiber diameter)		$k_{eff} = \epsilon^2 k_1 + (1 - \epsilon)^2 k_s + \frac{4\epsilon(1 - \epsilon)k_1 k_s}{k_1 + k_s}$	Ferrell et al., 1973; Soliman et al., 1970; Dul'Nev and Zarichny, 1974; Semena and Zaripov, 1977; Anonymous, 1979	Obtain experime[illegible]y or use manufacturer's data

TABLE 7.2 (*Continued*)

Minimum capillary radius			Permeability		
Equation	References	Accuracy Information	Equation	References	Accuracy information
$r_c = 1/2N$	Katzoff, 1967; Phillips and Hindeman, 1969; Tien and Sun, 1971; Roberts and Feldman, 1972; Goodacre, 1974	Eighty percent of the data lie within 0.8–1.2 times the value given by the equation	$K = t_w^2/12$		Equations apply strictly to these shapes with smooth surfaces in place of screens and are then accurate to within ±5%. The effect of screens is uncertain but is expected to be small
			$K = \phi t_w^2 \epsilon$	Anonymous, 1966	
			$K = \phi b t_w \epsilon$		
$r_c = 1/2N$	Ferrell et al., 1973; Keser et al., 1973; Roberts and Feldman, 1972; Freggens, 1969 Langston and Kunz, 1969; Chun, 1972; Birnbreier and Gammel, 1973; Hoogendoorn et al., 1973	Eighty percent of the data lie within 0.5–1.5 times the value given by the equation.	$K = d^2\epsilon^3/[122(1-\epsilon)^2]$	Ferrell et al., 1973; Keser, 1973 Roberts and Feldman, 1972; Freggens, 1969 Langston and Kunz, 1969; Chun, 1972; Birnbreier and Gammel, 1973; Marcus, 1972	Equation from Ref. (Marcus, 1972) Eighty percent of the data from other derivatives lie within 0.5–2.0 times the value given by the equation
$r_c = 0.21d$	Keser, 1973; Goodacre, 1974; Freggens, 1969; Langston and Kunz, 1969; Chun, 1972; Birnbreier and Gammel, 1973; Finlay et al., 1976; Eastman and Ernst, 1978	Eighty percent of the data lie within 0.7–1.3 times the value given by the equation.	$K = d^2\epsilon^3/[150(1-\epsilon)^2]$	Keser, 1973; Freggens, 1969; Langston and Kunz, 1969; Birnbreier and Gammel, 1973; Finlay et al., 1976; Eastman and Ernst, 1978; Marcus, 1972; Ferrell and Alleavitch, 1970	Equation from Ref. (Marcus, 1972) Eighty percent of the data from other derivations lie within 0.8–1.7 tim the value given by th ati on equation
$r_c = d/[2(1-\epsilon)]$	Ferrell et al., 1973; Goodacre, 1974; Freggens, 1969; Langston and Kunz, 1969	Eighty percent of the data lie within 0.5–1.5 times the value given by the equation.	$K = C_1[(y^2-1)]/[y^2+1]$ where $y = 1 + [C_2 d^2 \epsilon^3]/(1-\epsilon)^2$ $C_1 = 6.0 \times 10^{10}\ m^2$ $C_2 = 3.3 \times 10^3\ m^{-2}$	Ferrell et al., 1973; Anonymous, 1979; Goodacre, 1974; Freggens, 1969; Langston and Kunz, 1969 Semena et al., 1975	Eighty percent of the data lie within 0.75–1.25 times the value given by the equation

[a] In the sketches, the axis of the heat pipe and therefore the direction of the fluid flow are normal to the plane of the paper unless otherwise stated.

[b] These wicks are positioned so that the layers follow the contours of the inner surface of the heat pipe wall.

with heat pipes in which the liquid and vapor flows occur within the same channel can be greatly reduced. Third, with the majority of the fluid located in an external artery, heat transfer in the evaporator and condenser takes place across a relatively thin film of liquid in the circumferential wall grooves, thereby increasing the heat transfer coefficient. While somewhat different in shape, the basic principle of operation of the other advanced designs is the same: Separate the circumferential distribution and axial liquid flow to maximize the capillary pumping and reduce the liquid pressure drop.

7.4 CONTAINER OR CASE MATERIALS

Careful consideration should be given to the selection of the container or case material for heat pipes. Various factors that should be considered are the operating temperature range of the proposed device, compatibility with the wicking structure and working fluid, internal operating pressure, evaporator and condenser sizes and shapes, and the possibility of external corrosion. As mentioned previously, formation of NCGs typically limits the operational lifetime of heat pipes and thermosyphons. Degradation or contamination of the container or case material can lead to chemical reactions with the working fluid and, hence, generation of NCGs. These chemical reactions have a whole host of other potential problems, including the formation of corrosive films, which tend to increase the thermal resistance and decrease the surface tension and wetting angle; pitting of the internal surface of the container material, which may result in the formation and development of enhanced nucleation sights and hence premature boiling of the working fluid; and corrosion on the internal surface of the pipe, which may lead to increased thermal resistance and, hence, increased overall resistance of the heat pipe.

In addition to the compatibility problem associated with heat pipes it must be remembered that heat pipes and thermosyphons are in fact "unfired pressure vessels" and as a result must be designed to meet the appropriate pressure vessel codes. These codes (ASME, 1965) typically require that the maximum stress at any given temperature be less than one-fourth of the ultimate tensile strength at that temperature. This design criterion presents some rather unique problems with cryogenic heat pipes since, unlike most heat pipes, the maximum temperature to which cryogenic heat pipes are typically exposed occurs during storage or transportation, unlike high or moderate temperature heat pipes, which experience maximum temperatures during operating conditions. Because it is the internal to external pressure difference that is important, the environmental pressure must also be considered during the evaluation process of the container, particularly for spacecraft or satellite applications where the external pressure may be near zero absolute.

Basic requirements of the heat pipe case include a container capable of maintaining a leak-proof seal and structural integrity throughout the entire pressure range to which the heat pipe will be exposed. Possible materials include pure metal alloys such as aluminum, stainless steel, or copper; composite materials, either metal or carbon composite; or for higher temperature applications, refractory materials or linings to prevent corrosion.

7.5 PRELIMINARY DESIGN CONSIDERATIONS

Chemical reactions between the working fluid and the wall or wicking structure or decomposition of the working fluid may lead to the formation of NCGs or problems associated with corrosion. Because of the detrimental effect of NCGs on heat pipe performance, careful consideration must be given to the selection of working fluids and wicking and wall materials in order to prevent the occurrence of these problems over the operational life of the heat pipe. The effect of NCG formation has been discussed earlier and may result in either decreased performance or total failure. Corrosion problems can lead to physical degradation of the wicking structure since solid particles carried to the evaporator wick and deposited there will eventually reduce the wick permeability (Barantsevich et al., 1987).

Basiulis et al. (1976) conducted extensive compatibility tests with several combinations of working fluids and wicking structures, the results of which are summarized in Table 7.3, (Dunn and Reay, 1982) along with other investigations by Basiulis and Filler (1971). Other more recent investigations such as those performed by Zaho (1987), in which the compatibility of water and mild steel heat pipes were evaluated; Roesler et al. (1987), in which stainless steel, aluminum, and ammonia combinations were evaluated; and Murakami et al. (1987), in which a statistical predictive technique for evaluating the long term reliability of copper water heat pipes was developed—provide additional insight into the compatibility of various liquid–material combinations that might be used in the thermal control of electronics equipment. Most of the data available are the result

TABLE 7.3 Working Fluid, Wick, and Container Compatibility Data

Material	Water	Acetone	Ammonia	Methanol	Dow-A	Dow-E	Freon-11	Freon-113
Copper	RU	RU	NU	RU	RU	RU	RU	RU
Aluminum	GNC	RL	RU	NR	UK	NR	RU	RU
Stainless steel	GNT	PC	RU	GNT	RU	RU	RU	RU
Nickel	PC	PC	RU	RL	RU	RL	UK	UK
Refrasil	RU	RU	RU	RU	RU		UK	UK

Source: From Dunn and Reay, 1982).

Abbreviations:

RU, recommended by past successful usage; RL, recommended by literature; PC, probably compatible; NR, not recommended; NU, not used; UK, unknown; GNC, generation of gas at all temperatures; CNT, generation of gas at elevated temperatures when oxide present.

of accelerated life tests. Although a majority of it is based upon actual test results, care should be taken to ensure that the tests in which the data were obtained are similar to the application under consideration. Such factors as thermal cycling, mean operating temperature, and so on, must also be included.

In addition to NCG considerations, several other aspects must be considered in the preliminary design of heat pipes and thermosyphons. Among these are the source–sink interface conditions. As discussed in Chapter 6, the overall thermal resistance of a heat pipe system is made up of many components. However, in many applications the magnitude of the temperature gradients occurring at the heat source–heat sink interfaces are of the same magnitude as the end-to-end temperature gradient. In these cases it is extremely important to give careful consideration to the method or technique by which the heat pipe is attached to the heat source and heat sink. Figure 7.11 illustrates two attachment techniques frequently utilized in the evaporator. These consist of a grooved attachment technique or the use of saddle clamp, illustrated in Fig. 7.11. Other techniques that can be used either in conjunction with these or as stand-alone attachment techniques include the use of thermally conductive adhesives and soldered, brazed, or bolted techniques. In the condenser portion of heat pipes, the most common types of attachment techniques include the use of the parallel pipe flow heat exchanger illustrated in Fig. 7.12*a* or the finned configuration illustrated in Fig. 7.12*b*, which is often utilized in free or forced convection situations with air as the coolant.

Several recent investigations have been conducted to measure or predict

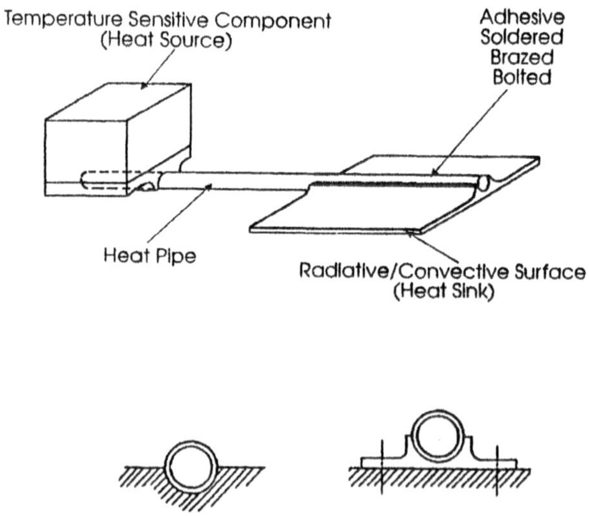

Figure 7.11 Evaporator attachment techniques.

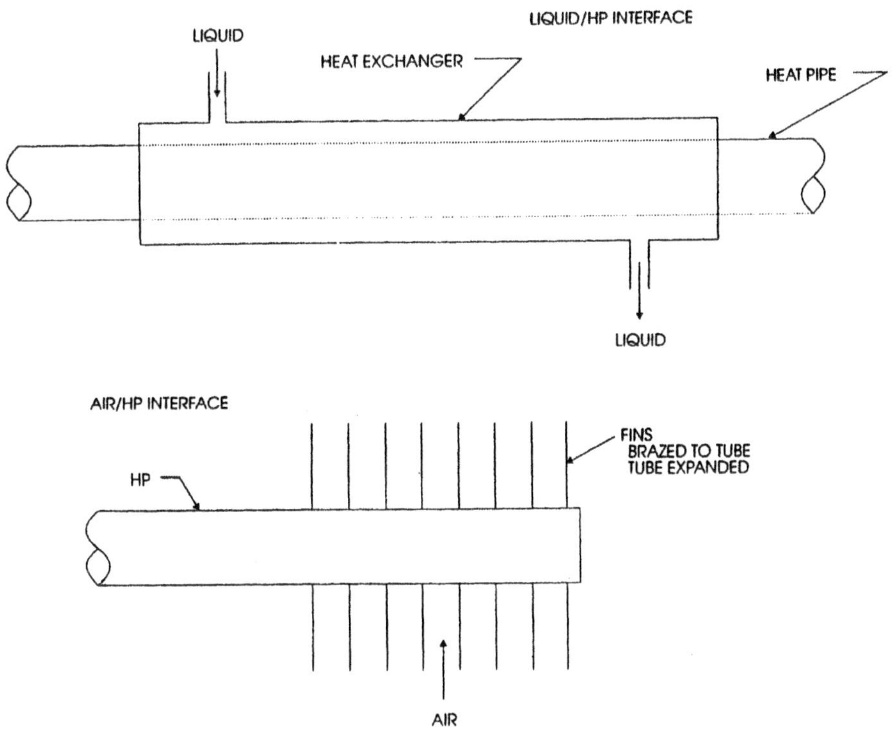

Figure 7.12 Condenser cooling schemes.

the temperature drop occurring at various interfaces. These include determination of the thermal contact conductance at similar metal-to-metal interfaces (Fletcher, 1988), dissimilar metallic interfaces (Stevenson et al., 1991), interfaces with anodized coatings (Peterson and Fletcher, 1990), and nonmetallic interfaces (Peterson and Fletcher, 1988).

7.6 HEAT PIPE FABRICATION

The first step in the fabrication process for any heat pipe or thermosyphon involves selection of the appropriate case, wick, and working fluid materials. While NCG generation and corrosion were specifically identified previously as potential sources of difficulty, these are only two of the factors to be considered when selecting heat pipe wicks, working fluids, and case materials. Others include wettability of the fluid–wick combination, strength-to-weight ratio, thermal conductivity, stability, and cost of fabrication. Once sufficient care has been taken to select materials that are compatible, care must be taken to ensure that proper fabrication procedures are employed.

7.6.1 Fabrication

Fabrication issues in heat pipe construction range from techniques for ensuring continuous connection of the wicking structure and maintaining adequate contact between the wick and the external case to issues involving the fabrication of internal grooves, arteries, or other more complicated configurations. Figure 7.13 illustrates a stainless steel heat pipe that utilizes a fibrous insulation held in contact with the case material by a coil spring. Because the resistance between the heat pipe case material and the wicking structure can make up a significant portion of the overall heat pipe thermal resistance, it is necessary to maintain good contact

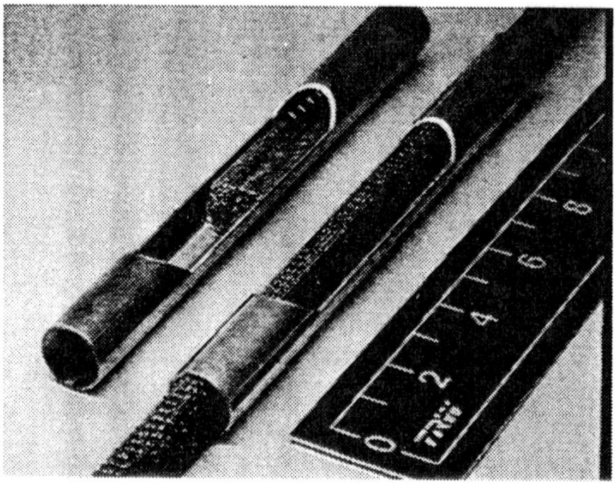

Figure 7.13 Fibrous insulation with spring retainer (courtesy of TRW).

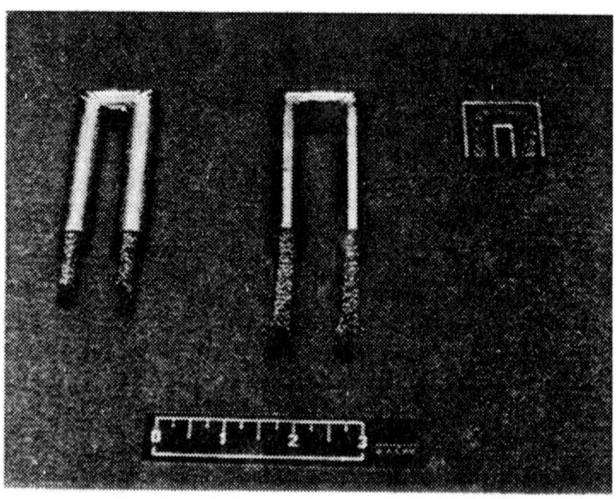

Figure 7.14 Corner mitering technique (courtesy of TRW).

between the two. This can be accomplished, as illustrated, through the use of coil springs or by utilizing other techniques involving spot welding, brazing, crimping, or thermally driven attachment techniques. In addition to maintaining good contact between the wicking structure and the heat pipe case material, proper operation requires that the wicking structure be continuous throughout the entire length of the heat pipe in most applications. Figure 7.14 illustrates a corner mitering technique utilized by TRW for a stainless steel pipe and a metallic fiber wicking structure.

7.6.2 Cleaning

The particular cleaning procedure to be employed for a given heat pipe is dependent upon the case material, wicking structure, and working fluid combination being used. There are numerous references available that describe various cleaning procedures for a wide variety of combinations. A majority of these procedures all contain the same basic steps: initial cleaning to remove any debris, metal filings, and so on; a chemical cleaning process to remove any water, oils, or films on the case or wick materials; a series of final rinse processes to remove any remaining solvent or chemical etchant; and in some cases a vacuum bakeout procedure in which the heat pipe is heated and evacuated to remove any dissolved gases. Because these processes are material combination specific, it is helpful first to discuss in generic terms the various alternatives and then to identify some specific procedures for use with the most common working fluid–material combinations encountered in electronic thermal control applications. As discussed by Brennan and Kroliczek (1979), a variety of methods can be used to clean heat pipes and thermosyphons; including solvent cleaning, alkaline cleaning, acidic etch or pickling, passivation, ultrasonic cleaning, and vacuum bakeout.

Solvent cleaning is the process by which oil grease or other contaminants can be dissolved using either organic or inorganic solvents. The cleaning process can occur after the heat pipe has been assembled or prior to assembly. If done prior to assembly, however, care must be taken to ensure that the parts are not exposed to contact with human skin as the oils on the surface will contaminate the various parts. The parts can be immersed, sprayed, or wiped with the solvent and can be accomplished at room temperature but is typically more effective if done at slightly elevated temperatures. In extreme cases where the parts are either too large to be immersed or of significantly complex geometry to make solvent application difficult, a vapor degreasing process can be used. Regardless of the application utilized, it is important that the solvent is removed in the liquid form so that dissolved oils, for example, can be carried away.

Alkaline cleaning can also be used to remove oil residue or other contaminants from the case or wicking materials. As was the case for solvent cleaning, the alkaline substance can be applied to the case or wicking material either by immersing the heat pipe or by spraying on the alkaline solution. As discussed by Brennan and Kroliczek (1979), the cleaning effectiveness of alkaline compounds is the result of "builders," carbonates, phosphates, silicates, or hydroxides, and as a result, no universal cleaner exists for all types of materials.

A third alternative cleaning procedure is the use of an acidic etchant or pickling solution. These methods while cleaning oils, grease, and other contaminants, are particularly useful for removing the oxide layer that may have formed on the surface of the case or wicking materials. Acidic etching and acid pickling can be differentiated by the severity of the process. Acid etching processes typically refer to processes designed to remove surface films, while pickling refers to processes designed to remove scale from metals (Brennan and Kroliczek, 1979).

Passivation is a chemical cleaning treatment typically utilized with stainless steels to remove or eliminate iron particles embedded in or near the surface during grinding or machining processes. In addition to dissolving these particles, the oxidizing action promotes the corrosion resistance of the stainless steel. The most common technique for passivation involves the application of a mild solvent to remove oils and greases remaining from the machining or extrusion processes followed by the application of a nitric acid solution to dissolve the iron or steel particles.

Copper with Water, Acetone, or Methanol

The use of solvents such as acetone or methanol as a working fluid typically does not require complicated or involved cleaning processes. However, if water is to be used, care must be taken to ensure good wetting. One simple cleaning procedure is to use a mild acidic solution as a cleaning agent in conjunction with vigorous brushing. This should be followed by several thorough flushing and rinsing cycles with water or the working fluid to be utilized.

For configurations that do not lend themselves to brushing, the following alternative procedure has been utilized with good success:

1. Soak and rinse the surfaces with an acetone solution to remove oil, grease, and so on.
2. Soak in an alkaline solution of 50% phosphoric acid, 50% nitric acid solution at room temperature for 15 min.
4. Rinse with a pure working fluid and let soak at 50°C for 1 h.
5. Repeat with a final rinse with a pure working fluid and then dry thoroughly with clean, dry air or nitrogen.

Aluminum with Ammonia, Acetone, or Freons

Because of its high conductivity, resistance to oxidation, and light weight, aluminum is the material of choice for most electronic applications. As a result, extensive investigations have been conducted to determine the optimum cleaning procedure (Brennan and Kroliczek, 1979). The results indicate that the following procedure works well for a wide variety of aluminum alloys:

1. Soak and brush the surfaces with a wire brush and trichloroethane.
2. Rinse with cold trichloroethane.
3. Immerse in a nonetching alkaline cleaner such as Ridoline No. 53 (Amchem Products Co.), Oakete No. 164 (Oakete Products Co.), or Kelite spray white (Kelite Corp.) for 5 min at slightly elevated temperatures.
4. Follow with a 2-min tap water rinse.
5. Immerse in chemical deoxidizer (i.e., nitric acid, 10–20% by volume) at room temperature.
6. Rinse with anhydrous isopropyl alcohol.

Between each successive step, the parts should be dried thoroughly with clean filtered air except after the final rinse, in which case dry nitrogen should be used. This procedure has been used with both 6061 and 6063 alloys.

Stainless Steel (300 series) with Water, Ammonia, or Acetone

The best methods for cleaning stainless steel materials is to utilize the passivation techniques outlined above to remove any iron or steel residue remaining from the fabrication process. This can be accomplished using the following procedure:

1. Soak and brush the surfaces with a wire brush and acetone to remove oil, for example.
2. Flush parts with acetone.
3. Soak in an alkaline solution of 15% nitric acid, 5% hydrochloric acid, and 80% water at 58°C for 15 min.
4. Soak in a 15% nitric acid, 85% water solution at 65°C for 15 min.
5. Dry thoroughly with clean, dry air or argon.

Although this procedure can be used when combinations of stainless steel materials are employed, it is recommended only for single-alloy systems.

An alternative cleaning method has been described by Dunn and Reay (1982). The procedures are similar, with the exception that the second soak utilizes a solution of 50% nitric acid and 5% hydrochloric acid and demineralized water is used for the rinse. The major difference is the recommendation that the parts be placed in a furnace at 400°C and heated in atmospheric air. This heating process volatizes or decomposes any remaining grease and the resulting carbon is given off as carbon dioxide.

Carbon Steel with Naphthalene or Toluene

Several cleaning techniques for use with carbon steel materials and ammonia, naphthalene, or toluene are recommended. Although numerous proprietary cleaning solutions may be commercially available, only the basic acid / alkaline methods are presented here, since the proprietary solutions require testing on a case-by-case basis and are only derivatives of basic acid / alkaline cleaning methods. These four methods are as follows:

Method 1

1. Soak the parts in a solution of pure acetone for approximately 30 min.
2. Follow this with a vigorous brushing with a wire brush and acetone for several minutes.
3. Rinse with water at room temperature.

Method 2

1. Soak the parts in an acid pickle of HCl (19% by volume) for approximately 3 min at 25°C.
2. Rinse with water.
3. Dry with air or dry nitrogen.

Method 3

1. Soak the parts in an alkaline solution (45 g / L NaOH) for approximately 3 min at 90°C.
2. Follow this with an acid pickle in HCl (19% by volume) for approximately 3 min at 25°C and finally a water rinse.
3. Rinse with water.
4. Dry with air or dry nitrogen.

Method 4

1. Soak the parts in an acid dip (HNO_3, 12% by volume) for 4.5 min at 25°C.
2. Follow this with a water rinse.
3. Dry with air or dry nitrogen.

For all four cleaning procedures, heat pipe materials should be immersed in the appropriate solution for a specified period of time and rinsed thoroughly with water after removal. During immersion, the sample should be agitated gently to encourage uniform application of the acid / alkaline solution. This process should be repeated for each successive solution for the different cleaning procedures identified above until each cleaning sequence has been completed.

All of these processes have been shown to improve the wetting and the static wicking height of the various working fluids, and while this increase will not directly translate into increases in the performance limit, it can be expected to significantly improve the overall heat transport capacity.

7.6.3 Assembly Procedures

Figure 7.15 illustrates a typical heat pipe manufacturing flow chart and identifies the different steps required for the various components. As illustrated, the process begins by first obtaining the proper materials from stock and ensuring that their purity and properties are adequate for the application at hand. Before the heat pipe can be assembled it is necessary to perform several operations on the various components: the case/container must be fabricated to the appropriate length, size, and shape through various machining operations, which may include threading, extrusion, and so on and then the heat pipe case must typically undergo a preliminary cleaning process to eliminate any oil residue or impurities

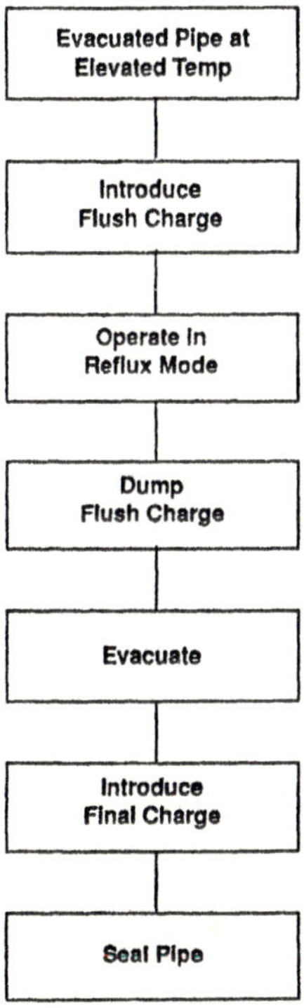

Evacuation and Charging Flow Chart

Figure 7.15 Heat pipe manufacturing flow chart.

remaining from the machining steps. The wicking material must also undergo a preliminary cleaning procedure to eliminate oils, surface films, or oxide layers that may have formed during the fabrication or storage of the material. The wick must then be fabricated and cut to the appropriate size and shape. Once the initial shaping and cleaning processes have been completed, the wicking structure can be inserted into the heat pipe case. Typically, this is followed by a secondary cleaning process consisting of a mild acidic etch (see Section 7.6.2) followed by several rinsing steps. At this point the heat pipe case can be closed and sealed with the exception of a fill tube.

A wide variety of techniques have been utilized to produce the vacuum tight seal required for heat pipe construction. These typically include the use of a plug or end cap fitted into the end of the heat pipe and brazed using a copper–silver eutectic alloy, a welding process (Tig or E-beam), or a simple silver solder process. All of these processes have been shown to produce leak tight seals with a leak rate of less than 1×10^{-9} std cm^3/s of helium. However, in the case of soldering or brazing, care must be taken to ensure that all of the flux has been removed from the joint.

Often it is helpful, at this point in the fabrication procedure, to do a pressurized verification test to determine if the joints, enclosures, and/or seams of the heat pipe case are appropriately sealed. This is best done by pressurizing the case with a small amount of helium and then using conventional leak testing equipment to check the case for leaks.

7.7 CHARGING PROCEDURES

The first step in the charging procedure or process is to determine the amount of working fluid required for a particular system. An approximation of the required charge can be made by estimating volume occupied by the wicking structure plus the volume of any arteries, grooves, and so on. During this estimation process, care should be taken to ensure that the specific volume of the working fluid utilized for these calculations is obtained for a temperature at or near the anticipated operating temperature. Failure to do so may result in a heat pipe that is properly charged at room temperature but at the required operating temperature has an insufficient amount of working fluid to function properly. In addition to utilizing the fluid properties at the correct temperature, consideration should be given to how any errors in the estimation of the amount of working fluid required would ultimately affect the operation. For example, in a 1g environment, a slight excess charge will typically result in the formation of a small puddle in the lower portion of the heat pipe. During periods of high heat transfer, this puddle may be swept toward the condenser or, in a worst case scenario, may result in premature entrainment due the vapor flowing over the free liquid surface. In a microgravity or zero-gravity environment, a slight overcharge may accumulate in the condenser, causing a partial blockage similar to that described in Chapter 6. In either case, the result of a slight overcharge will result in a small reduction in performance. Undercharge, however, regardless of the gravitational field, can result in premature dryout of the evaporator wick and total failure. For this reason, it is quite common for heat pipe manufacturers to spend a considerable amount of time calculating the amount of charge required and then to deliberately overcharge the heat pipe anywhere from 10 to 20%.

Once the amount of working fluid required is known, it is necessary to introduce the working fluid into the heat pipe in such a manner that only the pure solid, liquid, or vapor exists. This can be accomplished in one of four ways:

1. using an evacuation and back-filling technique,
2. using a liquid fill and vapor generation technique,
3. using a solid fill and sublimation technique, and
4. charging using a supercritical vapor technique.

Most common among these for the low to moderate temperature working fluids typically utilized in thermal control of electronic devices and equipment is the evacuation and back-fill technique. In this method the heat pipe is fabricated, cleaned, rinsed, and then evacuated. Once this has been accomplished, the working fluid goes through a multiple distillation process whereby the working fluid is distilled from one container to another and finally into the heat pipe itself. This process helps to ensure that no impurities or dissolved atmospheric gases are introduced into the heat pipe by the working fluid.

Figures 7.16 and 7.17 illustrate the apparatus typically utilized in this process. As shown, the charging system consists of a series of four valves, two fluid chambers, and the heat pipe. The four valves are placed in the system to control flow of the liquid and the regions being evacuated. Valve 1 is located between the two flasks, valve 2 is located between flask 1 and the tee connecting the heat pipe and vacuum pump, valve 3 is located on the line to the vacuum system, and the final valve, valve 4, is located between the tee and the heat pipe. The process begins by heating the working fluid in flask 2 while valve 2 is closed and valves 1, 3, and 4 are open, thus distilling the fluid from flask 2 to flask 1. During this process a vacuum is drawn on the heat pipe, that is, valves 3 and 4 are open. Once the correct amount of working fluid has been distilled from flask 2 to flask 1, valves 1 and 3 are closed and valve 2 is opened. The fluid in flask 1 is then distilled into the pipe. After all the fluid has been vaporized, valve 4 is closed and the pipe can be weighed to check the quantity of working fluid.

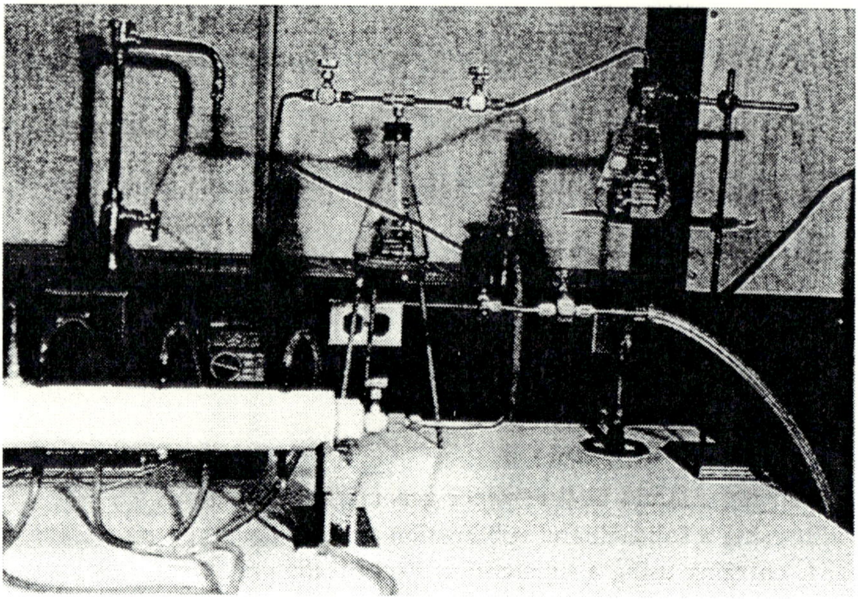

Figure 7.16 Double distillation charging facility.

The second and third methods are considerably less complicated and involve similar procedures for fabricating, cleaning, and rinsing the heat pipe. However, in these procedures, the working fluid is literally poured into the case for liquids (method 1) or placed in the case for solids (method 2). The heat pipe is then heated and evacuated to remove any air or NCGs present. While these methods have proven quite successful in numerous applications, particularly for situations where the working fluid either is in a solid form or is more dense than air at room temperature, the first method is preferred due to decreased probability of contamination and better NCG control.

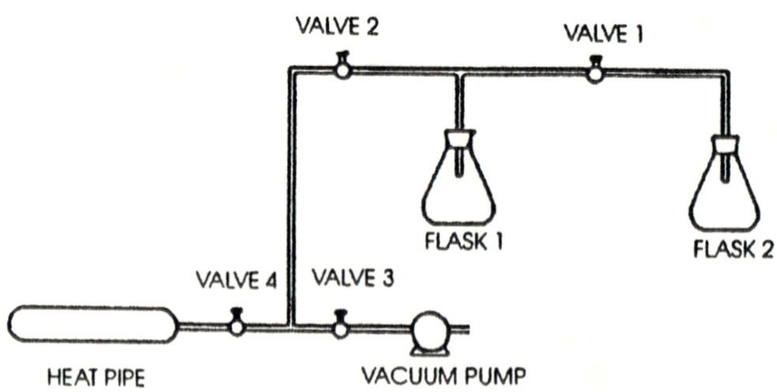

Figure 7.17 Schematic of double distillation charging facility.

The fourth and final method is charging using a supercritical vapor technique. This technique is particularly applicable for use with cryogenic working fluids or for the charging of micro–heat pipes, which are extremely sensitive to flooding. Although it is possible to utilize the priming / charging processes developed for conventional heat pipes for these types of applications, some problems with the control of the working fluid inventory or NCGs may occur. For this reason, it is necessary to carefully control the amount of working fluid present in these heat pipes. Although this process was developed specifically for use with micro–heat pipes (Peterson and Duncan, 1993), it can also be applied to larger, more conventional heat pipes. In this process, the individual micro–heat pipes or the entire micro–heat pipe array is placed in a high pressure chamber, which is then evacuated to a pressure of approximately 10^{-5} mm Hg. After evacuating the chamber, a predetermined amount of working fluid is added and the chamber is heated to a point above the critical temperature of the working fluid. In these conditions, the working fluid is in the supercritical state and exists entirely as a vapor, which, barring any gravitational or temperature variations, is at a relatively uniform density, evenly distributed throughout the individual micro–heat pipes and the chamber. The individual heat pipes or the entire heat pipe array is then sealed while inside the chamber, using an ionic or ultraviolet bonding process. When the chamber is opened and the temperature reduced, a portion of the working fluid condenses within the individual microchannels. By controlling the temperature of the pressure vessel and the quantity of the working fluid added after the initial evacuation process, the final amount of fluid in each heat pipe can be precisely controlled.

Regardless of the charging method utilized, it may be necessary to drain the working fluid and recharge the heat pipe after a short period of operation to ensure complete cleaning. Once this has been done, the heat pipe can be operated for a short period of time and then any NCGs that may remain may be evacuated through the fill tube by opening the valve to the vacuum pump momentarily. After venting this NCG, usually referred to as "burping the pipe," the vent or fill tube can be pinched off and the heat pipe can undergo final verification for leaks. In many applications, it may be possible to include a small amount of helium gas with the working fluid. This allows the use of conventional leak detection equipment to determine the integrity of the case / container material.

7.8 RELIABILITY AND LIFE TESTS

In the early stages of development, a number of investigations were conducted to determine the reliability of various types of heat pipes and thermosyphons. Many of these, such as those conducted by Basiulis and Filler (1971), focused on the reliability of various types of material combinations in the intermediate operating temperature range. Stainless steel heat pipes with water as the working fluid were evaluated by Busse et al. (1973). These investigations found evidence of large amounts of hydrogen generation at temperatures of approximately 250°C. Several techniques were used to mitigate this phenomenon, including the addition of large amounts of oxygen. The hydrogen generation occurred quite rapidly, and for the case of stainless steel 316, normal preventative measures, such as outgasing or passivation procedures, were ineffective in preventing the generation of hydrogen (Dunn and Reay, 1982).

Long duration life tests conducted on copper water heat pipes have been performed by numerous investigators. Most notable among these are the investigations of Kreed et al. (1973), which have indicated that this combination with proper cleaning and charging procedures can produce heat pipes with expected life spans of tens of years. Other tests (Dunn and Reay, 1982) have indicated similar results for copper–acetone and copper–methanol combinations. However, in these latter cases, care must be taken to ensure the purity of the working fluid, wick structure, and case materials.

If adequate care is taken, heat pipes have been shown to be highly reliable in a number of situations. One such example is the thermal print head heat sink illustrated in Fig. 7.18. This cooling system utilizes a single U-shaped copper–water heat pipe approximately 30 cm long. The print head is 12.7 cm × 3.2 cm × 5.8 mm and generates a total power of approximately 50 W. The original design for this application used three separate heat pipes but was later modified to use a single U-shaped pipe. The thermal print head to be cooled in this application was used to print commercial bar codes and the heat pipe design serves two functions: first, to maintain the print head at uniform temperature conditions, dissipating the heat uniformly over the entire length to prevent thermal distortion; and second, to remove heat from the thermal print head to a forced convective air stream. This particular device has been manufactured since 1985 and has reported lifetimes of approximately 2.5 million hours with no reported failures.

A summary of long term life tests for different heat pipe working fluids, wicks, and case materials is illustrated in Table 7.4. Much of the early test data contained in this table were originally summarized by Chisholm (1971) and Brennan and Kroliczek (1979) while some of the more recent information was the result of a summary assembled by Toth and Meyer (1992).

Heat pipe failure can be classified into several distinct categories. These include early failure occurring in the 0–1000 time frame or longer term failures occurring after successful periods of operation in excess of 10,000 h. Heat pipe failure in relatively short time frames is typically the result of either leaks or fundamental incompatibilities. The leaks usually result in failure in the 0–100-h time frame and are the result of problems associated with the fabrication process. The second primary reason for short term

Figure 7.18 Thermal printhead heat sink (courtesy of Thermacore, Inc.).

failure of heat pipes is the result of basic material incompatibilities of the materials and working fluid. These problems again are typically associated with inadequate fabrication processes and allow foreign substances such as water in a steel vessel, residual acids from cleaning, or residual flux from soldering processes to remain in the pipes after charging (Toth and Meyer, 1992). Longer term degradation, that is, heat pipe failure over time frames in excess of 10,000 h, usually results from leaks caused by external corrosion or long term incompatibilities resulting in the generation of NCGs.

Long term reliability can be ensured through careful inspection and preparation processes, including the following:

- Laboratory inspection to ensure high purity (i.e., OFHC copper) is used for the case, end caps, and fill tubes.
- Appropriate inspection procedures are used to ensure that the wicking material is made from high quality materials.

TABLE 7.4 Heat Pipe Life test Data

Working fluid	Container	Wick	Remarks	Reference
Ammonia with 0.05% water	Aluminum alloys: ADI 99.3%, aluminum	Axial groove	*Ammonium chromate additive was 1.2–1.5% of working fluid mass;* 8000 + hours of operation showed satisfactory operating characteristics and the same surface state was observed if no additives had been introduced	Barantsevich et al., 1990
Ammonia with 0.05% water	Aluminum alloys: ADI 99.3%, aluminum	Axial groove	*Diethanolamine additive* was 1.2–1.5% of working fluid mass; same results as above	
Ammonia with 0.05% water	Aluminum alloys: ADI 99.3%, aluminum	Axial groove	*Ammonium chromate mixture with ammonium nitrate was 1.2–1.5% of working fluid mass;* good results on corrosion resistance but small gas generation degrades performance of heat pipe	
Ammonia with 0.05% water	Aluminum alloys: ADI 99.3%, aluminum	Axial groove	*Ammonium nitrate additive was 1.2–1.5% of working fluid mass;* best inhibitor noted by: 1. Performance stability 2. Minimum noncondensible gas formation 3. State of internal metallic surface	
Ammonia	Alumin alloy: 6061-T6	Aluminum 5056, 82.7% porosity	Substantial amounts of aluminum and magnesium from pipe and wick alloy material have been dissolved in working fluid; variable amounts of noncondensible gas was found; age = 12–13 yr	Antoniuk and Lapinski, 1989
Ammonia	Aluminum 6061	Aluminum 5056	Condenser and evaporator regions of the heat pipes showed extensive erosion and corrosion; life on tests = 7–10 yr	Lapinski and Antoniuk, 1991
Ammonia	Stainless steel: X10 CrNiTi 189	Stainless Steel X5 CrNi 189 plus eight-shaped artery	Testing since 1975; no indication of noncondensible gas plug	Roesler, et al.
Ammonia	Aluminum alloy: AlMgSi 0.5	Axial groove, different dimensions	Four years of testing: 1. Inhomogeneous pattern in temperature distribution in condenser region. 2. Performance during stationary tests is much higher (50W) than other heat pipes (10W–20W)	

TABLE 7.4 *(Continued)*

Working fluid	Container	Wick	Remarks	Reference
Ammonia	Aluminum alloy: AlMgSi 0.5	Axial groove, different dimensions	9.5 years of testing: exhibited large temperature drop over middle condenser region proving gas plug noncondensible	
Dowtherm E		Stainless steel, copper	15,000 h at 300°C.	Basiulis and Dixon 1969
Indium silver	Rhenium	Tungsten mesh	1000 h at 2000°C; considerable mass transfer Considerable mass transfer at 300 h	Grover et al., 1968
Potassium	Nickel	—	16,000 h at 600°C	Chilsholm, 1971
Potassium	Berylium	—	1200 h at 750°C; no failure but recrystallization and grain growth had occurred	
Lead	Niobium	—	3 h at 1600°C; intercrystalline corrosion	Busse et al., 1965
Lithium, Li–13Ca	Deoxidized Nb-Zr	—	Both 1000 h at 1500°C; Zr loss	Busse et al., 1969
Lithium	SGD-TA	—	1000 h at 1500°C; no Zr loss	
Lithium	310 S	310 S screen, mesh 60	*Experiment*: 1. Crack at end cap weld after 1165 h, maximum temperature 860°C 2. Cracks through end cap after 1125 h, maximum temperature 860°C 3. Circumferential crack near downstream end evaporator after unknown time, maximum temperature 900°C Test performed in a vacuum	Kaufman and Tower, 1985
Lithium	Hastalloy X	Hastalloy X metal fiber	*Experiment*: 1. Lithium attack of wick, evaporator blockage after 448 h, maximum temperature 910°C 2. Lithium attack of wick, cracks at end cap weld after 754 h, maximum temperature 940°C 3. Lithium attack of wick, evaporator blockage after 1145 h, maximum temperature 960°C Test performed in a vacuum	
Lithium	Haynes 188	Haynes 188 metal fiber	*Experiment*: 1. Radial crack adjacent to seam weld after 997 h, maximum temperature 980°C 2. Cracks in end cap weld and seam weld after 273 h, maximum temperature 840°C 3. No apparent failure after 27286 h, maximum temperature 985°C Test performed in a vacuum	

TABLE 7.4 (*Continued*)

Working fluid	Container	Wick	Remarks	Reference
Lithium	Hastalloy B	Hastalloy B metal fiber	*Experiment:* 1. Severe lithium attack of wick and end cap after 606 h, maximum temperature 960°C 2. Severe lithium attack of wick and end cap after 552 h, maximum temperature 970°C 3. Severe lithium attack of wick, end cap, and wall after 435 h, maximum temperature 980°C Test performed in a vacuum	Kaufman and Tower, 1985
Silver	Tungsten	Tungsten Mesh	1000 h at 1900°C; negligible mass transfer	Grover, 1969
Sodium	304 L	304 L screen Mesh 200 Mesh 100	*Experiment:* 1. No failure after 29,187 h, maximum temperature 890°C 2. No failure after 23,505 h, maximum temperature 910°C Test performed in a vacuum	Kaufman and Tower, 1985
Sodium	304 L	310 S screen, mesh 200	Experimental corrosion of evaporator wall at 1200 and 4070 h, maximum temperature 900°C	
Sodium	310 S	310 S screen, mesh 60	*Experiment:* 1. no failure after 25,417 h, maximum temperature 885°C 3. External corrosion due to heater over temp after 1492 h, maximum temperature 925°C 3. No failure after 16,874 h, maximum temperature 880°C	
Sodium	Hastalloy B	Hastalloy B, metal fiber	*Experiment:* 1. External corrosion at evaporator end weld after 1186 h, maximum temperature 1000°C 2. External corrosion at evaporator end weld after 547 h, maximum temperature 910°C 3. No failure after 1145 h, maximum temperature 1020°C	
Water	Stainless steel	Stainless steel mesh	Rapid deterioration in condenser performance over first 31 days due to hydrogen generation	Schwartz, 1969
Water	OFHC copper	Wickless	Successful tests at 150°C for 129,267 h with copper / silver braze	Toth and Meyer, 1992
Water	OFHC copper	Phosphor bronze screen	Successful tests at 65°C for 85,630 h with copper / silver braze	
Water	OFHC copper	Sintered copper powder	Successful tests at 100°C for 75,166 h with copper / silver braze	
Water	OFHC copper	Phosphor bronze screen	Successful tests at 75°C for 99,318 h with copper / silver braze	
Water	OFHC copper	Swaged groove	Successful tests at 55°C for 85,630 h with copper / silver braze	

- Inspection and distillation procedures have been established to ensure that the working fluid is of consistently high purity.
- Fabrication in a clean environment to ensure or eliminate the presence of oils, vapors, and so on.
- Use of clean solvents during the rinse process (Toth and Meyer, 1992).

In an effort to obtain a better understanding of the effects of long term exposure to elevated temperatures and repeated thermal cycling on heat pipes and thermosyphons, several investigations utilizing accelerated life tests have been performed. While not nearly as well understood as those used in the electronics industry, these tests have been moderately successful in predicting the effect of long term exposure to elevated temperature on the generation of NCGs and other activation processes such as corrosion, oxidation, and diffusion. These tests typically utilize an Arrhenius model to predict the response parameter F of the form

$$F = Ce^{-A/kT} \tag{7.5}$$

where C = constant
A = reaction activation energy
k = Boltzmann constant
T = absolute temperature

Baker (1979) used this model and experimental results obtained from the Jet Propulsion Laboratory to predict the rate and amount of hydrogen gas generated over a 20-year lifetime for a stainless steel heat pipe with water as the working fluid. The gas generation rate was described as

$$\dot{m}(t, T) = f(t)F(T) \tag{7.6}$$

The results of this process, when plotted for several different temperatures, allow the mass of hydrogen generated to be predicted as a function of time.

Anderson (1973) used oxidation theory to predict the passivating film growth along with the methods outlined in equation 7.6 obtain reaction activation energies of 8.29×10^{-20} and 10.3×10^{-20} J for stainless steel–water and nickel–water combinations, respectively. While these results compared favorably with those of Baker (Dunn and Reay, 1982) and tended to confirm the results, very little additional work has been done in this area since then.

In summary, experience has demonstrated that mechanical cleaning of the case and wicking structure with an appropriate solvent combined with an acidic etch and vacuum bakeout under elevated temperatures will produce heat pipes free of contaminant that will experience negligible performance degradation (less than 5%) over a product lifetime of 10 years. As discussed by Toth and Meyer (1992), it is difficult to calculate a mean time to failure (MTTF) for heat pipes, thermosyphons, and other similar devices due to the relatively small amount of data that exist on actual products in operation. A conservative estimate of MTTF, however, of 135,000 h has been made. This is a significant life expectancy when compared to electrical fans of the type used in many electronic components of 80,000–100,000 h (Toth and Meyer, 1992).

Worldwide production of heat pipes specifically designed for applications involving the thermal control of electronic components or devices was in excess of one million heat pipes per year in 1992. These heat pipes were used in a wide variety of applications ranging from consumer electronics to industrial equipment (Toth and Meyer, 1992).

NOMENCLATURE

A Reaction activation energy (J)
C Constant
F Functional relationship
g Gravitational acceleration (m/s^2)
k Boltzmann constant
m Mass (kg)
$\dot{m}$ Mass flow rate (kg/s)
N_l Liquid figure of merit (W/m^2)
P Pressure (Pa)
r Radius (m)
T Temperature (K)
z Height (m)

Greek Symbols

ρ Density (kg/m^3)
λ Latent heat of vaporization ($J/kg \cdot K$)
μ Viscosity ($N-s/m^2$)
σ Surface tension (N/m)

Subscripts

A Substance A
B Substance B
v Vapor
l Liquid

REFERENCES

Adamson, A. W., 1990, *Physical Chemistry of Surfaces*, 5th ed., Wiley, New York.

Alario, J., Brown, R., and Otterstadt, P., 1984, "Space Constructable Radiation Prototype Test Program," Paper No. 84-1793, American Institute of Aeronautics and Astronautics, Washington, DC.

Anderson, W. T., 1973, "Hydrogen Evolution in Nickel-water Heat Pipes," Paper No. 73-726, 8th AIAA Thermophysics Conference, American Institute of Aeronautics and Astronautics, Washington, DC.

Anonymous, 1966, "Friction Losses for Fully-Developed Flow in Straight Pipes," Data Item No. 66027, Engineering Sciences Data Unit, London, UK.

Anonymous, 1979, Product Data from Brunswick Corporation, Miami, FL.

Antoniuk, D. and Lapinski, R. J., 1989, "Aging Effects on Aluminum/Ammonia Heat Pipes and Resultant Impact on System Performance," *AIAA 24th Thermophysics Conference*, June, Buffalo, New York.

ASME Boiler and Pressure Vessel Committee, 1965, *ASME Boiler and Pressure Vessel Code – Section VIII Unfired Pressure Vessel*, American Society of Mechanical Engineers, New York.

Baker, E., 1979, "Prediction of Long Term Heat Pipe Performance from Accelerated Life Tests," *AIAA Journal*, Vol. 11, No. 9, September.

Barantsevich, V. L., Barakove, L. V., and Tribunskaja, I. A., 1987, "Investigation of the Heat Pipe Corrosion Resistance and Service Characteristics," Proc. 6th Int'l. Heat Pipe Conf., Grenoble, France, pp. 188–193.

Basiulis, A., and Dixon, J. C., 1969, "Heat Pipe Design for Electron Tube Cooling," Paper No. 69-HT-25, American Society of Mechanical Engineers, New York.

Basiulis, A., and Filler, A., 1971, "Operating Characteristics and Long Life Capabilities for Organic Fluid Heat Pipes," Paper No. 71-408, 6th AIAA Thermophysics Conference, Tullahoma, TN, AIAA.

Basiulis, A., Prager, R. C., and Lamp, T. R., 1976, "Compatibility and Reliability of Heat Pipe Materials," Proc. 2nd Int'l. Heat Pipe Conf., Bologna, Italy, pp. 357–372.

Birnbreier, H., and Gammel, G., 1973, "Measurement of the Effective Capillary Radius and the Permeability of Different Capillary Structures," Paper No. 5-4, Proceedings of the First International Heat Pipe Conference, Stuttgart, West Germany.

Bohdanski, J., and Schins, H. E. J., 1967, "The Surface Tension of Alkali Metals," *Journal Inorganic Nuclear Chemistry*, Vol. 29, pp. 2173–2179.

Brennan, P. J., and Kroliczek, E. J., 1979, "Heat Pipe Design Handbook", Contract Report No. NAS5-23406, National Aeronautics and Space Administration. Washington, D.C.

Busse, C. A., Caron, R., Geiger, F., and Pötzschke, M., 1965, "Performance Studies on Heat Pipes," IEEE—First International Conference on Thermionic Electrical Power Generation, London.

Busse, C.A., Geiger, F., and Strub, H., 1969, "High Temperature Lithium Heat Pipes", Second International Conference on Thermionic Electrical Power Generation, Stresa, Italy, May, pp. 495–506.

Busse, C. A., Campanile, A., and Loens, J., 1973, "Hydrogen Generation in Water

Heat Pipes at 250°C," Paper No. 4-2, First International Heat Pipe Conference, Stuttgart, Germany, October.

Chi, S. W., 1976, "Heat Pipe Theory and Practice, McGraw-Hill, New York.

Chiakovsky, V., and Simonov, G., 1978, "Complex Investigation of Characteristics and Processes in Artery-grooved Heat Pipes," Proceedings of the Third International Heat Transfer Conference, Palo Alto, CA, pp. 4246–433.

Chisholm, D., 1971, *The Heat Pipe*, Mills and Boon, London, England.

Chun, K. R., 1972, "Some Experiments on Screen Wick Dry-Out Limits," *Transactions of the ASME, Series C, Journal of Heat Transfer*, Vol. 94, pp. 46–51.

Dean, D. J., 1976, "An Integral Heat Pipe Package for Microelectronic Circuits," Proceedings of the Second International Heat Pipe Conference, Bologna, Italy, pp. 481–502.

Dul'Nev, G. N., and Zarichny, Y. P., 1974, "Thermal Conductivity of Blends and Compound Materials," *Energiya* (USSR) Vol. 19, pp. 135–146, (in Russian).

Dunn, P. D., and Reay, D. A., 1982, *Heat Pipes*, 3rd ed., Pergamon, Oxford.

Eastman, G. Y., and Ernst, D. M., 1978, "High Performance, High Temperature Heat Pipes," Proceedings from the Third International Heat Pipe Conference, Palo Alto, CA, pp. 268–273.

Feldman, K. T., 1976, The Heat Pipe: Theory, Design and Applications, Technology Application Center, Univ. of New Mexico, Albuquerque, New Mexico.

Feldman Jr., K. T., and Berger, M. E., 1973, "Analysis of a High-Heat-Flux Water Heat Pipe Evaporator," Technical Report No. ME-62(73) ONR-012-2, University of New Mexico, Albuquerque, NM.

Ferrell, J. K., and Alleavitch, J., 1970, "Vaporization Heat Transfer in Capillary Wick Structures," *Chemical Engineering Progress Symposium Series*, Vol. 66, No. 102, pp. 82–91.

Ferrell, J. K., Alexander, E. G., and Piver, W. T., 1973, "Vaporization Heat Transfer in Heat Pipe Wick Materials," *AIAA Progress in Astronautics and Aeronautics: Thermal Control and Radiation*, Vol. 31, C. L. Tien, (Ed.), MIT Press, Cambridge, MA, pp. 3–18.

Finlay, I. C., Loy, B., and Cree, D. G., 1976, "Manufacture and Performance of a Water Filled Heat Pipe with a Sintered Copper Wick," Proceedings of the Second International Heat Pipe Conference, Bologna, Italy, pp. 93–101.

Fletcher, L. S., 1988, "Recent Developments in Contact Conductance Heat Transfer," *ASME Transactions, Journal of Heat Transfer*, Vol. 110, No. 4b, pp. 1059–1070.

Freggens, R. A., 1969, "Experimental Determination of Wick Properties for Heat Pipe Applications," Proceeding from the Fourth Intersociety Energy Conversion Conference, Washington, DC, pp. 888–897.

Goodacre, J. B., 1974, "Study of Heat Pipe Wick Structures—Phase 2," Final Technical Report on Contract No. K/GW31a/135, Marconi, United Kingdom.

Gorring, R. L., and Churchill, S. W., 1961, "Thermal Conductivity of Heterogeneous Materials," *Chemical Engineering Progress*, Vol. 57, No. 7, pp. 53–59.

Grover, G. M., Kemme, J. E., and Keddy, E. S., 1968, "Advances in Heat Pipe Technology," Second International Conference on Thermionic Electrical Power Generation, Stresa, Italy, May.

Hoogendoorn, C. J., Nio, S. G., and Hardy, H. J. J., 1973, "Permeability Studies on Wire Screens and Grooves," Paper No. 5-3, Proceedings of the First International Heat Pipe Conference, Stuttgart, West Germany.

Katzoff, S., 1967, "Heat Pipes and Vapor Chambers for Thermal Control of Spacecraft," Paper No. 67-310, American Institute of Aeronautics and Astronautics, Washington, DC.

Kaufman, W. B., and Tower, L. K., 1985, "Compatibility of Sodium and Lithium in Superalloy Heat Pipes," Final Report for Period Jan. 1, 1984–Dec. 31, 1984, Aero Propulsion Laboratory, Air Force Wright Aeronautical Laboratory, Wright Patterson Air Force Base, Ohio (Report No. AFWAL-TR-85-2006).

Kesser, D., 1973, "Experimental Determination of Properties of Saturated Sintered Wicks," Paper No. 6-1, Proceedings of the First International Heat Pipe Conference, Stuttgart, West Germany.

Kreed, H., Kroll, M., and Zimmermann, P., 1973, "Life Test Investigations with Low Temperature Heat Pipes," Paper No. 4-1, First International Heat Pipe Conference, Stuttgart, West Germany, October.

Langston, L. S., and Kunz, H. R., 1969, "Liquid Transport Properties of Some Heat Pipe Wicking Materials," Paper No. 69-HT-17, American Society of Mechanical Engineers, New York.

Lapinski, R. J., and Antoniuk, D., 1991, "Characterization of Aging Mechanisms in Aluminum/Ammonia Heat Pipes," AIAA Paper No. 91–1361, AIAA 26th Thermophysics Conference, June, Honolulu, Hawaii.

Marcus, B. D., 1972, "Theory and Design of Variable Conductance Heat Pipe," Report No. CR-2018, National Aeronautics and Space Administration, Washington, D.C.

Marto, P. J., and Peterson, G. P., Chapter 4, 1988 " Application of Heat Pipes to Electronics Cooling," in *Advances in Thermal Modeling of Electronic Components and Systems*, A. Bar-Cohen and A. D. Kraus (eds.), Hemisphere Publishing Corporation, New York, NY, pp. 283–336.

Murakami, M., and Arai, K., 1987. "Statistical Prediction of Long-term Reliability of Copper Water Heat Pipes From Acceleration Test Data," Proc. 6th Int'l. Heat Pipe Conf., Grenoble, France, pp. 2194–199.

Ochterbeck, J. M., and Peterson, G. P., 1992, "Freeze/Thaw Characteristics of a Copper–Water Heat Pipe: Effects of Noncondensible Gas Charge," *AIAA Journal of Thermophysics and Heat Transfer*, Vol. 7, No. 1, pp. 127–132.

Peterson, G. P., and Babin, B. R., 1989, "Analytical and Experimental Investigation of Miniature Heat Pipes—Phase II," Wright Research and Development Final Report No. WRDC-TR-89-2067, Wright Research and Development Center, Wright-Patterson Air Force Base, OH.

Peterson, G. P., and Duncan, A. B., 1993, *Micro Heat Pipe Priming and Charging Processes*, Pending U.S. Patent pending, August 1, 1993.

Peterson, G. P., and Fletcher, L. S., 1988, "Evaluation of the Thermal Contact Conductance Between Mold Compounds and Heat Spreader Material," *ASME Transactions, Journal of Heat Transfer*, Vol. 110, No. 4, pp. 996–999.

Peterson, G. P., and Fletcher, L. S., 1990, "Measurement of the Thermal Contact Conductance and Thermal Conductivity of Anodized Aluminum Coatings," *ASME Transactions, Journal of Heat Transfer*, Vol. 112, No. 3, pp. 579–585.

Phillips, E. C., and Hindeman, J. D., 1969, "Determination of Properties for Capillary Media Useful in Heat Pipe Design," Paper No. 69-HT-18, American Society of Mechanical Engineers, New York.

Ponnappan, R., 1985, "Double Wall Artery Heat Pipe," Final Report No. AFWAL-TR-85-2041. Wright-Patterson Airforce Base, Dayton, OH.

Roberts, C. C., and Feldman Jr., K. T., 1972, "Predicting the Performance of Heat Pipes with Partially Saturated Wicks, Paper No. 72-WA/HT-38, American Society of Mechanical Engineers, New York.

Roesler, S., Heine, D. and Groll, M., 1987, "Life Testing With Stainless Steel/Ammonia and Aluminum/Ammonia Heat Pipe," Proc. 6th Int'l. Heat Pipe Conf., Grenoble, France, pp. 211–216.

Schwartz, J., 1969, "Performance Map of the Water Heat Pipe and the Phenomenon of Noncondensible Gas Generation," Paper No. 69-HT-15, American Society of Mechanical Engineers, New York.

Semena, M. G., Kostornov, A. G., 1975, "Investigation of the Structural and Hydrodynamic Characteristics of the Wicks of Heat Pipes," *High Temperature*, Vol. 13, No. 1, pp. 137–140.

Semena, M. G., and Zaripov, V. K., 1977, "Influence of the Diameter and Length of Fibers on Material Heat Transfer of Metal Fibre Wicks of Heat Pipes," *Thermal Engineering*, Vol. 24, No. 4, pp. 69–72.

Soliman, M. M., Graumann, D. W., and Berenson, P. J., 1970, "Effective Thermal Conductivity of Dry and Liquid-saturated Sintered Fiber Metal Wicks," Paper No. 70-HT/SpT-40, American Society of Mechanical Engineers, New York.

Stevenson, P. F., Peterson, G. P., and Fletcher, L. S., 1991, "Thermal Rectification in Similar and Dissimilar Metal Contacts," *ASME Transactions, Journal of Heat Transfer*, Vol. 113, No. 1, pp. 30–36.

Swanson, L. W., 1990, *Heat Pipes in Heat Transfer and Fluid Flow Data Books*, Section 507.5, Genium Publishing, New York.

Tien, C. L., and Sun, K. H., 1971, "Minimum Meniscus Radius of Heat Pipe Wick Materials," *International Journal of Heat and Mass Transfer*, Vol. 14, pp. 1853–1855.

Toth, J. E., and Meyer, G. A., 1992, Heat Pipes: Is Reliability an Issue?" IEPS Conference, Austin, TX, September 28.

Udell, K. S., and Jennings, J. D., 1984, A Composite Porous Heat Pipe," Proc. 5th Int'l. Heat Pipe Conf., Tskuba, Japan, pp. 41–47.

Vansant, J. H., and Malet, J. R., 1975, "Thermal Conductivity of Some Heat Pipe Wicks," *Letters in Heat and Mass Transfer*, Vol. 2, pp. 199–206.

Zaho, R. D., Zhu, Y. H., and Liu, D. C., 1987, "Experimental Investigation of the Compatibility of Mild Carbon Steel and Water Heat Pipes," Proc. 6th Int'l. Heat Pipe Conf., Grenoble, France, pp. 200–204.

Chapter

8

HEAT PIPE TESTING AND ANALYSIS

8.1 STEADY-STATE TESTING OF HEAT PIPES

Steady-state testing and evaluation procedures for heat pipes and thermosyphons are typically designed to measure the maximum heat transport capacity, the end-to-end temperature drop, the thermal resistance, and/or the effect of variations in the local gravitational acceleration on the overall thermal performance. For low or moderate temperature heat pipes of the type typically encountered in the thermal control of electronics, all of these measurements can be accomplished using a single, relatively simple test facility and straightforward test procedures. Figure 8.1 illustrates a steady-state test facility of the type that would normally be used for low or moderate temperature heat pipes. As shown, the test facility consists of four principal subsystems: the test pipe, a heat source, a heat sink, and a temperature measurement system. The test facility can be constructed for use in an air or vacuum environment, but should be adjustable so that the evaporator section of the heat pipe can be elevated above or below the condenser to determine the effect of the gravitational body force on the transport capacity.

8.1.1 Steady-State Test Facility

Because of the simplicity and accuracy with which the thermal power applied to the test pipe can be measured, electrical resistance heating is the most common type of heating system used. Other methods, including induction heating, heated constant temperature baths, and radiation, have also been used successfully. Regardless of the method, what is needed is a means by which the temperature and power input into a specific region of

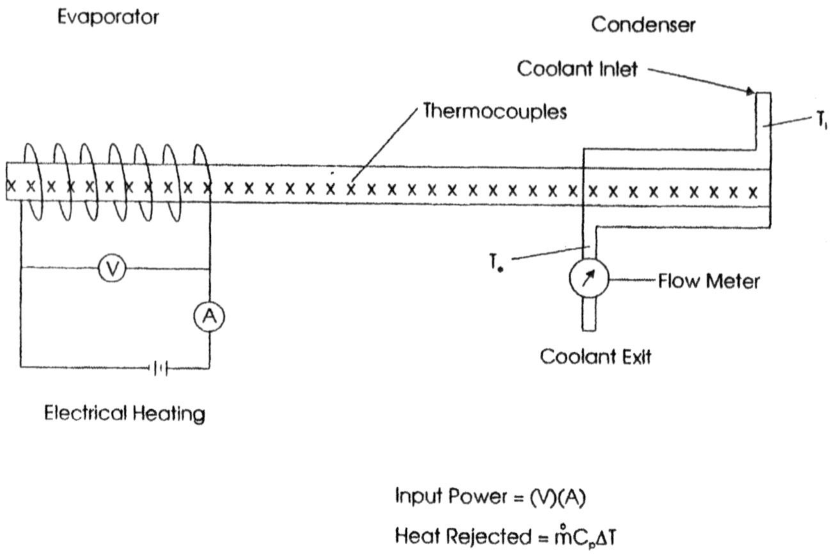

Figure 8.1 Test apparatus for low to moderate temperature heat pipes (Dunn and Reay, 1982).

the heat pipe can be carefully monitored and controlled. For the case of electrical resistance heating, the electrical energy input to the heat pipe or Joule heating can be measured either by measuring the input voltage and current or by measuring the voltage and using the known electrical resistance of the heating element. The electrical power input to the system can be controlled by means of a simple variac.

In addition to a means for adding power to the evaporator section of the heat pipe, a system is also required by which the heat can be rejected. Again, there are several alternatives available, including radiation, free convection to the surrounding atmosphere, latent heat rejection to a spray cooling system, and rejection to a liquid-cooled heat exchanger. The latter of these typically presents the most attractive alternative for heat pipes designed to operate in the low to moderate temperature range, due to the accuracy with which the temperature of the condenser portion can be controlled. Measurements of the amount of heat rejected in this type of system can also be made quite easily by performing a sensible heat balance across the coolant inlet and exit.

If the evaporator, condenser, and adiabatic section are well insulated, the total amount of heat transported by the heat pipe, or the *axial heat transport capacity*, can be estimated by averaging the input power and the heat rejected.

To determine the transport characteristics, it is also necessary to evaluate the axial and radial temperature distributions along the length of the heat pipe. These measurements are typically made on the external case of the test article, but can also be made internally if compatible materials and leak-proof seals are used. Thermocouples, in conjunction with a computerized data acquisition system, are frequently used to monitor the input power, heat rejected, and temperature distribution throughout the tests. While the temperature gradients along the length of a heat pipe may be quite small, it is important to note that in most heat pipe test procedures, the absolute temperature measurements are not as critical as the relative temperature differences. For this reason, the use of thermocouples with a relatively large uncertainty (±0.5°C) when compared with the temperature gradients being measured, can yield results that are quite acceptable. However, care should be taken to ensure that at any given temperature all of the thermocouples will yield the same value. This can be accomplished by verifying the thermocouples and data acquisition system at the freezing point and atmospheric boiling point of water to ensure that, for a given condition, they yield similar results prior to installation on or in the test pipe. Previous tests have shown that temperature differences can be measured to within an accuracy of ±0.05°C using this technique. When combined with the placement errors of the thermocouples, this value can result in overall experimental uncertainties for the temperature measurement that are quite acceptable.

8.1.2 Steady-State Test Procedure

Typical test procedures consist of setting a constant temperature bath for the condenser region at a specified temperature and incrementing the input power to the evaporator in predetermined increments. By observing the axial temperature distribution along the outside surface of the heat pipe, the maximum transport capacity can be determined. Figure 8.2 illustrates a typical axial temperature profile for various input powers. As shown, increasing heat fluxes result in gradual increases in both the evaporator and condenser regions of the heat pipe. In the evaporator portion, the temperature distribution may be uniform or may exhibit some slight reduction in temperature at the end of the evaporator due to axial losses.

As the input power is gradually incremented, the rapid increase in the temperature in the evaporator portion becomes apparent. This increase in temperature is an indication of partial dryout of the evaporator wick and defines the so-called *capillary limit* discussed in Chapter 3. Continued increases in the input power will result in continued operation with a gradual progression of the temperature spike in the evaporator. The progression of this spike is usually associated with the location of the dryout point. It is important to note that while the initial presence of a temperature spike in the evaporator is usually an indication of the onset of dryout and signifies that the heat pipe has reached the maximum transport capacity as defined by the capillary limit, the heat pipe will continue to

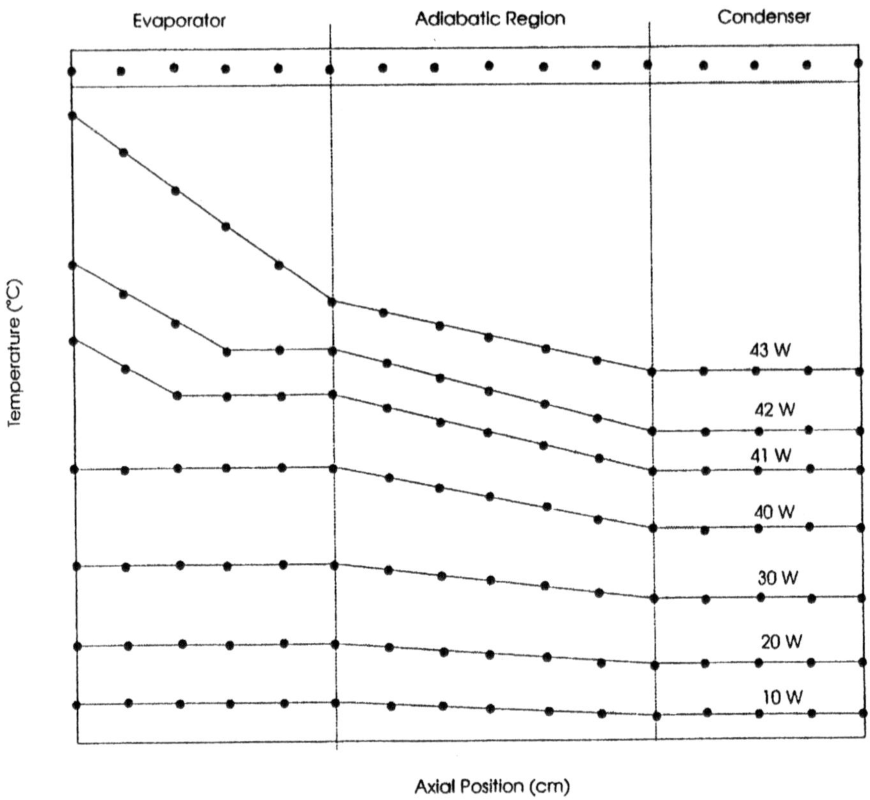

Figure 8.2 Axial temperature profile for evaporator dryout with increasing heat transport levels.

operate. Progression of the dryout point throughout the evaporator will continue with increased heat loads until the entire evaporator portion of the heat pipe has completely dried out, at which point the heat pipe is assumed to have reached complete or total failure.

The effect of variations in the local gravitational acceleration can be determined by conducting similar tests. However, in these tests the procedure typically involves presetting the condenser coolant temperature and the input power to the evaporator and slowly incrementing the tilt angle to elevate the evaporator slightly above the condenser level. Again, during this procedure, the axial temperature distribution is observed to identify that point at which the onset of evaporator dryout is apparent. In this way, a graph similar to that illustrated in Fig. 8.3 can be produced that shows the effect of elevation on the maximum axial transport capacity. Using the measured elevation difference between the evaporator and condenser, the effect of variations in the local gravitational acceleration can

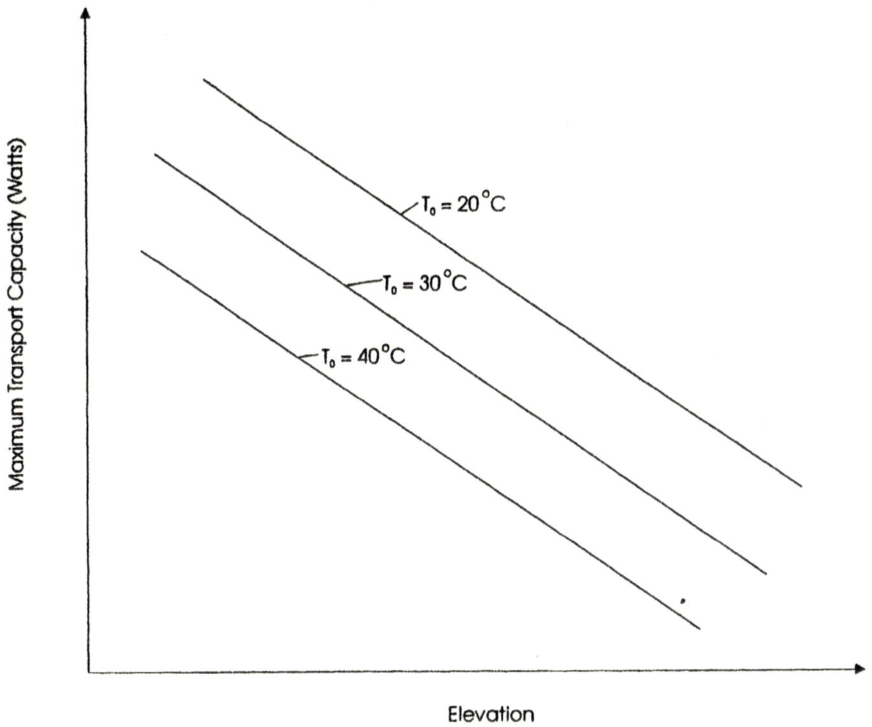

Figure 8.3 Effect of variations in the adverse tilt on the maximum transport capacity of heat pipes.

be estimated. In this figure, the effect of elevation on the maximum transport capacity is illustrated for three different operating temperatures.

Control of the operating temperature or mean adiabatic vapor temperature can be accomplished by combining increases in input power and reductions in the condenser coolant temperature. Figure 8.4 illustrates the effect each of these has on the relationship between power and temperature. As shown, increases in power alone result in increases in operating temperature while reductions in the temperature of the cooling fluid result in reductions in the mean adiabatic vapor or operating temperature. Through a combination of these two factors, the maximum transport capacity can be determined.

To do this, an operating temperature is established and the input power and coolant temperature are adjusted until the mean adiabatic vapor temperature, as determined by thermocouples located on the outer wall of the heat pipe in the adiabatic section, reach the desired value. By slowly increasing the power and decreasing the coolant temperature, the transport capacity can be measured and increased incrementally while maintaining a

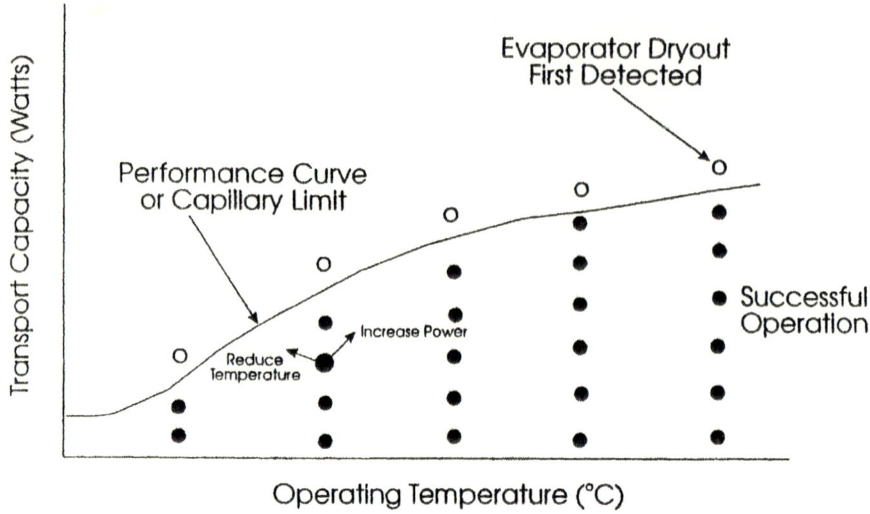

Figure 8.4 Development of a thermal performance map as a function of operating temperature.

constant operating temperature. This results in a series of points along a constant operating temperature line where successful operation has been measured. Continual increases in the transport capacity will eventually result in evaporator dryout (as determined by the axial temperature distribution illustrated in Fig. 8.2). The maximum transport capacity is then located at some point between the last successful operating power (the solid symbols) and the point at which initial dryout of the evaporator occurs (the open symbols). If this process is repeated over a range of operating temperatures, a curve that describes the maximum transport capacity as a function of operating temperature can be developed experimentally, as shown in Fig. 8.4. Tests such as those described herein can provide significant information about the operational characteristics and performance limits of heat pipes and thermosyphons and can be used to compare with and validate the predictive techniques described earlier in Chapters 3–5.

8.2 TRANSIENT STARTUP CHARACTERISTICS OF HEAT PIPES

The transient behavior of heat pipes is of particular interest during sudden changes in the thermal load or during initial startup. In both of these situations, the startup characteristics are strongly dependent upon the initial state of the working fluid and the operating temperature range. The low or moderate temperature heat pipes of the type most commonly used in electronics applications typically start up from a state where the wick is either fully or partially saturated with working fluid and the bulk of the fluid exists in a saturated state. In contrast, heat pipes with cryogenic working fluids must at some point start up from a supercritical state, and liquid metal heat pipes typically start up from a frozen state.

A majority of the transient studies have focused on the behavior of high temperature heat pipes. Tolubinsky et al. (1978) have investigated the transient behavior of sodium and potassium heat pipes with instantaneous power additions, and Bystrov et al. (1982) conducted experiments on vacuum and gas-loaded sodium heat pipes. A significant decrease in the startup time was noted by Ivanovskii et al. (1982) for increasing amounts of gas loading in heat pipes. Sockol (1973) presented an analytical model and predicted that startup is not initiated until the vapor pressure of the working fluid is of the same order of magnitude as the pressure of the initial gas charge. Transient analytical and experimental investigations for gas-loaded heat pipes have been presented by Bystrov and Goncharov (1983) and by Ponnappan et al. (1990). In both cases, the addition of NCGs was found to enhance the transient response and restart capabilities.

8.2.1 Normal Startup Characteristics

In one of the earliest analytical papers written on heat pipes and thermosyphons, Cotter (1967) identified three fundamental startup modes: uniform startup, frontal startup, and frontal startup with NCGs. These three startup modes are illustrated in Fig. 8.5, where the numbers indicate increasing time from startup. In the first of these Fig. 8.5*a*, uniform startup, occurs when the vapor density is high at ambient temperatures. In this situation the initial vapor density is high enough that a uniform or nearly uniform temperature exists prior to and throughout the startup period. This condition typically occurs during startup of heat pipes with cryogenic working fluids. In situations where the initial vapor density is low, as represented by Fig. 8.5*b*, *frontal startup* occurs. In this mode compressible flow effects may be important and sonic velocities are likely to occur during the initial phases due to large axial temperature variations in the vapor. In cases where NCGs are present, the characteristic startup curves will be similar to those illustrated in Fig. 8.5*c*. As illustrated, the movement of the vapor results in the NCG being swept toward the condenser. As the average operating temperature (adiabatic vapor temperature) increases, the NCG front is slowly compressed, blocking a smaller and smaller portion of the condenser.

8.2.2 Frozen Startup

In general, the fundamental principles involved in the steady-state and transient operation of heat pipes are relatively well understood. However, several phenomena commonly found in heat pipe operations have not

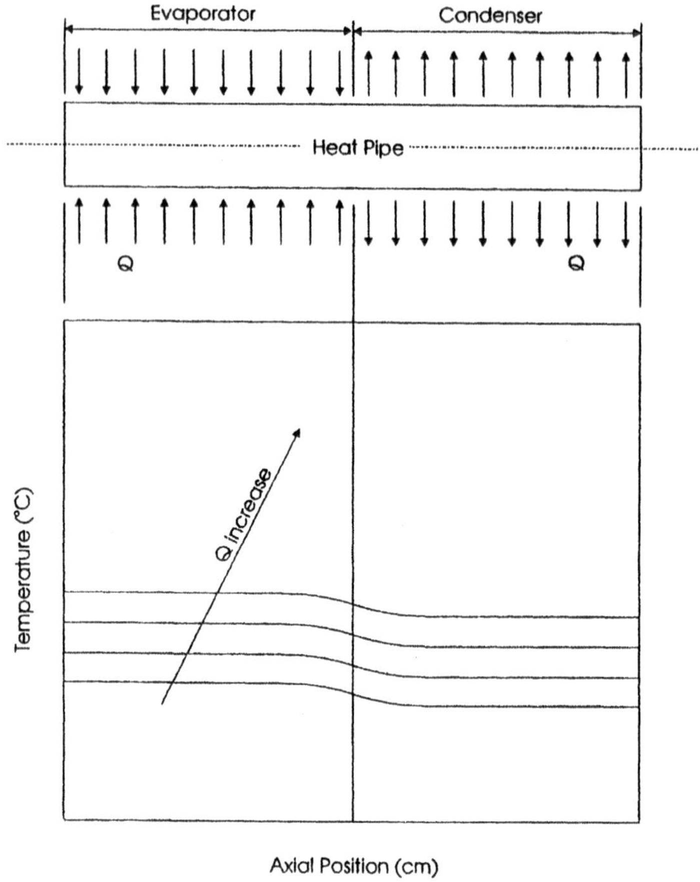

(*a*)

Figure 8.5 Fundamental start-up modes of heat pipes: (*a*) uniform startup (high vapor densities at ambient temperatures); (*b*) frontal startup (low vapor densities at ambient temperatures); (*c*) frontal startup (non-condensible gases present) (Chi, 1976).

been fully investigated, including the transient processes of freezing and restart from the frozen state. Experimental data exist primarily from investigations conducted on liquid metal heat pipes, as these designs must at one point always start from the frozen state and largely ignore the effect of the initial frozen working fluid configuration on restart. Although heat pipes that utilize working fluids such as water and ammonia would normally not freeze during standard operation or manufacturing, several situations arise where the freeze–thaw behavior is important (Ochterbeck and Peterson, 1993).

Three types of freezing that occur in heat pipes have been described by

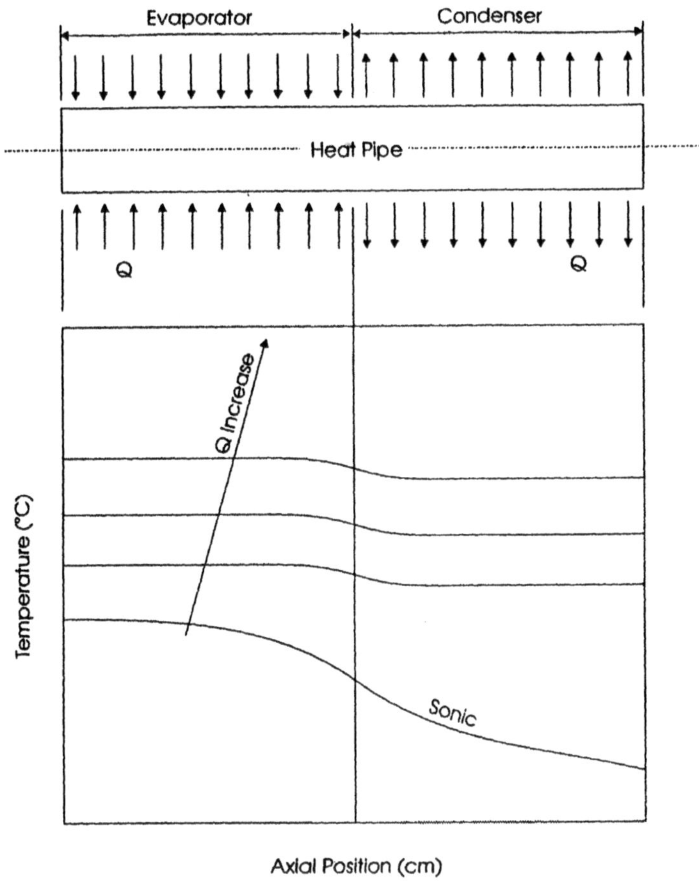

(*b*)

Figure 8.5 (*Continued*)

Antoniuk and Edwards (1990): suction freezeout, freezing blowby, and diffusion freezeout. Suction freezeout is the freezing of the fluid and corresponding depletion of the available liquid inventory in the active region of the heat pipe. Freezing blowby corresponds to a phenomenon occurring when a complete solid blockage of the vapor and liquid passages is thawed. During melting of the blockage a high pressure region (evaporator) exists on one side of the blockage and a low pressure region (condenser) exists on the other. When breakthrough of the blockage occurs, liquid is rapidly driven from the higher pressure evaporator into the lower pressure condenser region and may result in rapid dryout of the evaporator wicking structure. Diffusion freezeout, also discussed by Edwards and Marcus (1972), is the freezing of the working fluid that

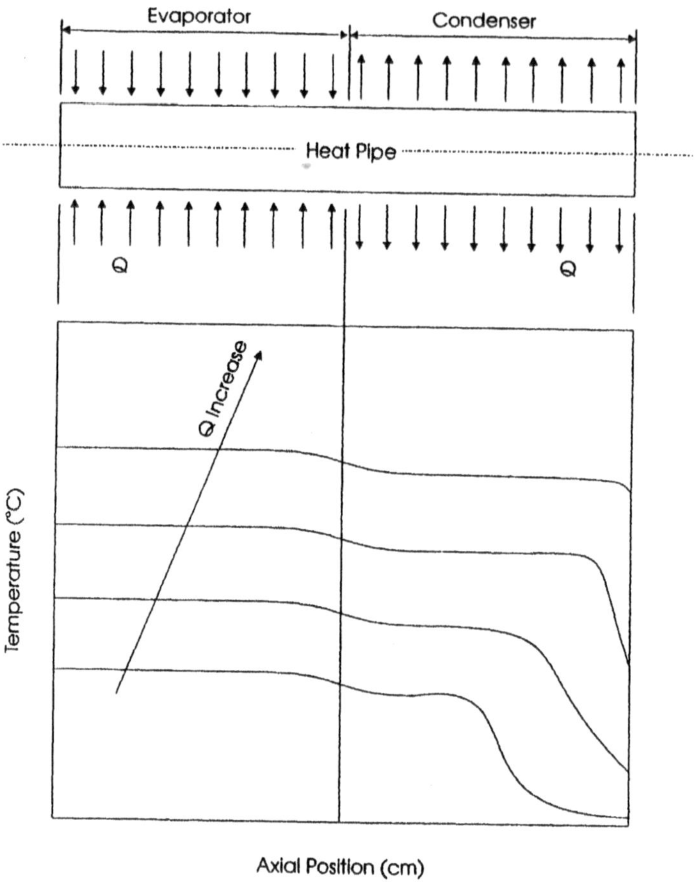

(*c*)

Figure 8.5 (*Continued*)

diffuses from the vapor located in the condenser region into the portion of the condenser blocked by NCGs.

The ability to restart a frozen heat pipe has been discussed by Chi (1976) and Ivanovskii et al. (1982) as being a direct function of the condenser heat removal rate, heat pipe geometry, working fluid, and condenser-to-evaporator-length ratio. Both sources also noted that the addition of NCGs would aid in the startup due to a progression of the melt front along the length of the heat pipe. It was hypothesized that the amount of NCG required for frontal heating and successful startup corresponded to the amount necessary to achieve a NCG pressure in excess of the working fluid saturation pressure at the melting temperature. Neither source, however, discusses the process of the working fluid freezing, and both assume an essentially uniformly distributed working fluid in the discussions of frozen startup.

Experimental investigations of frozen startup of liquid metal heat pipes have been conducted by several investigators, including Merrigan et al. (1985) for a molybdenum–lithium heat pipe with varying radiation heat rejection rates during freezing of the working fluid. In this investigation, uneven rates were found to hinder startup due to partial dryout of the evaporator during the freezing process. Also, successful restart was found to be directly dependent on the heat rejection rate during startup.

With regard to room temperature working fluids of the type more commonly encountered in electronics applications, several investigations have been conducted. Deverall et al. (1970) conducted an experimental investigation of frozen startup with a water heat pipe. Evaporator dryout was observed for an initially frozen heat pipe at −65°C and an evaporator heat addition of 10 W. Restart and isothermal operation were achieved only after the fluid in the condenser had melted and rewet the evaporator. A temperature gradient of approximately 100°C from the evaporator to the condenser was required to completely thaw the heat pipe. Abramenko et al. (1986) performed an experimental investigation using an aluminum–ammonia heat pipe with radiation heat rejection, thus resulting in conditions similar to those found in spacecraft thermal control systems. Startup from the frozen state was obtained only after the radiative sink temperature was increased, thus reducing the heat rejection rate at the condenser.

Two recent numerical investigations have been presented that examine the startup of a frozen heat pipe. The numerical model of Bowman (1990) was compared with experimental data obtained for a copper–water heat pipe and a molybdenum–lithium heat pipe. A more extensive finite element numerical model has been presented by Jang et al. (1990) in which both the free molecular flow and continuum flow regimes were examined. This model was used to simulate the startup of a frozen stainless steel–sodium heat pipe, but was not compared with any experimental data.

8.2.3 Compressibility and Startup from Supercritical State

Startup from a supercritical state occurs most frequently for heat pipes utilizing cryogenic working fluids and as a result is not often encountered in electronic applications. Perhaps of more importance are the compressibility effects during startup. While compressible flow effects can occur in all three temperature ranges of heat pipes, they are most common in heat pipes starting up from a supercritical or frozen state. Figure 8.6 illustrates how the vapor compressibility can affect the mass flow rate and axial pressure distribution when the condenser region of the heat pipe is suddenly reduced and held constant. As shown, the mass flow rate behaves as would be expected, as does the pressure distribution during

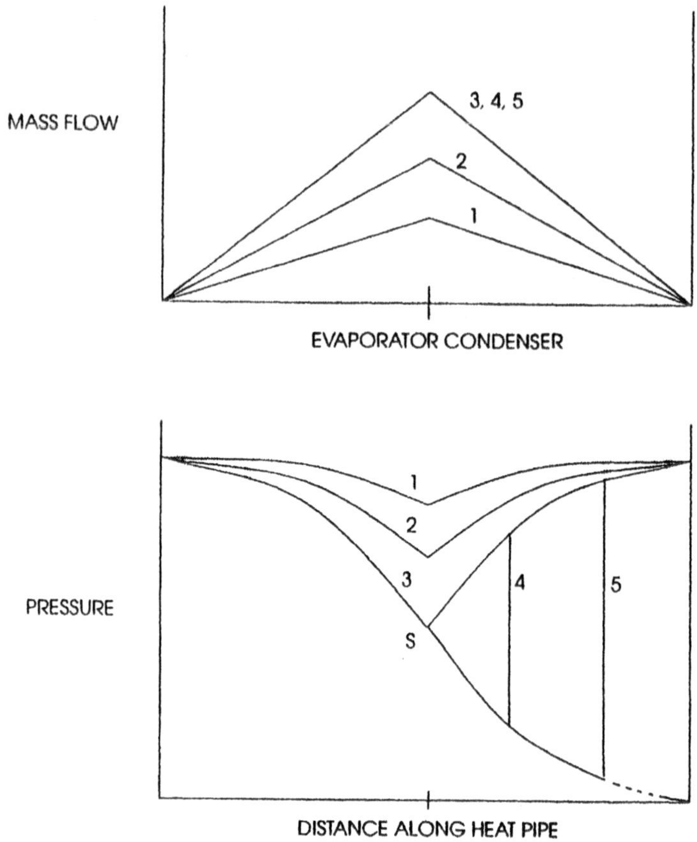

Figure 8.6 Compressibility effects during start-up as a function of time.

time intervals 1–3. However, with increasing time it is possible for the vapor flow rate to increase until sonic velocity occurs, as illustrated by time intervals 4 and 5. As the pressure in the condenser continues to decrease due to decreased temperature, supersonic flow and the accompanying shock wave may also occur. This so-called sonic limit has been discussed previously in Chapter 4.

Difficulties are often encountered in attempting to startup heat pipes from a condition where a portion or all of the working fluid is in either the supercritical or frozen state. While these are not typical phenomena encountered in the low or moderate temperature heat pipes used in the electronics industry, they may occur in liquid metal heat pipes or in extremely low temperature applications. These difficulties are primarily a result of the low vapor density and the rapid shift from a viscous limiting condition to one in which sonic or supersonic flow occurs. Figure 8.7 illustrates the axial temperature variation for a stainless steel–water heat

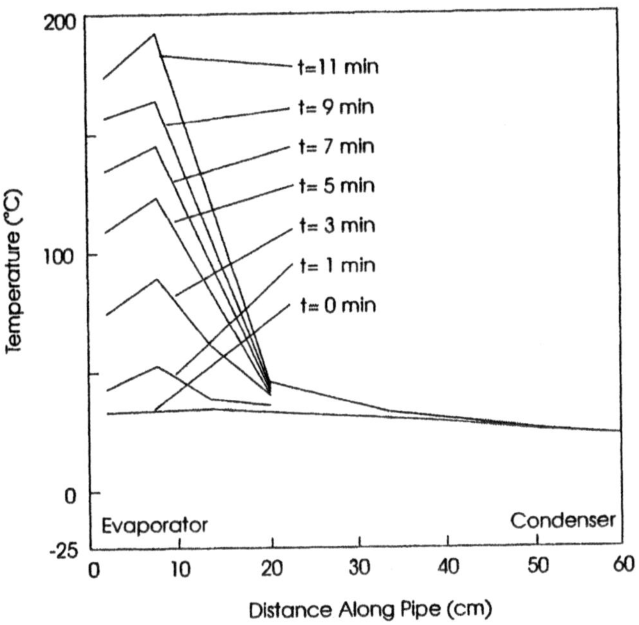

Figure 8.7 Time dependent temperature profiles along a heat pipe starting up from the frozen state (Shlosinger, 1969)

pipe (Shlosinger, 1969). As shown, the condenser portion of the heat pipe is in a frozen state and remains frozen 11 min after startup in spite of increases in the evaporator temperature of nearly 200°C. The result of this process is vapor velocities that approach or exceed the sonic velocity for the conditions inside of the pipe.

Several methods have been proposed by which the problems associated with startup from a supercritical state can be reduced or eliminated. Among these are the use of multiple fluids such as alcohols or ethylene glycol solutions, which solidify at the normal operating temperatures, or the use of secondary heat pipes with a working fluid designed to operate at higher temperatures, contained within the principal heat pipe to assist in startup (Shlosinger, 1969).

REFERENCES

Abramenko, A. N., Kanonchik, L. E., and Prokhorov, Y. M., 1986, "Start-up Dynamics of an Arterial Heat Pipe from the Frozen or Chilled State," *Journal Engineering Physics* (translation of *Inzhenerno-Fizicheskii Zhurnal*), Vol. 51, No. 5, pp. 1283–1288.

Antoniuk, D., and Edwards, D. K., 1990, "Depriming of Arterial Gas-controlled Heat Pipes," Proc. 7th Int'l Heat Pipe Conf., Minsk, USSR, May.

Bowman, W., 1990, "Transient Heat-Pipe Modeling, the Frozen Start-Up Problem," Paper No. 90-1773, AIAA/ASME 5th Joint Thermophysics and Heat Transfer Conference, Seattle, WA, June, American Institute of Aeronautics and Astronautics, Washington, D.C.

Bystrov, P. I., and Goncharov, V. F., 1983, "Starting Dynamics of High-Temperature Gas-filled Heat Pipes," *High Temperature (USA)*, Vol. 21, No. 6, pp. 927–936.

Bystrov, P. I., Goncharov, V.F., Kharchenko, V. N., and Shul'ts, A. N., 1982, "Transient Heat and Mass Transfer in Liquid-Metal Heat Pipes," *Heat Transfer-Soviet Research*, Vol. 14, No. 3.

Chi, S. W., 1976, *Heat Pipe Theory and Practice*, Hemisphere, Washington.

Chisholm, D., 1971, *The Heat Pipe*, Mills and Boon, London, England.

Cotter, T. P., 1967, "Heat Pipe Startup Dynamics," USAEC Ref. LA-DC-9026, Conf. 671045-2, Los Alamos Scientific Laboratories, University of California, Los Alamos, NM.

Deverall, J. E., Kemme, J. E., and Florschuetz, L. W., 1970, "Sonic Limitations and Startup Problems of Heat Pipes," Report No. LA-4518, Los Alamos National Laboratory, Los Alamos, NM.

Dunn, P. D., and Reay, D. A., 1982, *Heat Pipes*, 3rd ed., Pergamon, Oxford.

Edwards, D. K., and Marcus, B. D., 1972, "Heat and Mass Transfer in the Vicinity of the Vapor-Gas Front in a Gas-loaded Heat Pipe," *ASME Journal of Heat Transfer*, Vol. 94, May, pp. 155–162.

Ivanovskii, M. N., Sorokin, V. P., and Yagodkin, I. V., 1982, *The Physical Principles of Heat Pipes*, Clarendon, Oxford.

Jang, J. H., Faghri, A., Chang, W. S., and Mahefkey, E. T., 1990, "Mathematical Modeling and Analysis of Heat Pipe Start-up From the Frozen State," *ASME Journal of Heat Transfer*, Vol. 112, August, pp. 586–594.

Merrigan, M. A., Keddy, E. S., and Sena, J. T., 1985, "Transient Heat Pipe Investigation for Space Power Systems," Report No. LA-UR-85-3341, Los Alamos National Laboratory, Los Alamos, NM.

Ochterbeck, J. M., and Peterson, G. P., 1993, "Freeze/Thaw Characteristics of a Copper–Water Heat Pipe: Effects of Non-Condensible Gas Charge," *AIAA Journal of Thermophysics and Heat Transfer*, Vol. 7, No. 1, pp. 127–132.

Ponnappan, R., Boehman, L. I., and Mahefkey, E. T., 1990, "Diffusion-controlled Startup of a Gas-loaded Liquid-Metal Heat Pipe," *AIAA Journal of Thermophysics*, Vol. 4, No. 3, July, pp. 332–340.

Shlosinger, A. P., 1969, "Heat Pipe Devices for Space Suit Control," Report No. CR-1400, Contract No. NAS 2-3817, National Aeronautics and Space Administration, Washington, DC.

Sockol, P. M., 1973, "Startup Analysis for High-Temperature Gas-loaded Heat Pipes," Report No. TM X-2840, NASA-Lewis Research Center, Cleveland, OH, July.

Tolubinsky, V. I., Shevchuk, E. N., and Stambrovsky, V. D., 1978, "Study of Liquid-Metal Heat Pipe Characteristics at Start-up and Operation Under Gravity," Proceedings of the 3rd International Heat Pipe Conference, Palo Alto, CA, pp. 274–282.

Chapter

9

HEAT PIPE APPLICATIONS IN THE ELECTRONICS INDUSTRY

9.1 INTRODUCTION

The thermal control of electronic components has one principal objective, to maintain relatively constant component temperature equal to or below the manufacturer's maximum specified service temperature. Investigations have demonstrated that an increase in temperature of as little as 10°C can reduce the reliability of some systems by as much as 50% (Nelson et al., 1978a). For this reason, new thermal control schemes must be capable of eliminating hot spots within the electronic devices, removing heat from these devices, and dissipating this heat to the surrounding environment. Figure 9.1 compares the thermal acceleration factor, defined as the ratio of the failure rate at temperature T and at 70°C, with the mean chip operating temperature. This figure illustrates the relationship between the reliability of bipolar digital components and their operating temperature and clearly demonstrates the need for a well-controlled thermal environment (Peterson, 1992).

Thermal control schemes to remove heat from individual devices and systems include the traditional means of free and forced gaseous and liquid convection as well as conduction and radiation or combinations thereof (Goodling et al., 1987). Several strategies have been developed for controlling and removing the heat generated in these devices and systems. These include advanced air cooling schemes, such as the one in the IBM 4381 midrange processor (Biskhorn et al., 1984), the Hitachi SiC RAM module (Okutani et al., 1984), and the Mitsubishi high thermal conduction module (Kohara et al., 1985); direct cooling without phase change, such as

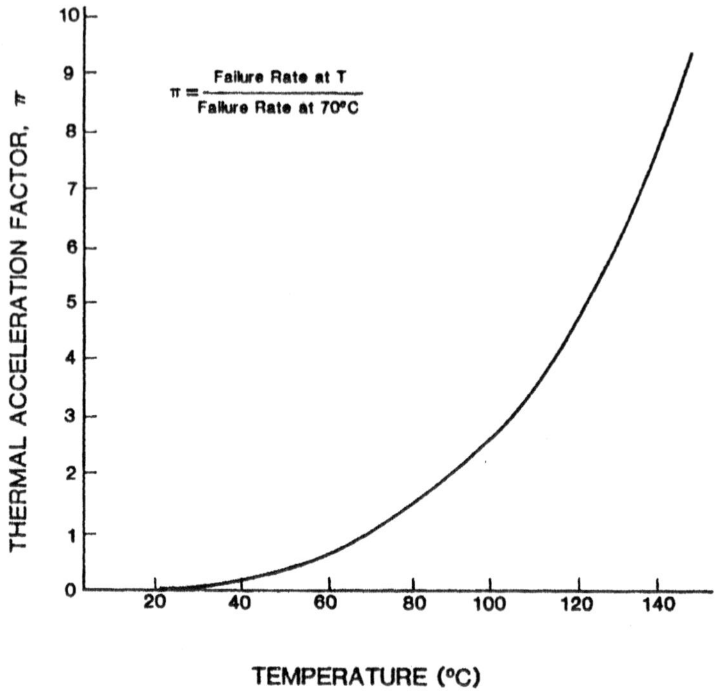

Figure 9.1 Effect of temperature on the failure rate of bipolar digital devices. (Reprinted with permission of McGraw-Hill Publishing Company.)

that in the Cray-2 supercomputer (Danielson et al., 1986), or with phase change, such as the liquid encapsulated module (Moran and Oktay, 1982), where the electronic package is submerged in a liquid pool; and indirect cooling schemes, such as the IBM thermal conduction module (TCM) (Chu et al., 1982) in the IBM 3081 / 3090 processor or the liquid-cooled module (LCM) (Watari and Murano, 1985) in the NEC SX. Other more advanced techniques proposed include miniature thermosyphons (Wayner and Parks, 1985) or free-falling liquid films (Mudawwar et al., 1987).

Although air cooling is the best understood and most frequently used technique, it is limited in the heat removal rate by the convection coefficient. Direct cooling methods are capable of attaining extremely high heat flux levels, but they present problems with contamination and are extremely expensive. Although an issue of considerable discussion, indirect cooling strategies appear to be the best near-term solution for the thermal control of advanced computer architectures, and as noted by Incropera (1986), "the most fruitful approach to enhancing the performance of indirect cooling technologies is likely to be one which reduces the thermal path between the electronic packages and the [cooling fluid]."

Heat pipes, because of their high thermal conductivity, provide an essentially isothermal environment with very small temperature gradients between the individual components. Hence, they are an acceptable alternative to the large, bulky aluminum or copper fin structures of complex geometries that are currently the industry standard. The high heat transfer characteristics, the ability to maintain constant evaporator temperatures under different heat flux levels, and the diversity and variability of evaporator and condenser sizes make the heat pipe an effective device for the thermal control of electronic components. This review of applications includes recent advances and developments that affect the implementation of heat pipes in the thermal control of electronic devices. The applications fall into three categories: indirect, where the heat pipe is placed in contact with the component or device and serves the same function as a heat sink; direct, where the device is an integral part of the heat pipe and / or is in direct contact with the working fluid; and system level heat pipes, where a heat pipe is used to control the temperature in equipment cabinets or systems.

Ruttner (1978) summarized the application of heat pipe thermal control systems in the cooling of electronic components developed before 1977. Therefore, this work focuses on more recent applications. Marto and Peterson (1988) reviewed the operating principles, modeling, design, and testing of heat pipes as applied to electronic applications and provided a comprehensive review of specific applications, as have several other reviews (Peterson, 1988a, 1992).

9.2 INDIRECT HEAT PIPE THERMAL CONTROL

Because of the high effective conductivity of heat pipes compared to that of conventional heat sinks, heat pipes have been proposed and selected for thermal control of individual components, series of components, and entire printed wire boards. The simplest heat pipe heat sinks are cylindrical with a copper or aluminum case and water, acetone, or methanol as the working fluid. Using this configuration, heat can be removed from power transistors, thyristors, or individual chips. These components are often mounted on the evaporator portion of the pipe and attached mechanically. A series of fins attached to the condenser end of the heat pipe provides the mechanism for heat rejection to a coolant, either through free or forced convection to a gas or a liquid.

An example of such a device is illustrated in Fig. 9.2, which shows a thermal electric heat sink that utilizes a 30.5-cm copper–water heat pipe. The application was originally developed for medical applications involving the removal of heat in a tissue-embedding system. The thermal electric cooler illustrated on the evaporator portion of the heat pipe is approximately 4.25 cm square and has a total power of 75 W. The heat pipe is used to cool the hot side of the thermal electric cooler and allows the cool side of

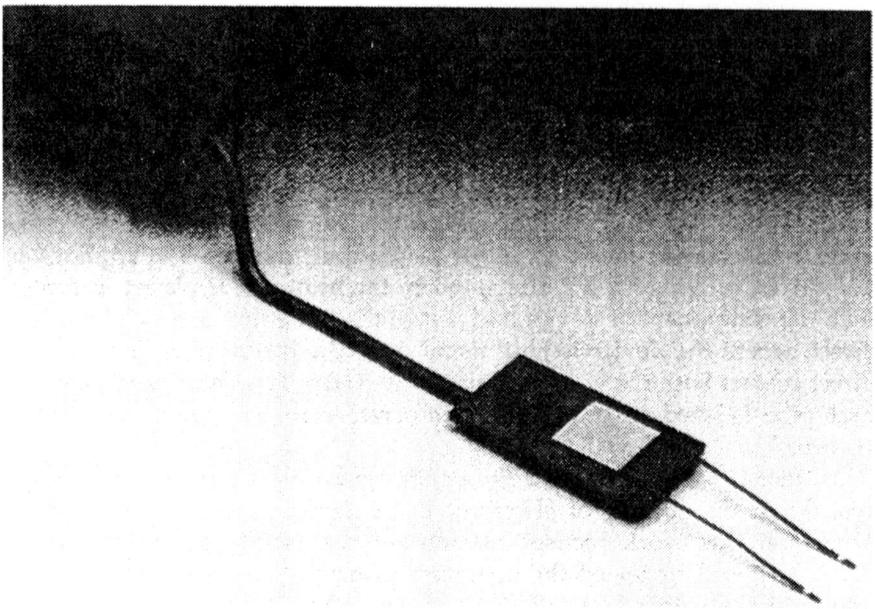

Figure 9.2 Thermoelectric heat sink using heat pipes and forced convection (courtesy of Thermacore, Inc.).

the thermal electric cooler to be maintained at uniform temperature. Heat transported to the condenser region is removed by forced air convection.

Murase et al. (1982) and Yoshida et al. (1984) reported on a more conventional design termed a "heat kicker." This device consisted of a small, 12.7–15.88-mm-outside-diameter copper–water heat pipe with aluminum fins attached to one end, as illustrated in Fig. 9.3, or both ends depending on the location of the evaporator. Figure 9.4 illustrates the maximum heat transfer rate as a function of the vapor temperature where the heat source is at one end of the heat pipe. In Fig. 9.5 the heat source is in the center with heat-rejecting fins at either end. In the former case, the wick was constructed from 150-mesh copper screen while the centrally heated version had a rectangular spiral-grooved wick. The effect of the tilt angle on the thermal performance of the screen-wick heat pipe is shown in Fig. 9.4, where it is apparent that changes in tilt can significantly affect the heat transport capacity. For example, in a horizontal orientation at a vapor temperature of 80°C, the heat pipe can transport about 500 W. If the heat pipe is tilted 5° with the evaporator end down (i.e., gravity assisted), the heat transport capacity increases to nearly 600 W. However, when the heat pipe is tilted 5° in the opposite direction, with the condenser end down, the heat transport capacity decreases dramatically, to about 300 W. Similar

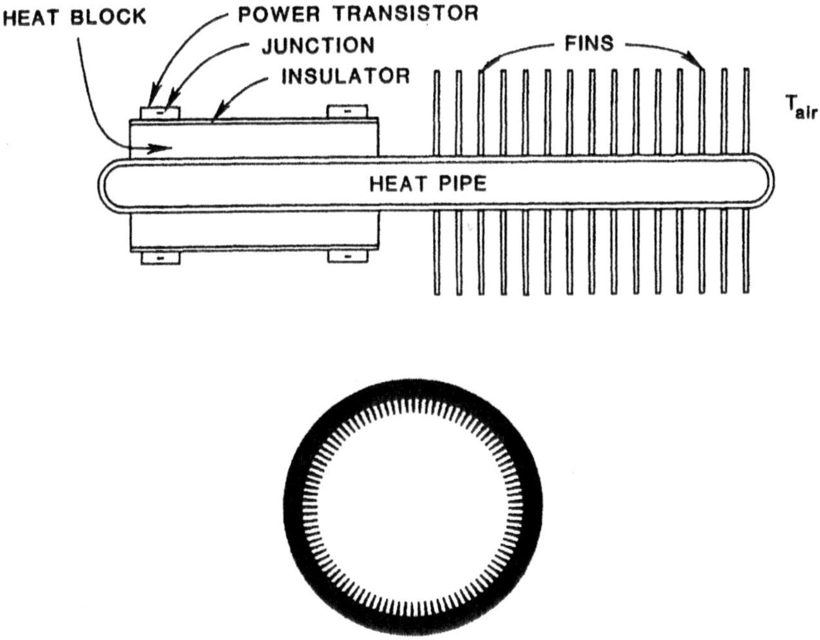

Figure 9.3 Heat pipe heat sink for power transistors (Murase et al., 1982).

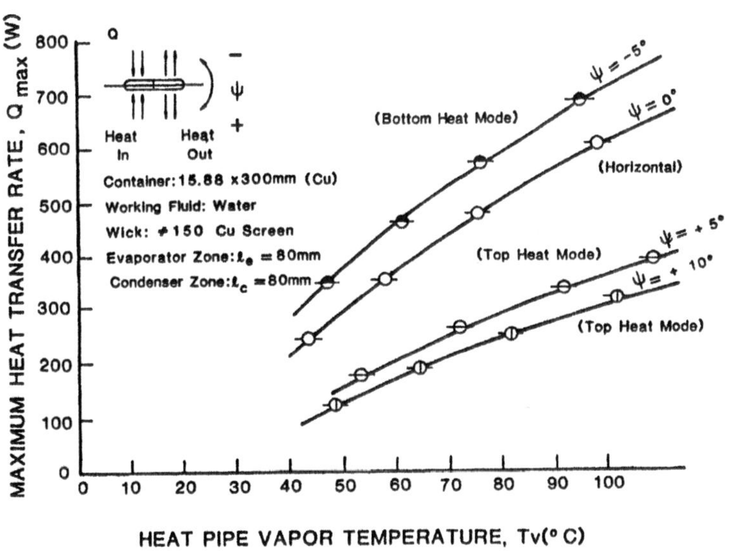

Figure 9.4 Maximum heat transfer characteristics (Screen wick) (Murase et al., 1982).

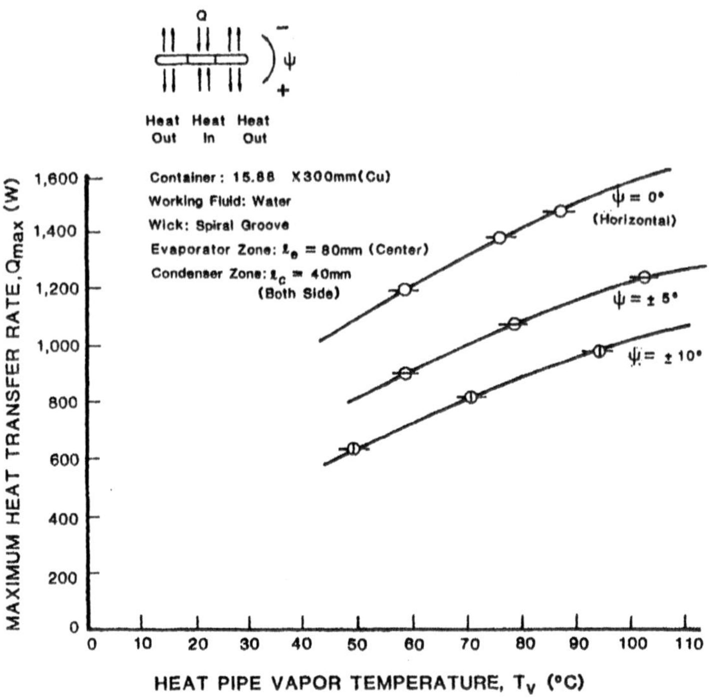

Figure 9.5 Maximum heat transfer characteristics (Grooved wick) (Murase et al., 1982).

curves for the centrally heated, grooved heat pipe are presented in Fig. 9.5. With this wick, the heat pipe thermal transport capacity at the same vapor temperature of 80°C is about 1400 W when horizontal and about 1100 W when inclined ±5°. Clearly, the performance of this heat pipe is far superior to the screen-wick heat pipe, which exhibits a larger liquid flow resistance than the axial grooves. Murase et al. (1982) showed that as a heat sink in an audio amplifier system, the grooved heat pipe has a lower thermal resistance than an extruded aluminum heat sink. This improved performance can lead to a 50% savings in weight, as shown in Fig. 9.6, which illustrates the overall thermal resistance as a function of weight.

The use of multiple, parallel heat pipes to cool high power thyristors has also been demonstrated by Murase et al. (1987). Several heat pipes were mounted in parallel in a single heater block with a single set of heat-rejecting fins. The heat pipes were made of copper / groove-wick / water, with an outside diameter of 15.9 mm and a length of 1200 mm. The aluminum fins were placed at a pitch of 8 mm to be suitable for natural convection. A black coating was used to enhance the effects of radiation. This concept was later expanded to a heat pipe heat sink for high capacity semiconductor devices (diodes, thyristors, GTO thyristors, or SI thyristors)

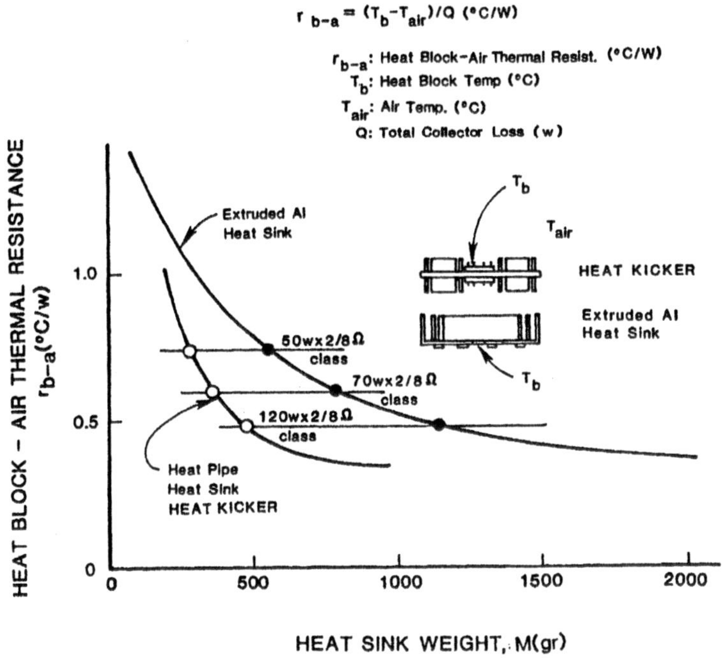

Figure 9.6 Performance as a function of weight (Murase et al., 1982).

as applied to power controllers for large sized motors and railway substations. Other standardized thermal control devices are now commercially available for thermal loads of 250–2500 W (Murase et al., 1986).

Figure 9.7 illustrates a finned heat pipe design for the cooling of high power semiconductor devices (Ptacnik and Polasek, 1988). The heat pipes shown are constructed of copper with a sintered powder wick on the end and axial grooves on the inner longitudinal surface. Two working fluids were evaluated, water and methanol. The pipe was 0.1 m in length and 0.064 m in diameter, and had a maximum performance level of nearly 2000 W when equipped with 0.1×0.15-m rectangular fins. Two similar devices are shown in Figs. 9.8 and 9.9. In Fig. 9.8, heat is conveyed through a metal block perpendicular to the longitudinal axis of the pipe. The thermal resistance of the single- and double-pipe configurations are shown in Figs. 9.10 and 9.11, respectively. In Fig. 9.5 a thyristor cooling assembly used for a motor controller is cooled by a series of four vertical heat pipes. The total power of the device is approximately 3 kW and heat is removed from the condenser by forced convection. The finned stack is approximately 20 cm tall and 15 cm square.

Malik et al. (1984) investigated another device for cooling a thyristor by mounting it on a commercially available heat pipe. The thyristor had a rated value of 150 A and a maximum allowable junction temperature of

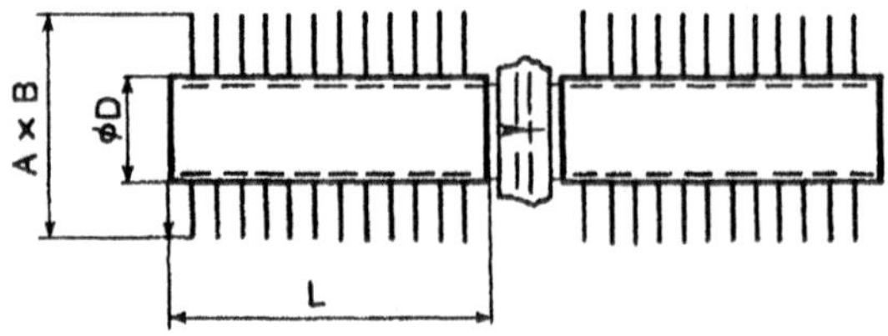

BASIC PARAMETERS:		
Tube material		copper
Type of wicks		grooved, sintered
Working Fluid		Water, Methanol
Tube Length	L (m)	0.1
Tube Diameter	ϕD (m)	0.032 0.04 0.064
Fin Dimensions	A x B (m)	0.125 x 0.125 0.1 x 0.1 0.1 x 0.125 0.1 x 0.15
Performance Range	$\dot{Q}$ (W)	500 - 2000

Figure 9.7 Finned high power semiconductor coolers (Ptacnik and Polasek, 1988).

120°C. It was mounted on a copper–water heat pipe 300 mm long and 25.4 mm in diameter. The transient thermal response of the heat pipe cooling system was far superior to the conventional methods of both free and forced convection. In addition, the thermal resistance, 0.20°C / W at an air flow rate of 217 liters / s, was greatly reduced under low power conditions and tended to remain constant as the power increased.

Other applications in this area include the bonding of integral heat pipes directly to large transcalent devices, which typically include 250-A rectifiers with blocking voltages of 1000 V, 100-A transistors, or 400-A thyristors (Adams et al., 1979).

Aside from just serving as highly efficient fins, heat pipes can serve other useful functions. In 1981, Wolf built and tested a device for vibration-sensitive electronics. This device, illustrated in Fig. 9.12, was capable of transporting up to 50 W at adverse inclinations of 2.0 cm and slightly over 100 watts at horizontal orientations. It had a source–sink separation distance of approximately 10 cm and a thermal footprint of 77 cm^2. In 1984, Miayazaki and Sasaki reported on the development of a heat pipe thermal switch for spacecraft electronics that also used the bellows concept. This device, which consisted of a space radiator, a baseplate, and a stainless steel bellows vessel (Fig. 9.13), had a grooved wicking structure. The thermal footprint was approximately 61 cm^2, and the transport capacity was 60 W in a horizontal orientation.

Figure 9.8 Heat pipe heat block (Ptacnik and Polasek, 1988).

Recently, Peterson (1987, 1988b) reported on the analysis and testing of a similar, although much smaller, device, referred to as a bellows type heat pipe for use with electronic components or multichip modules. This device, shown in Fig. 9.14, was used to provide a thermal conduction path between the surface of a semiconductor chip and a liquid-cooled cold plate. The advantages of this device can be seen by comparing the overall thermal resistance paths between the bellows heat pipe and a solid metal rod, which was as much as 10 times higher. In addition, as the thermal power increased, increases in the operating pressure would result in increases in the internal pressure and, hence, increases in the force applied to the contact, thereby reducing the contact resistance. Also, the inherent flexibility of the bellows structure compensates for misalignment, again reducing the contact resistance at the interface of the heat sink and the chip.

This concept, although capable of providing intermittent contact, was somewhat similar to one proposed by Eldridge and Peterson (1983) but never constructed (Fig 9.15). In this conceptual design, it was proposed

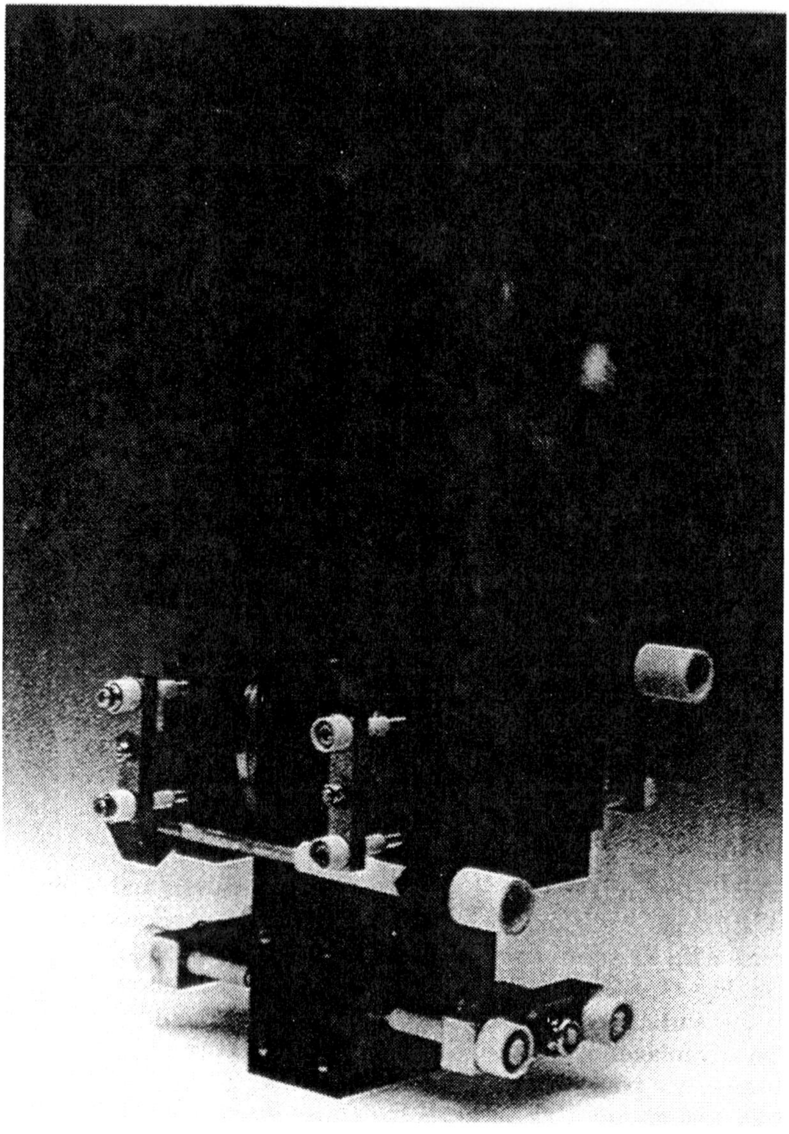

Figure 9.9 Thyristor cooling assembly used for motor controllers (courtesy of Thermacore, Inc.).

that the back of the integrated circuit chip be bonded to the evaporator portion of a heat pipe evaporator, which could be constructed from screen, sintered powder, or a series of axial grooves. A porous wicking material lining the inside circumference of the heat pipe would connect the evaporator and condenser, where heat would be dissipated by free

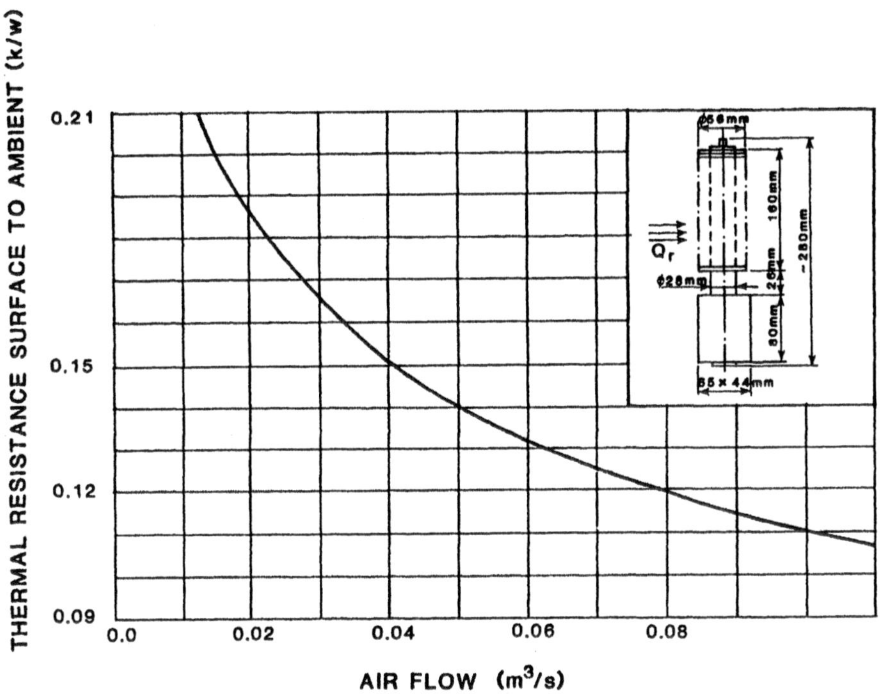

Figure 9.10 Heat pipe heat block performance, single pipe (Ptacnik and Polasek, 1988).

convection. In a later investigation, Babin and Peterson (1989, 1990) presented the results of prototype testing on a similar device designed to replace the thermal piston in the IBM thermal conduction module (TCM) (Chu et al., 1982). In this application, the bellows heat pipe illustrated in Fig. 9.16 would replace the metal piston of the TCM illustrated in Fig. 9.17. As seen by the comparison of the overall internal resistance R_{int} shown in Figs. 9.16 and 9.17, the resistance of the bellows heat pipe, R_{hp}, replaces the resistance of the metal housing, R_h, the resistance of the gap, R_{p-h}, and the resistance of the thermal piston, R_t. The contact resistance R_{c-p} and the chip resistance R_c remain the same for either case. The results of this investigation demonstrated the feasibility of this concept and identified several problem areas, including NCG formation and boiling limitations resulting from the small evaporator surface area.

In addition to using heat pipes to cool individual wafers or components, several methods have been proposed for cooling multiple arrays of devices or entire printed wiring boards. Bonding flat heat pipes to the back of printed circuit boards, as illustrated in Fig. 9.18, is a common method of employing two-phase heat transfer mechanisms. Research performed at the Institute for Mathematical Machines in Prague, Czech Republic, demon-

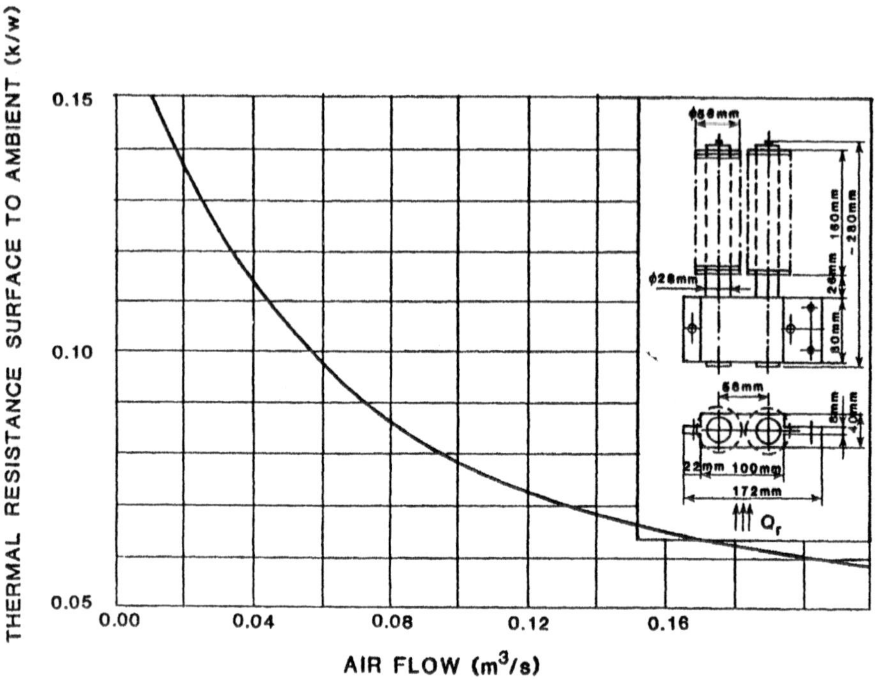

Figure 9.11 Heat pipe heat block performance, two pipes (Ptacnik and Polasek, 1988).

strated that flat heat pipes designed to rest underneath a standard dual-in-line package (DIP) are capable of removing and transporting significant amounts of heat from individual components (Zorbil and Stulc, 1984). A flat heat pipe with cross-sectional dimensions of 2.5 × 6.5-mm and 225 mm in length can dissipate up to 10 W while maintaining a temperature difference between the individual components of only 1°C. A more compact 1.2 × 6.5-mm heat pipe with the same configuration transported 7 W of power while maintaining an isothermal bandwidth of 5°C. Both of these were flat heat pipes with sintered copper powder wicks and water as the working fluid. The components were mounted on the heat pipe surface, as illustrated in Fig. 9.19.

A separate but similar investigation has been reported in the literature. This investigation involves the placement of the hot components directly on small, flat heat pipes (Token, 1986). Figure 9.20 compares the thermal performance of a flat stainless steel heat pipe 1.02 mm thick, 4.01 mm wide, and 178 mm long with both copper and aluminum strips of similar dimensions. As illustrated, the heat pipe yields a maximum temperature rise of approximately 7°C while the copper and aluminum yield 42 and 97°C, respectively. Of greater importance, however, is the uniform tem-

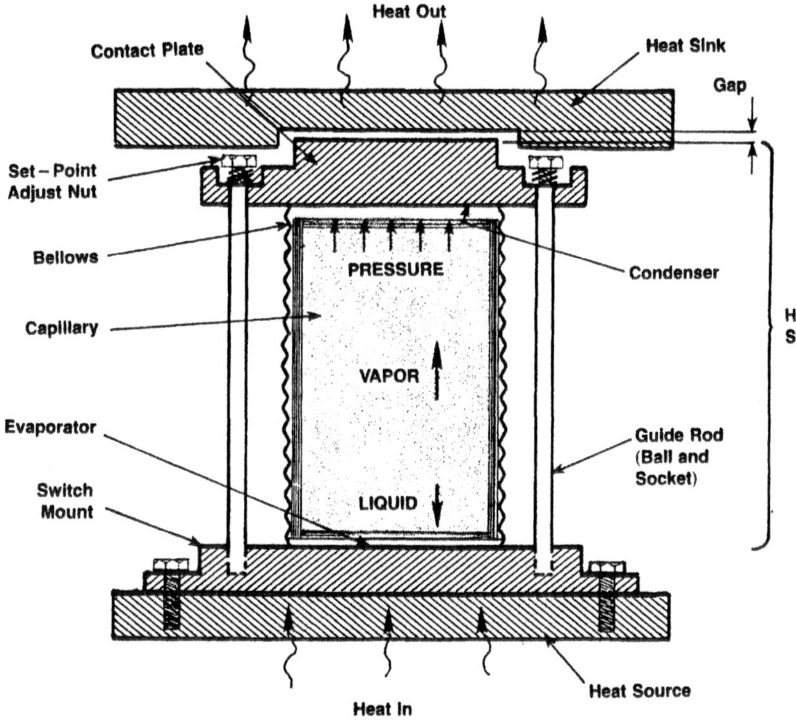

Figure 9.12 Heat pipe thermal switch (Wolf, 1981).

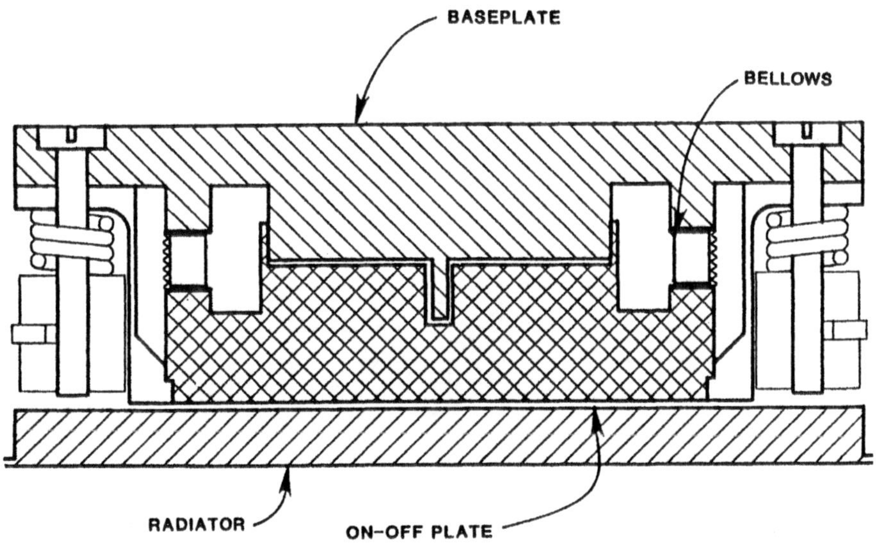

Figure 9.13 Thermal switch for space applications (Miayazaki and Sasaki, 1984).

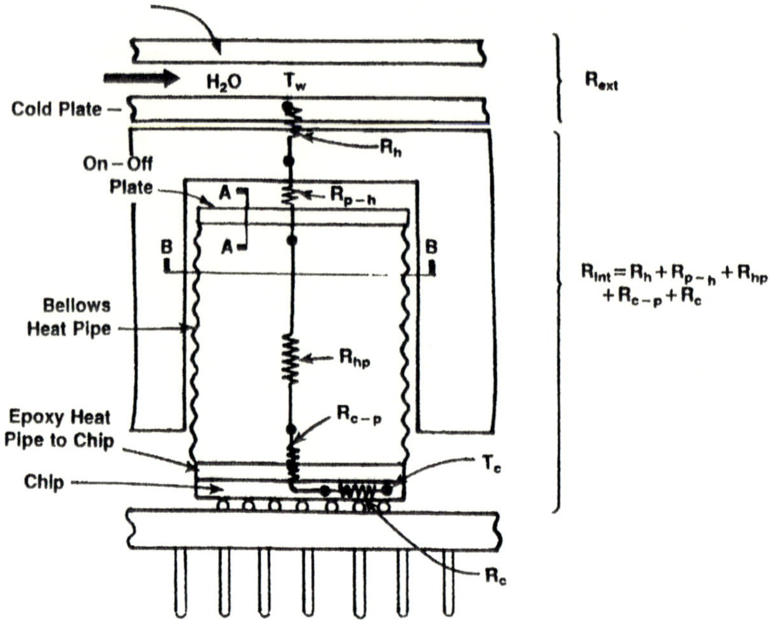

Figure 9.14 Bellows heat pipe (Peterson, 1986).

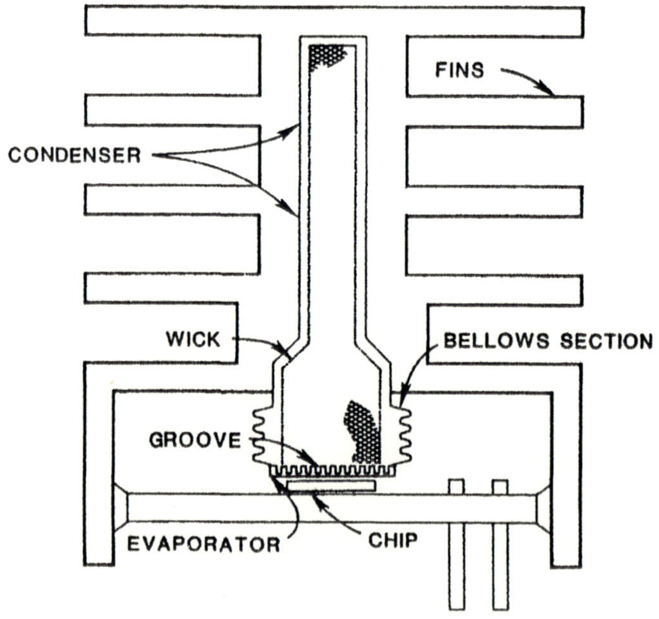

Figure 9.15 Heat pipe cooling of semiconductor chip (Eldridge and Peterson, 1983).

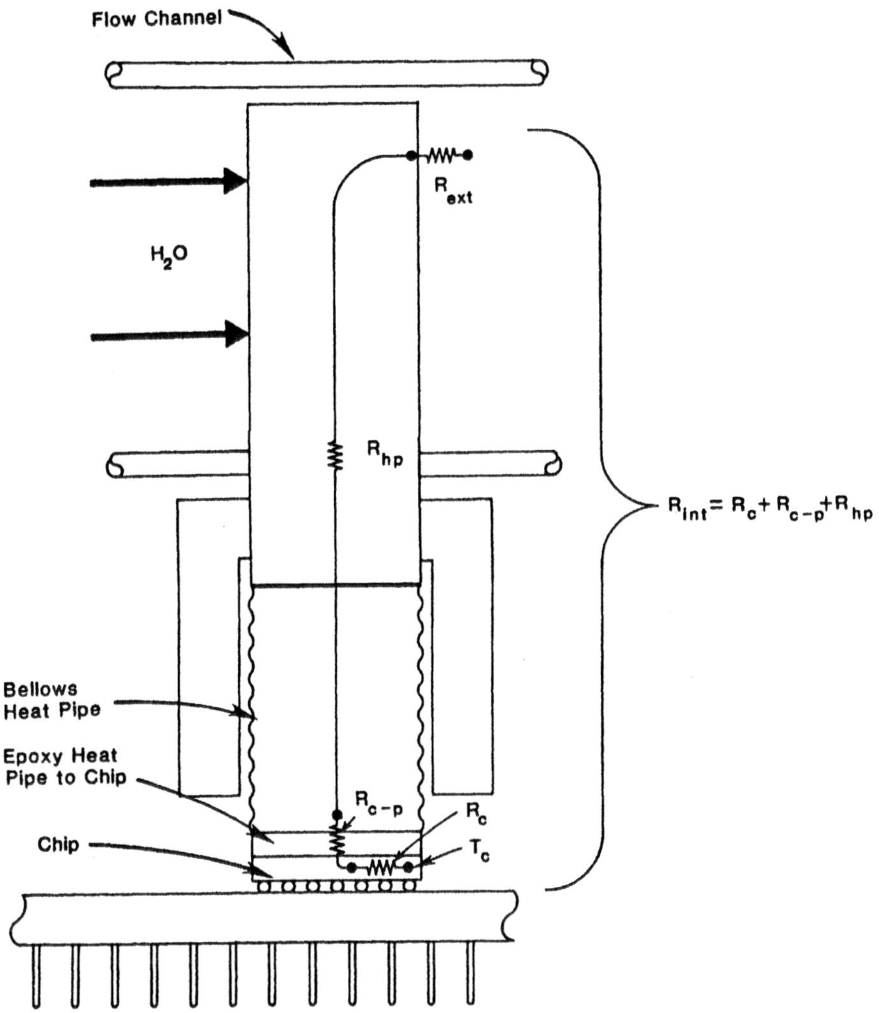

Figure 9.16 Flexible bellows heat pipe for use in multichip modules (Babin and Peterson, 1990).

perature across the face of the heat pipe caused by minimizing the temperature variations between the components.

Zorbil et al. (1988) conducted further investigation expanding the earlier work to include an experimental study on three different sizes of heat pipes. Figure 9.21 illustrates the general shape and dimensions of the test pipes. A trial printed wiring board was constructed using eight type 1 heat pipes 230 mm long. Seven ceramic packages were mounted on each pipe and the board was assembled. Each package was supplied with power sufficient to generate a total of 8.7 W. A maximum temperature of 60°C was

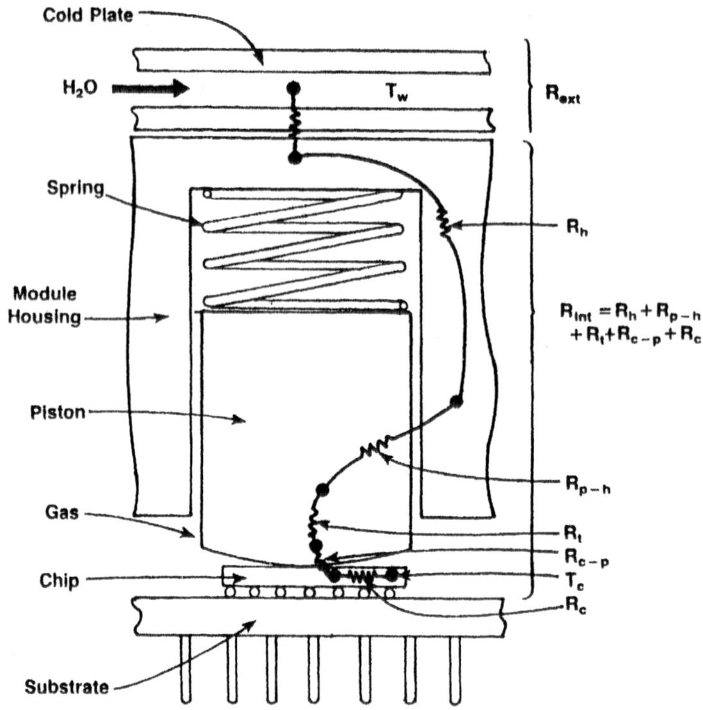

Figure 9.17 IBM Thermal Conduction Module piston (Chu et al., 1982).

obtained at the interface of the heat pipe and the ceramic packages before the onset of dryout occurred.

Murase et al. (1987) also proposed the use of multiple, parallel heat pipes to cool a series of high power thyristors. In this investigation, several heat pipes were mounted in a common heater block. These heat pipes, mounted in parallel, conducted heat to a finned array where the heat was removed by natural convection. The heat pipes, constructed from copper with water as the working fluid and with an outside diameter of 15.9 mm and a length of 1200 mm, were capable of transporting significant quantities of power.

Basiulis et al. (1987) proposed that the entire printed wiring board could function as a heat pipe and reported on two separate board cooling schemes, illustrated in Fig. 9.22, by which this could be accomplished. The first of these was to embed a series of flat heat pipes within the walls of the wiring board; the second utilized the entire wiring board as the vapor chamber. The first scheme (i.e., individual embedded heat pipes) provides the option of using several heat pipes with different working fluids to accommodate nonuniform power dissipation requirements and reduce the number and temperature of localized hot spots.

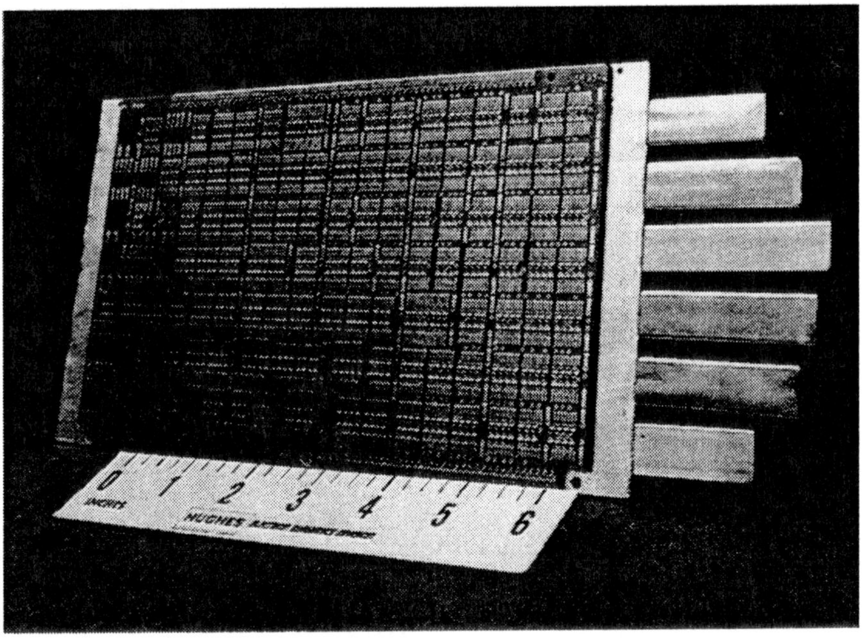

Figure 9.18 Flat heat pipes attached to a printed wiring board (courtesy of Hughes Aircraft Company).

Figure 9.23 illustrates two embedded heat pipe cold plates. These devices, which are approximately 15×15 cm and 15×30 cm, contain individual copper–water heat pipes embedded in the plate in a parallel array. The plates are approximately 3 mm thick and are designed to provide uniform temperatures over the entire surface of the cold plate. As illustrated in Fig. 9.24, the temperature is considerably more uniform for the embedded heat pipe system than for a pure aluminum cold plate, and the maximum temperature has been reduced by approximately 160%.

In a similar application, illustrated in Fig. 9.25, a flat plate with embedded heat pipes was designed for use with tactical jammer electronics. The function of the embedded heat pipe cold plate was to dissipate the heat uniformly over the entire bottom surface of the plate acting as a thermal spreader for the four square transistors located in the center of the board. Each transistor was approximately 1 cm square and had a power of 320 W. At maximum power, the temperature gradient between a point directly under the transistors to anywhere on the opposite surface of the cold plate was less than 15°C.

Because of the three-dimensional vapor flow, the second board cooling scheme proposed by Basiulis (1987), illustrated in Fig. 9.22, may result in more uniform surface temperatures. In this scheme, a sandwich type heat pipe is created by bonding copper or aluminum skins. This flat structure

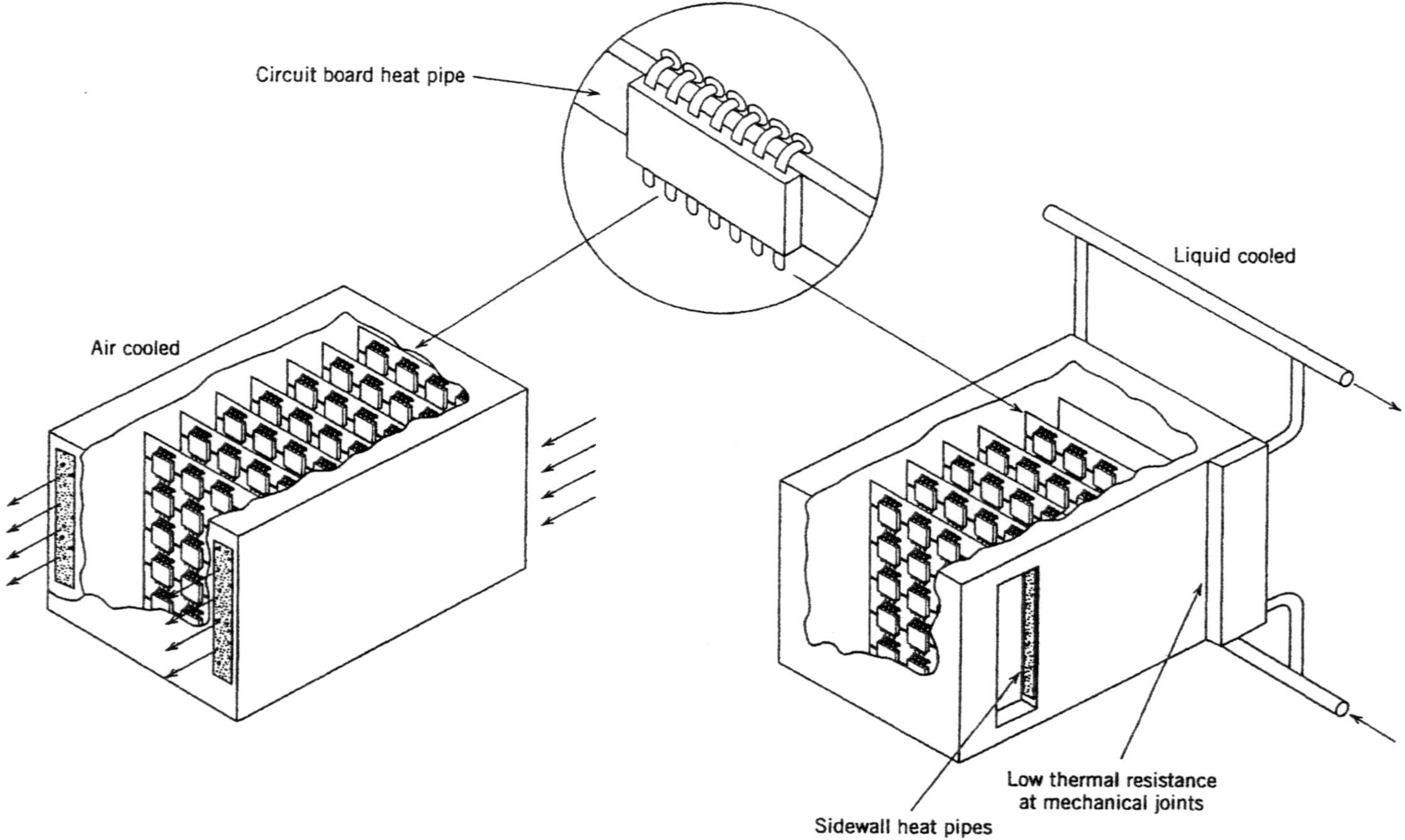

Figure 9.19 Heat pipe mounted DIPS (Zorbil et al., 1988).

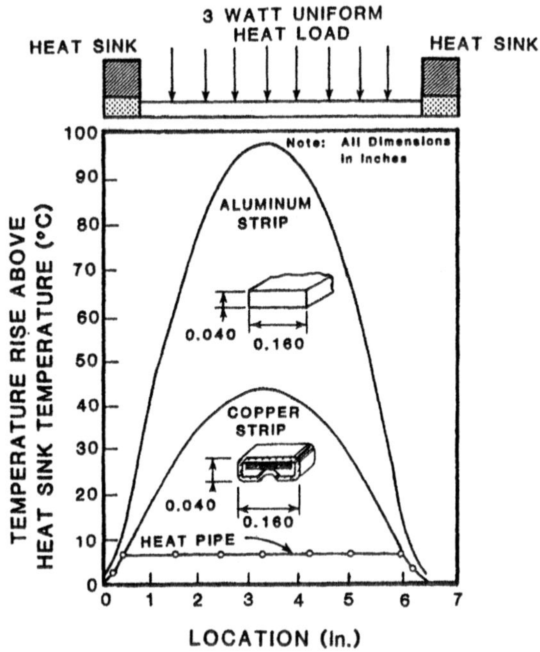

Figure 9.20 Comparison of temperature rise of heat pipe to metal conductors (Token, 1986).

provides a nearly isothermal condition that is over four times more efficient than a solid aluminum plate of the same size. Heat from the individual components vaporizes the working fluid, which can condense either in the condenser portion or in other regions where the heat flux is not as high, as shown in Fig. 9.26. This type of heat pipe is particularly useful in eliminating localized hot spots at individual components.

Ciekurs (1986) also reported an approach to solving hot spot problems with a flat-plate heat sink. In this work, it was noted that with this type of device, the evaporative cooling that occurs near a hot spot diffuses the heat uniformly over the heat sink face, aiding in the effectiveness of the heat rejection process. Additional advantages are a weight savings and an overall reduction in package size.

Adami and Yimer (1988) reported the results from an experimental investigation of such a device (Fig. 9.27) constructed from 1-mm copper sheets, 100-mesh copper screen, and water as the working fluid. Both steady-state and transient operations were evaluated, along with the effects of gravity, impurities, and amount of working fluid. The device operated with a high degree of stability between 30 and 95°C, and there was good agreement between the analytical model and the experimental results. At approximately the same time, Ogushi et al. (1988) conducted an in-

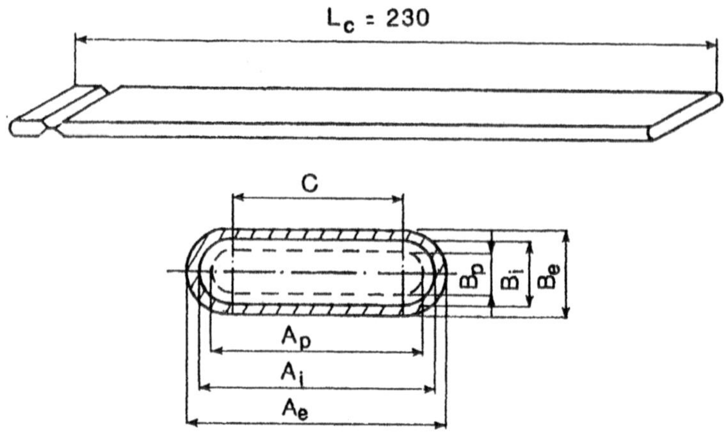

	Pipe semi product	A_e	A_i	A_p	B_e	B_i	B_p	C
		mm						
Type 1	ϕ4 x 0.5	5.0	4.0	3.6	2.3	1.3	0.9	2.7
Type 2	ϕ9.2 x 0.4	13.5	12.7	12.3	2.3	1.5	1.1	10.2
Type 3	ϕ15.8 x 0.4	23.4	22.6	22.2	2.3	1.5	1.5	21.0

Figure 9.21 Flat heat pipe to fit under DIPs (Zorbil et al., 1988).

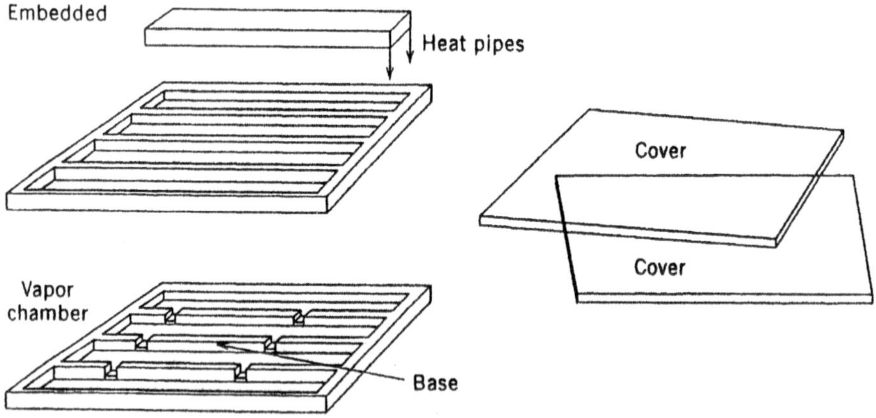

Figure 9.22 Avionics cooling scheme for thermal control of printed wiring boards (Basiulis et al., 1988).

Figure 9.23 Embedded heat pipe cold plates utilizing copper / water heat pipes (courtesy of Thermacore, Inc.).

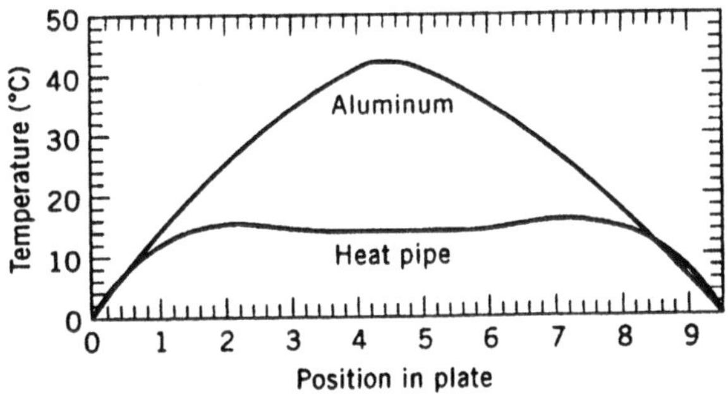

Figure 9.24 Comparison of heat pipe cold plate performance with an aluminum plate (courtesy of Thermacore, Inc.).

Figure 9.25 Heat pipe cold plate for removing heat from tactical jammer electronics (courtesy of Thermacore, Inc.).

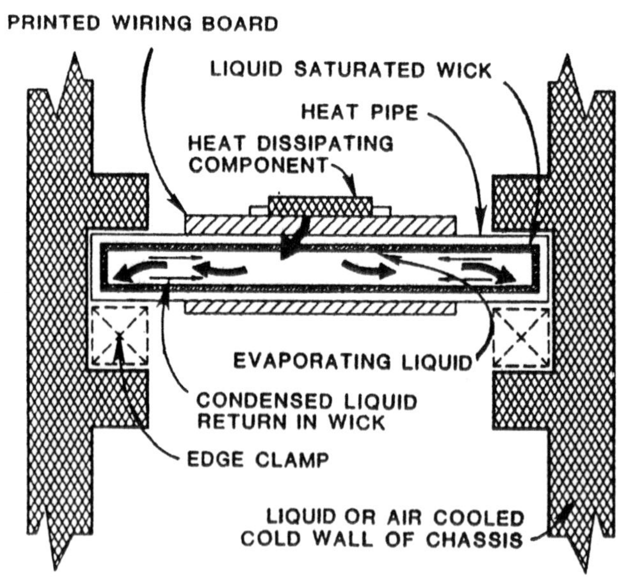

Figure 9.26 Edge-cooled heat pipe printed wiring board (Basiulis et al., 1987).

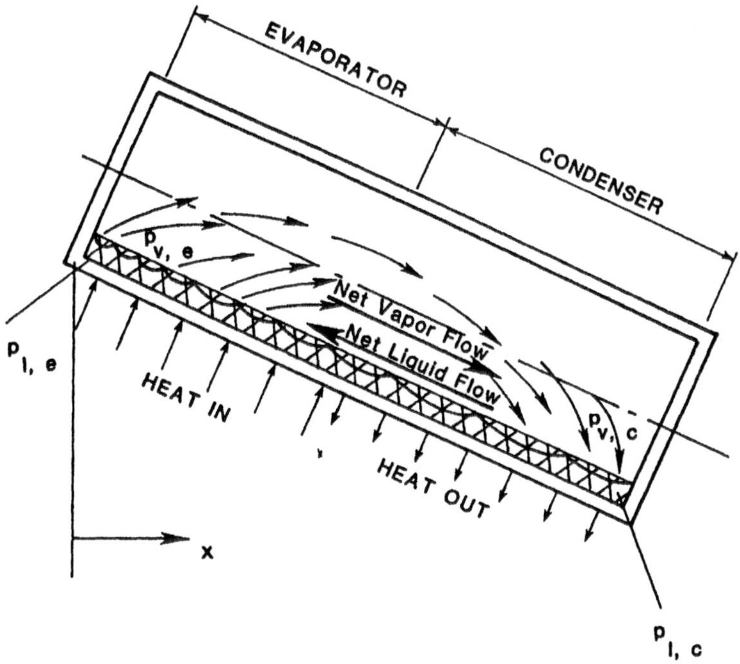

Figure 9.27 Flat plate heat pipe (Adami and Yimer, 1988).

vestigation on a similar device, shown in Fig. 9.28. This device had a vapor space height of 2.5 mm with axial grooves as the wicking structure to distribute the working fluid, methanol, or ammonia throughout the evaporator and a sharp-edged corner for communication between the top and bottom plates. Although some problems with slugging were encountered, the overall temperature drop was half that for an aluminum plate of the same dimensions when methanol was used and one-fourth that for ammonia. Figure 9.29 illustrates a comparison of the predicted and measured maximum heat transfer rate for methanol. As shown, the liquid slugging resulted in substantial decreases in the maximum heat transport capacity.

Nelson et al. (1978b) proposed a more intricate circuit card cooling scheme with a composite wick structure, as shown in Fig. 9.30. The circuit card heat pipe case was constructed from beryllium copper and the evaporator and condenser wicks were constructed from stainless steel screen. Using a sintered fiber, stainless steel artery, and methanol as the working fluid, a high capillary pumping pressure was accompanied by a low liquid flow resistance. This particular design is as yet too costly to manufacture compared to solid metal conductors, but in some applications it may be the only technically feasible solution. Ingenious, inexpensive fabrication techniques will serve to make these flat-plate heat pipes even more attractive.

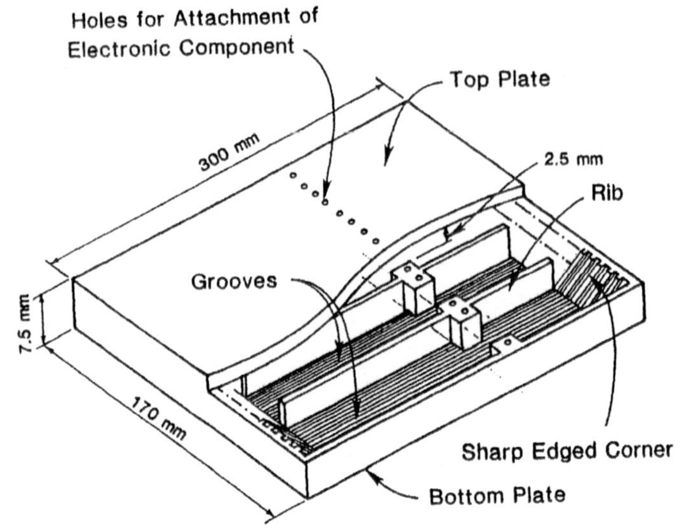

Figure 9.28 Flat plate type heat pipe with axial grooves and sharp edged corner for enhanced evaporator and condenser communication (Ogushi et al., 1988).

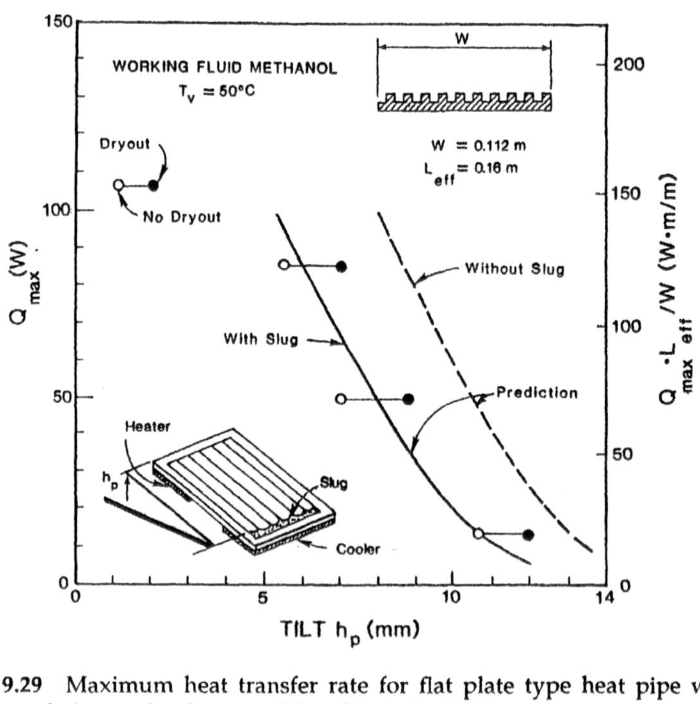

Figure 9.29 Maximum heat transfer rate for flat plate type heat pipe with axial grooves and sharp edged corner (Ogushi et al., 1988).

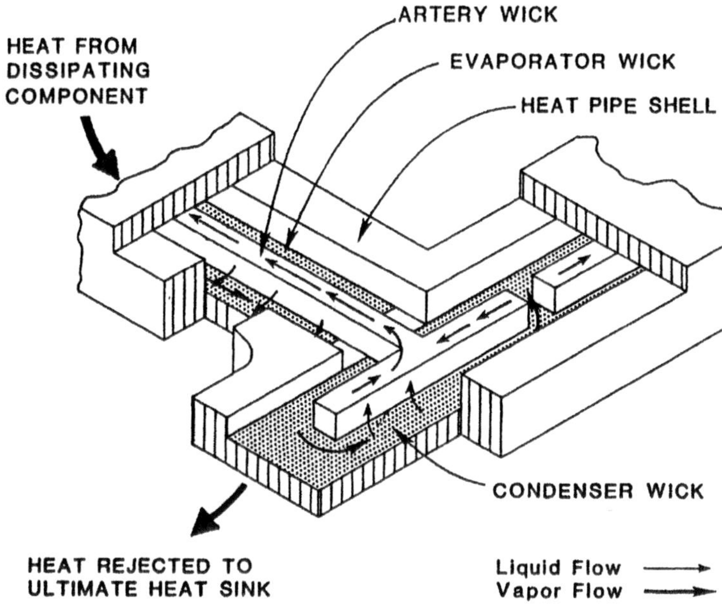

Figure 9.30 Card cooling scheme of Nelson et al. (1978a).

Although all of the previously discussed applications are relatively large compared to the size of most semiconductor devices, this need not be the case. As discussed earlier, Cotter (1984) proposed the micro-heat pipe concept, that is, a wickless heat pipe "so small that the mean curvature of the vapor–liquid interface is necessarily comparable in magnitude to the reciprocal of the hydraulic radius of the flow channel." The proposed device had an equilateral triangular cross section with side dimensions of approximately 20 μm and a length of 10–20 mm. The theoretical transport capacity of a single micro–heat pipe with these dimensions and an optimum amount of methanol as the working fluid was estimated to be approximately 0.03 W. Using this type of device in a solid array approximately 10% by volume, it was demonstrated theoretically that up to a few tens of watts per cubic centimeter of cooling could be provided.

In an application of this concept, a combined analytical and experimental investigation was conducted on a micro–heat pipe with a trapezoidal configuration (Peterson, 1988b; Babin and Peterson, 1990). Four trapezoidal test articles of the shape illustrated in Fig. 9.31 were manufactured by ITOH Research and Development Company. The application of these devices was to remove heat from a ceramic chip carrier, as shown in Fig. 9.32. The length of the test pipes was 57 mm, and the cross-sectional area was approximately 1 mm^2. The test pipes were constructed from both copper and sterling silver with ultrapure water as the working fluid. Preliminary results indicated that the liquid–vapor interface changes

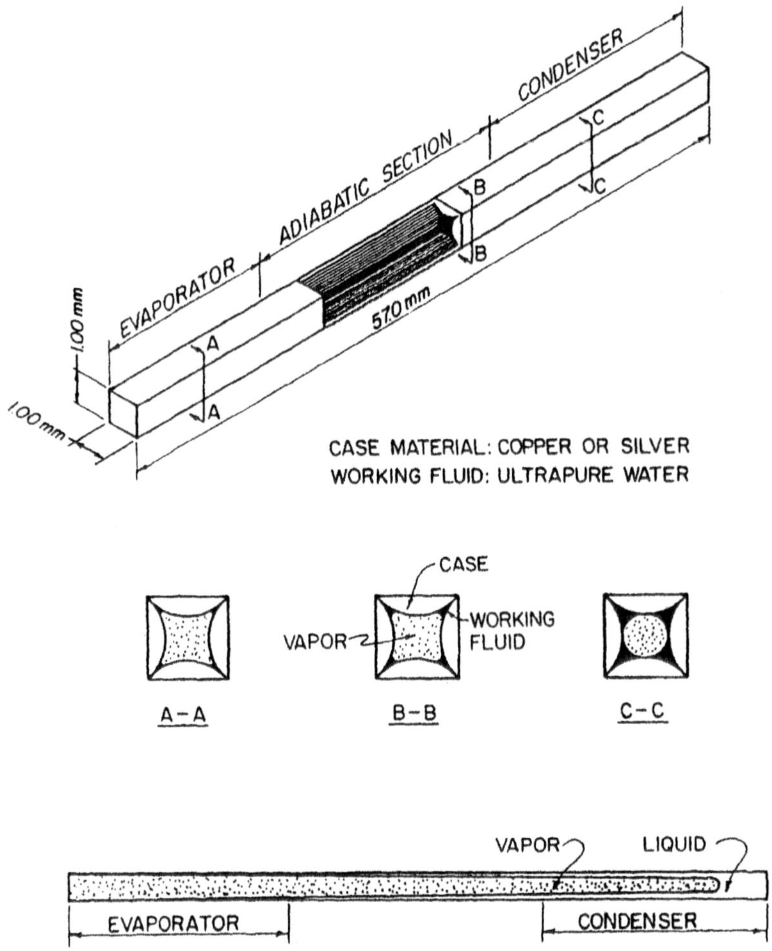

Figure 9.31 Trapezoidal micro heat pipe of Itoh Research and Development Co. (Babin et al. 1990).

continually along the pipe and that the heat pipe is very sensitive to the amount of charge, since proper wetted conditions must be maintained without flooding the microchannels. Figure 9.33 illustrates the principal limitations as predicted using the conventional steady-state modeling techniques outlined in Chapters 3 and 4. Because of the extremely small cross-sectional area, these devices could play an important role in future semiconductor and microelectronic device technologies where thermal loads are high and space is at a premium.

An extension of this concept is a finned array of heat pipes, shown in Fig. 9.34. This device was fabricated and tested for rejecting heat from high power telephone switching devices. The array consists of 16 micro–heat

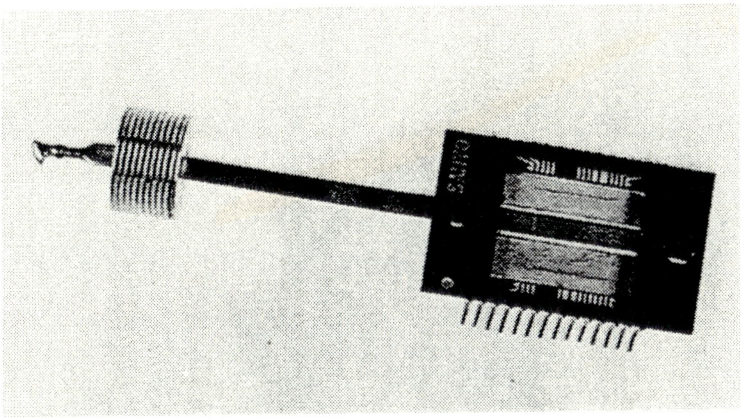

Figure 9.32 Micro heat pipe application (Peterson, 1988).

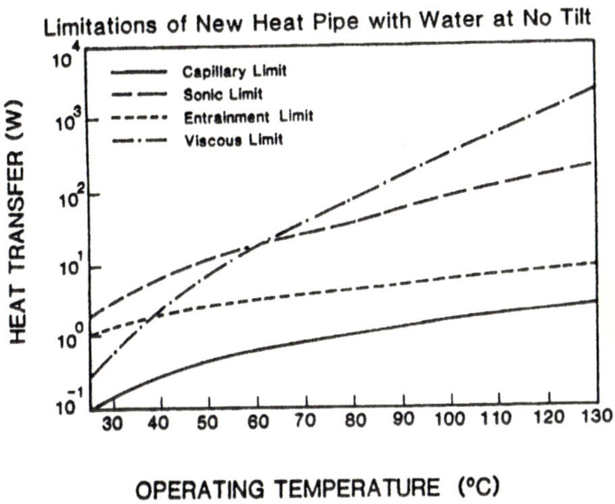

Figure 9.33 Micro heat pipe limitations (Peterson, 1988b).

pipes each approximately 3 mm in diameter and 25 mm in length with the capability of rejecting more that 4 W of thermal power.

9.3 DIRECT HEAT PIPE THERMAL CONTROL

Where the electrical power is high and the heat rejection requirements large, it may be necessary to control the temperature by immersing the devices in a dielectric fluid. Fluid near the saturation temperature typically

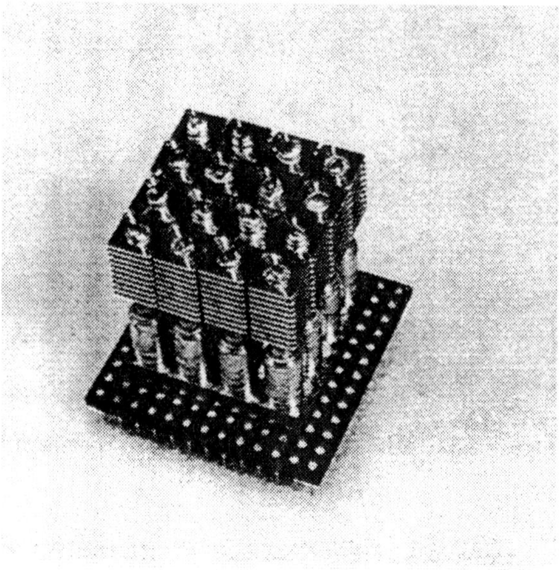

Figure 9.34 Micro heat pipe array (courtesy of Itoh Research and Development)

results in nucleate pool boiling and requires the use of a vapor space condenser. This two-phase loop (i.e., the boiling of the liquid, the condensation of the vapor, and the return of the condensate) is viewed as one form of a two-phase, closed loop thermosyphon (Marto and Peterson, 1988). Although Bergles et al. (1977) reviewed several aspects of direct liquid cooling, applications dealing with heat pipes and the heat pipe effect were omitted.

The generation of vapor bubbles imposes several problems on the thermal control of electronic devices (Marto and Peterson, 1988). First and most important among these is the critical heat flux, the maximum permissible level of the evaporator heat flux. Beyond this level, the vapor completely blankets the heat source and results in an increased temperature drop, leading to dryout and overheating. Second, the formation and collapse of vapor bubbles may generate dynamic forces on the chips and leads, creating high frequency mechanical vibration and subsequent failure. Third and finally, the presence of vapor bubbles may decrease the electric breakdown voltage of the dielectric fluid.

While not as potentially damaging as film boiling, nucleate pool boiling will result in an increased temperature drop. Two techniques have been investigated to reduce this temperature drop. The first is to make the device an integral part of the wick structure to ensure that fresh liquid always remains in contact with the heat source. The second is the direct evaporation (with no bubble nucleation) of a very thin liquid film.

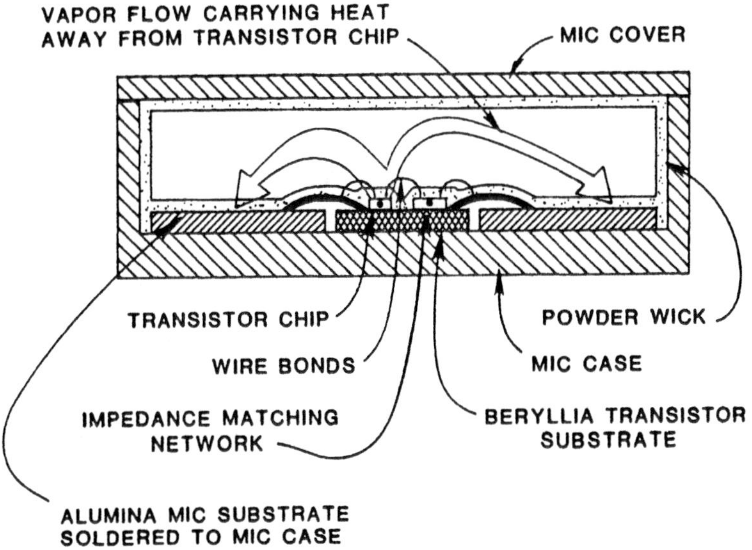

Figure 9.35 Cross-section of heat pipe cooled MIC RF transistor (Nelson et al., 1978a).

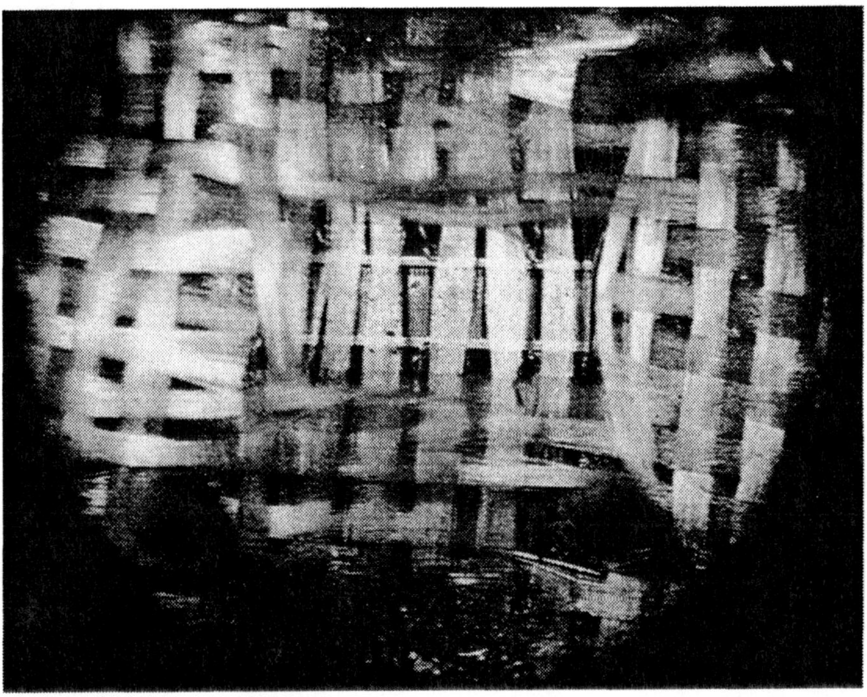

Figure 9.36 Fiber bundle wick applied to a high frequency power transistor (Nelson et al., 1978) (courtesy of Hughes Aircraft Company).

An example of the first of these two techniques was proposed originally by Dean (1976). Nelson et al. (1978a,b) expanded this concept and investigated the inclusion of electronic devices as an integral part of a heat pipe wick, as illustrated in Fig. 9.35. This approach overcame the problems of material performance and electrical compatibility and was a means to substantially reduce the junction temperature. Several types of wicking structures, including glass fiber bundles (Fiberglas strands 0.1 mm in diameter, spaced 0.5 mm apart to permit a high heat flux density) and powder wicks (U.S. Patent No. 4,047,198), were investigated (Sekhon et al., 1977). Difficulties were encountered in applying the fiber bundle wick to microcircuits because of the numerous lead wires. This required tedious hand placement of the fiber bundles. Figure 9.36 shows a photograph of a fiber bundle wick applied to a high frequency power transistor.

As an alternative, a high performance powder wick was developed and patented (Sekhon et al., 1977) that could be applied by mass production techniques. Figure 9.37 is a photograph of the powder wick applied to a high frequency transistor. Comparison of the thermal performance of these wicks to normal pool boiling reveals a substantial reduction in the thermal resistance. Figure 9.38 shows the reduction in junction temperature that

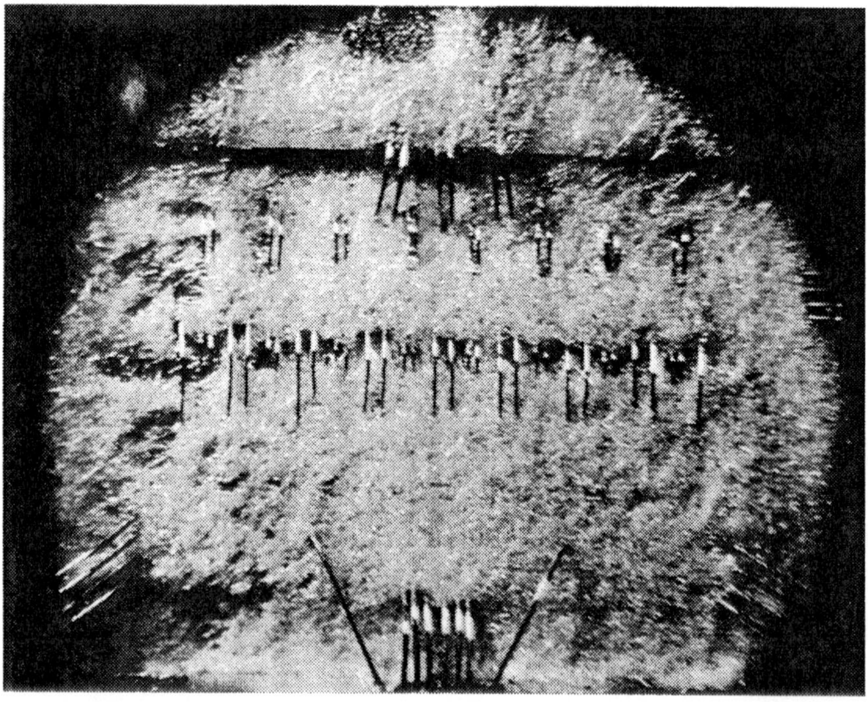

Figure 9.37 Powder wick applied to a 45 watt RF power transistor (Nelson et al., 1978) (courtesy of Hughes Aircraft Company).

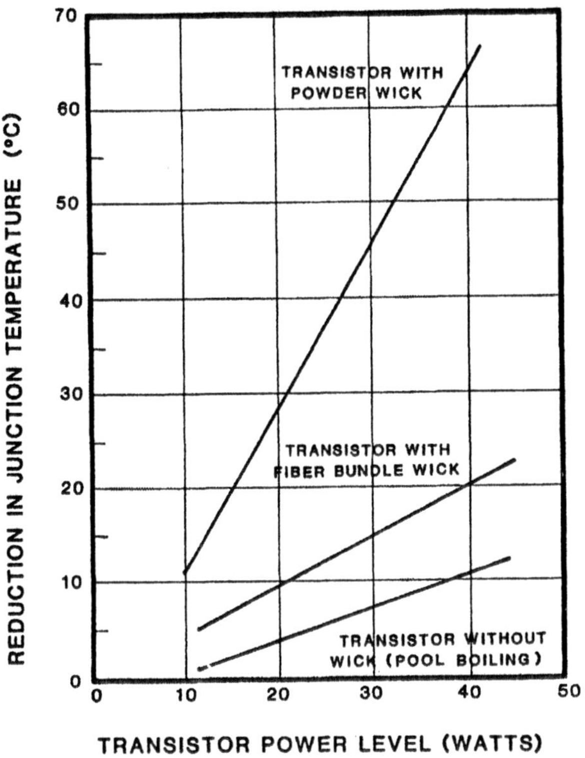

Figure 9.38 Comparison of reduction in junction temperature achieved by fiber bundle and powder wick and nonwicked transistors immersed in heat pipe fluid (Nelson et al., 1978a).

occurs when a conventionally bonded transistor and heat sink are exposed to two-phase cooling. Clearly, the powder wick heat pipe is far superior to either pool boiling or the fiber wick heat pipe, with the junction temperature reduced by 63°C at power levels of 40 W. Immersion cooling with pool boiling provided only an 11°C reduction, demonstrating the effectiveness of this cooling technique. In addition, it would be reasonable to utilize this technique to cool several semiconductor chips within a single case. Figure 9.39 illustrates a comparison of the thermal performance of this type of device with conventionally cooled semiconductor devices. As shown, the thermal performance of these types of devices is superior with junction-to-case resistances reduced by as much as 33% over a wide range of power levels.

Kromann et al. (1986) proposed a device (Fig. 9.40) similar to others mentioned in that a wicking structure promoted the flow of liquid to the heat source. Heat generated in the chip die was transferred to the liquid in the wick, where it was evaporated. Parallel channels within the cover

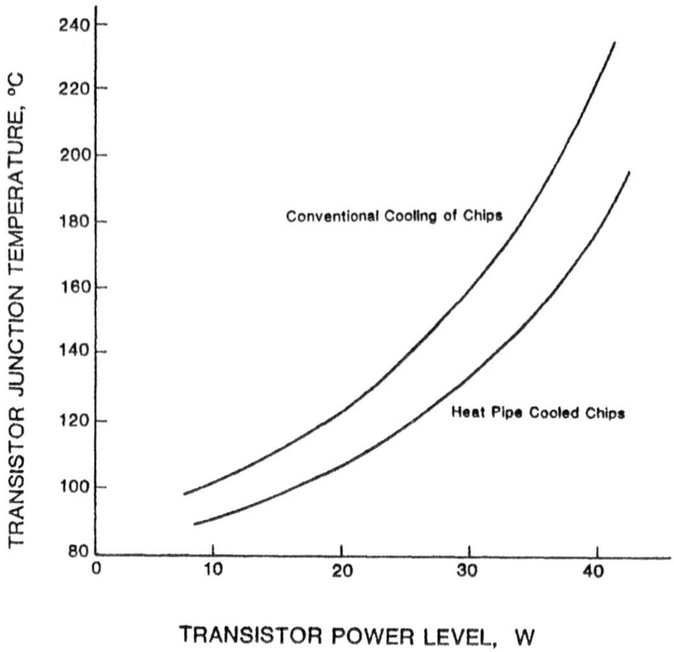

Figure 9.39 Cooling characteristics of heat pipe cooled chips and conventionally cooled chips (Nelson et al., 1978b).

provided a path for the liquid to flow from the condenser surface to the evaporator wick by gravitational forces. The wick consisted of eight layers of polyester cloth and the working fluid was pentane. Figure 9.41 compares the internal thermal resistance of this device at various die heat fluxes. As illustrated, the total package resistance is a strong function of the percentage of working fluid present. The best performance occurred when 52% of the volume was filled with working fluid. At this level, the total package resistance was approximately 0.2 W over a heat flux range of 10–27 W/cm^2.

Another method by which the temperature drop between the coolant and the heat source can be reduced is thin-film evaporation. In 1980, Andros and Shay (1980) proposed a closed miniature thermosyphon that utilized this technique. As illustrated in Fig. 9.42, heat added from a disc-shaped heat source vaporized liquid at the liquid–vapor interface. The vapor flowed across the bubble, where it condensed, rejecting heat and returning liquid to the evaporator by flow through the continuous liquid film. Wayner and Parks (1985) and Kiewra and Wayner (1986) have conducted fundamental investigations of this device and demonstrated that an evaporating thin circular liquid film in the shape of an extended meniscus combined with condensation within a small volume acts as an effective heat spreader. Experimental procedures to determine the op-

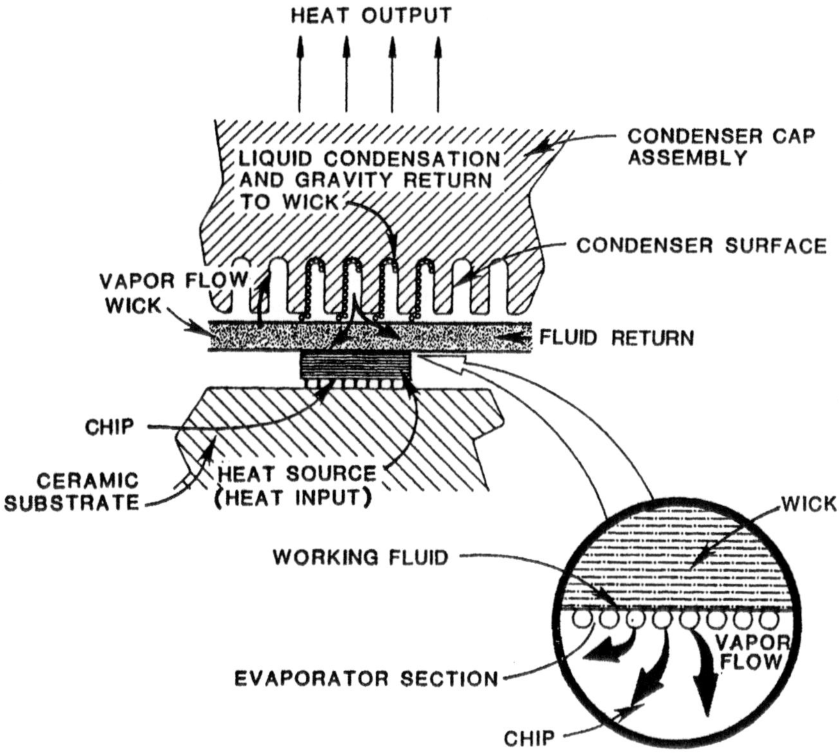

Figure 9.40 Integral heat pipe showing the fluid/vapor transport and the resulting heat transfer (Kromann et al., 1986).

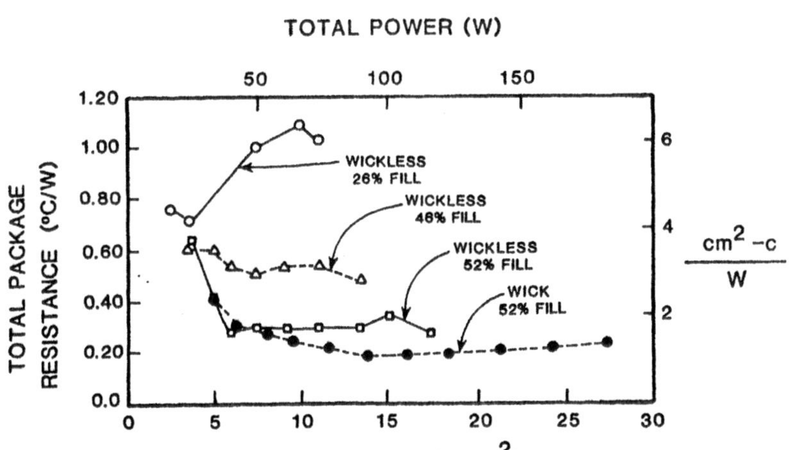

Figure 9.41 Total package thermal resistance (Kromann et al., 1986).

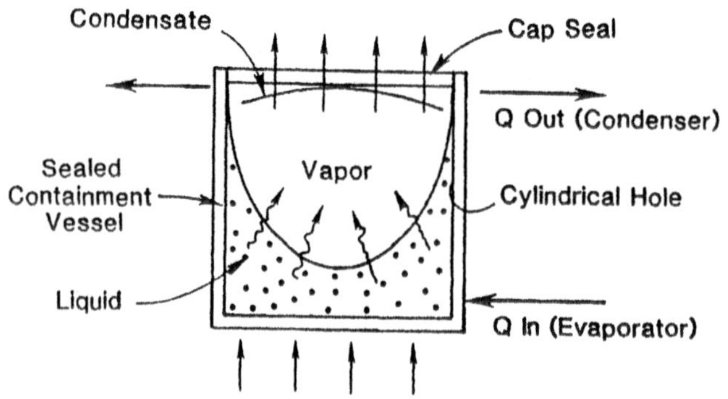

Figure 9.42 Thin film evaporation module (Andros and Shay, 1980).

timum parameters were developed to clarify the basic mechanisms of heat, mass, and momentum transfer where both body and interfacial forces were present. The effects of both NCGs and fluid flow in ultrathin films in the presence of evaporation were investigated.

9.4 SYSTEM LEVEL THERMAL CONTROL USING HEAT PIPES

The superior heat transfer characteristics of heat pipes make them ideal for the cooling of electronics in applications where the circuitry must be enclosed and isolated from dust, moisture, oil mist, or harsh environmental conditions. Since the point at which a heat pipe passes through an enclosure can be sealed easily, it is possible to obtain a closed, airtight system that prevents external contamination (Askalu and Carlen, 1984).

Johnson (1983) investigated heat pipes for the removal of heat from sealed enclosures. Shah and Giovannelli (1987) detailed design information for a heat pipe heat exchanger similar to the one illustrated in Fig. 9.43, including the important coupling of the heat pipe thermal performance to the heat exchange capability of the finned surfaces. Figure 9.44 illustrates a similar heat pipe cooling system designed to remove heat from existing cabinets. The system shown is approximately $16 \times 18 \times 38$ cm. It was designed to remove approximately 1200 W using copper–methanol heat pipes with a temperature rise of less than 20°C above ambient.

Figure 9.45*a* shows a design by Gerak et al. (1987) consisting of a heat exchanger with 8 rows of 17 finned copper heat pipes 16 mm in diameter with R12 as the working fluid. The measured power level as a function of the temperature drop is shown in Fig. 9.45*b*, demonstrating the depen-

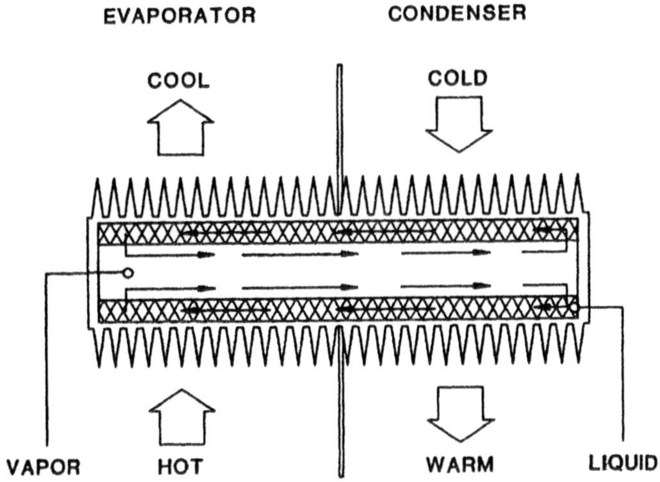

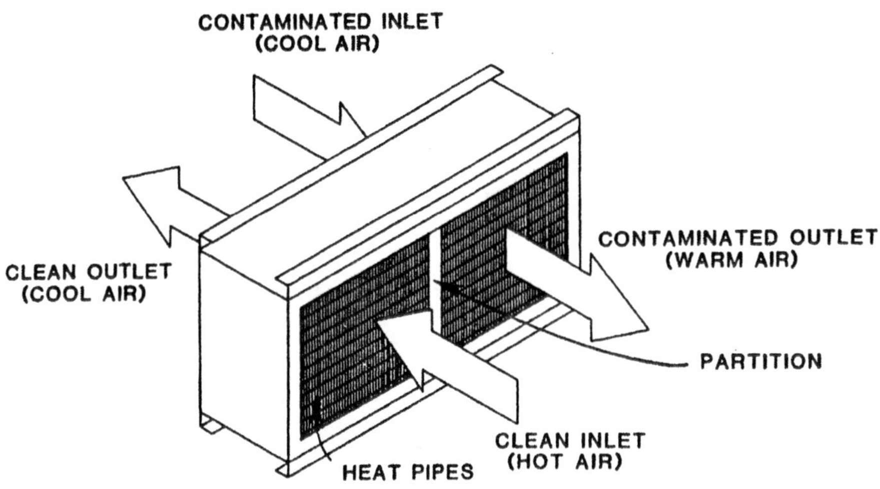

Figure 9.43 Heat pipe heat exchanger (Marto and Peterson, 1988).

dence of the heat transfer capability of the heat exchanger on the inlet air temperature difference. Jakes et al. (1988) described a heat exchanger equipped with a bank of finned thermosyphons for use in the cooling of a closed electrical cabinet. The heat pipe system reduced the heat exchanger volume by 65% without changing the velocity of the air flow. Both the

Figure 9.44 Heat pipe cabinet cooler (courtesy of Thermacore, Inc.).

calculation procedure and the methodology used to optimize the heat pipe dimensions and exchanger configuration are presented.

Figure 9.46 illustrates an application involving the use of two 6.25-mm diameter copper–water heat pipes designed for use in the thermal control of high power transistors. The mounting plate shown on the right was designed to accommodate three 25-W transistors. Heat generated at this mounting plate is conducted through an enclosed wall to the fins where it is rejected to the environment by natural convection. The overall length of the device is approximately 30.5 cm.

It is sometimes necessary to add heat to a specific system or device to maintain the desired operating temperature. Larkin successfully controlled the electronic circuitry in harsh environments with simple heat pipes (Larkin, 1984). The objective was to add heat to the electronic cabinets at remote weather stations. Figure 9.47 is a schematic of the device tested. The electronic equipment was in a cabinet on the surface and connected to a 3.65-long, 7.9-cm-diameter thermosyphon with one end buried in the

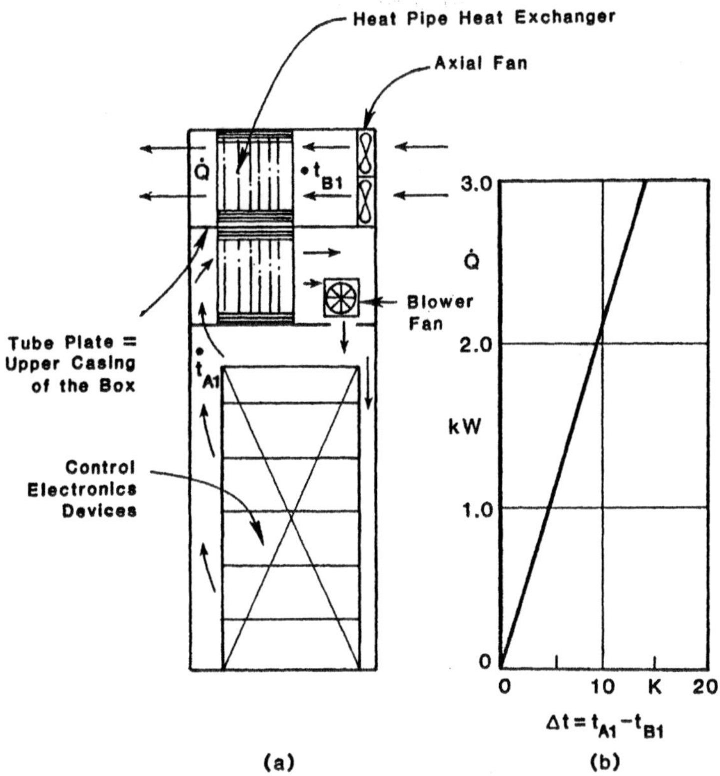

Figure 9.45 Heat pipe heat exchanger to cool electronic equipment in a sealed cabinet (Gerak et al., 1987).

ground. Power levels of approximately 30 W were maintained with 500 g of R22.

9.5 SUMMARY

Increasing the density of electronic circuits requires a more effective means of thermal control. Heat pipes, because of the vaporization and condensation process, have extremely high (effective) thermal conductivities and the capability of a nearly uniform temperature distribution in the evaporator. They are compact and lightweight and can eliminate localized hot spots at individual devices, printed wiring boards, or entire systems. Numerous applications over the past 20 years indicate that heat pipes enable higher density devices with higher reliability and lower cost and are an acceptable, proven alternative to more conventionally cooled systems.

Figure 9.46 Transistor cooling assembly using two heat pipes (courtesy of Thermacore, Inc.).

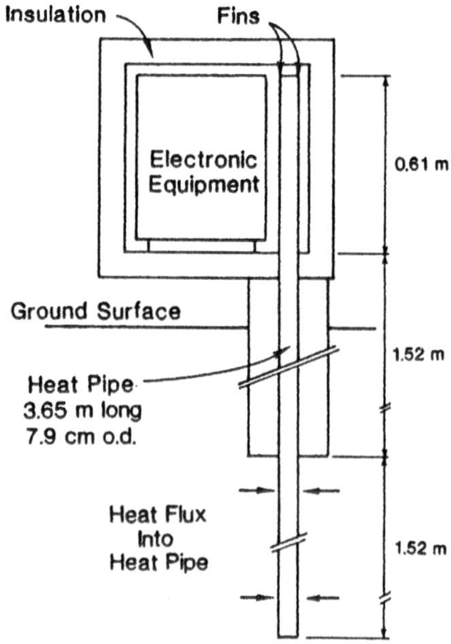

Figure 9.47 Heat pipe application in remote weather stations (Larkin, 1984).

REFERENCES

Adami, M., and Yimer, B., 1988, "Development and Evaluation of a Planar Heat Pipe for Cooling Electronic Systems," Proc. ASME Natl. Heat Trans. Conf., Houston, TX, pp. 499–507.

Adams, B. B., Kessler, S. W., and Reed, R. E., 1979, "Unique Liquid Cooled Solid State Power Device Development," Proc. 14th IAS Conf., San Francisco, CA, pp. 141–148.

Akalu, N. G., and Carlen, R. A., 1984, "Heat Pipe Links for Water-cooled Large Computer Circuit Assemblies," *IBM Technical Disclosure Bulletin*, Vol. 27, pp. 3551–3552.

Andros, F. E., and Shay, R. J., 1980, "Micro Bellows Thermo-Capsule," U.S. Patent 4,212,349.

Babin, B. R., and Peterson, G. P., 1989, "Modeling and Testing of a Flexible Bellows Heat Pipe for Cooling Isolated Heat Sources," Proc. ASME Natl. Heat Trans. Conf., Philadelphia, PA.

Babin, B. R., and Peterson, G. P., 1990, "Experimental Investigation of a Flexible Bellows Heat Pipe for Cooling Discrete Heat Sources," *ASME Journal of Heat Transfer*, Vol. 112, No. 3, pp. 602–607.

Babin, B. R., Peterson, G. P., and Wu, D., 1990, "Steady-state Modeling and Testing of a Micro Heat Pipe," *ASME Journal of Heat Transfer*, Vol. 112, No. 3, pp. 595–601.

Basiulis, A., Tanzer, H., and McCabe, S., 1987, "Heat Pipes for Cooling of High Density Printed Wiring Boards," Proc. 6th Int. Heat Pipe Conf., Grenoble, France, pp. 531–536.

Bergles, A. E., Chu, R. C., and Seely, J. H., 1977, "Survey of Heat Transfer Techniques Applied to Electronic Packages," Natl. Elect. Packg., and Prod. Conf., San Diego, CA, pp. 370–385.

Biskhorn, R. G., Horvath, J. L., and Hultmark, E. B., 1984, "Integral Cap Heat Sink Assembly for the IBM 4381 Processor," Proc. 4th Int. Electronic Packaging Soc. Conf., San Diego, CA, pp. 468–474.

Chu, R. C., Hwang, U. P., and Simons, R. E., 1982, "Conduction Cooling for an LSI Package: A One-dimensional Approach," *IBM Journal of Research and Development*, Vol. 26, pp. 45–54.

Ciekurs, P.V., 1986, "An Approach to Solving the Hot Spot Cooling Problem in Electronic Packaging," ASME Paper No. 86-WA / EEP-1.

Cotter, T. P., 1984, "Principles and Prospects of Micro Heat Pipes," Proc. 5th Int. Heat Pipe Conf., pp. 328–335. Quote on p. 328.

Danielson, R. D., Krajewski, N., and Brost, J., 1986, "Cooling a Superfast Computer," *Electronic Packaging and Production*, Vol. 14, No. 6, pp. 44–45.

Dean, D. S., 1976, "An Integral Heat Pipe Package for Microelectronic Circuits," Proc. 2nd Int. Heat Pipe Conf., Bologna, Italy, pp. 481–502.

Eldridge, J. M., and Peterson, K. E., 1983, "Heat Pipe Vapor Cooling Etched Silicon Structure," *IBM Tech. Disclos. Bull.*, Vol. 25, No. 8, pp. 4118–4119.

"Flat Heat Pipe Assemblies Keep PC Boards Cool," *Mach. Des.*, Vol. 71, No. 3 p. 272.

Gerak, A., Horvath, L., Jelinek, F., Stulc, P., and Zboril, V., 1987, "Examples of Heat Pipe Application in Chemical, Electrical and Other Industries," Proc. 6th Int. Heat Pipe Conf., Grenoble, France, pp. 522–530.

Goodling, J. S., Jaeger, R. C., Williamson, N. V., Ellis, C. D., and Slagh, T. D., 1987, "Wafer Scale Cooling Using Jet Impingement Boiling Heat Transfer," Paper No. 87-WA / EEP-3, American Society of Mechanical Engineers, New York.

Incropera, F. P., 1986, "Research Needs in Electronic Cooling," *Proc. of the Research Needs in Electronic Cooling Workshop*, sponsored by the National Science Foundation and Purdue University, Purdue University, West Lafayette, IN, p. 27.

Jakes, J., Vassicek, F., Polasek, F., and Oslejsek, O., 1988, "Cooling of Air in Cooled Electrical Machines by Heat Pipe Heat Exchangers," Proc. 3rd Int. Heat Pipe Symp., Tsukuba, Japan, pp. 290–302.

Johnson, R. D., 1983, "Cooling of Electrical and Electronic Equipment in Sealed Enclosures," *Electron Power*, Vol. 29, No. 4, pp. 332–336.

Kiewra, E. W., and Wayner, P. C., 1986, "A Small Scale Thermosyphon for the Immersion Cooling of a Disc Heat Source," *Heat Transfer in Electronic Equipment*, ASME Symposium Vol. HTD-Vol. 57, Bar-Cohen, A. (Ed.), American Society of Mechanical Engineers, New York, pp. 77–82.

Kohara, M., Nakao, S., Tsutsumi, K., Shibata, H., and Nakata, H., 1985, "High Thermal Conduction Module," Proc. 1985 IEEE Electric Components Conference, pp. 180–186.

Kromann, G. B., Hannermann, R. J., and Fox, L. R., 1986, "Two-Phase Internal Cooling Technique for Electronic Packages," *Heat Transfer in Electronic Equipment*, ASME Symposium Vol. HTD-Vol. 57, Bar-Cohen, A. (Ed.), American Society of Mechanical Engineers, New York, pp. 61–66.

Larkin, B. S., 1984, "Temperature Control of Electronic Systems Using Inexpensive Heat Pipes," *Research and Development of Heat Pipe Tech.*, JaTec Publishing, Tokyo, Japan, pp. 346–351.

Malik, P. R., Rao, K. S., and Venkateswarlu, P., 1984, "Thermal Response Studies on Heat Pipes—Thyristor Assembly and its Comparison with Conventional Cooling Systems," *Research and Development of Heat Pipe Technology*, Vol. 1, JaTec Publishing, Tokyo, Japan, pp. 352–357.

Marto, P. J., and Peterson, G. P., 1988, "Application of Heat Pipes to Electronics Cooling," *Advances in Thermal Modeling of Electronic Components and Systems*, A. Bar-Cohen and A. D. Kraus (Eds.), Hemisphere, Washington, DC, Chapter 4, pp. 283–336.

Miayazaki, Y., and Sasaki, T., 1984, "Thermal Switch for Space Applications," *Research and Development of Heat Pipe Technology.*, Vol. 1, JaTec Publishing, Tokyo, Japan, pp. 445–450.

Moran, K., and Oktay, S., 1982, "Cooling Concepts for IBM Electronic Packages," Proc. 2nd Annual Int. Electronic Packaging Soc. Conf., San Francisco, CA, pp. 120–140.

Mudawwar, I., Incropera, T. A., and Incropera, F. P., 1987, "Microelectronic Cooling by Fluorocarbon Liquid Films," Proc. Int. Symposium on Cooling Technology for Electronic Equipment.

Murase, T., Yoshida, K., Fujikake, J., Koizumi, T., and Ishida, N., 1982, "Heat Pipe Heat Sink "HEAT KICKER" for Cooling of Semi-Conductors," *Furukawa Review*, Tokyo, Japan, Vol. 2, pp. 24–33.

Murase, T., Endo, S., and Koizumi, T., 1986, "Heat Pipe Cooling System "POWER-KICKER" for Power Semiconductor Devices," *Furukawa Review*, Tokyo, Japan, Vol. 4, pp. 38–46.

Murase, T., Tanaka, S., and Ishida, S., 1987, "Natural Convection Type Long Heat Pipe Heat Sink "POWERKICKER-N" for the Cooling of GTO Thyristor," Proc. 6th Int. Heat Pipe Conf., Grenoble, France, pp. 537–542.

Nelson, L. A., Sekhon, K. S., and Fritz, J. E., 1978a, "Direct Heat Pipe Cooling of Semiconductor Devices," Proc. 3rd Int. Heat Pipe Conf., Palo Alto, CA, pp. 373–376.

Nelson, L. A., Sekhon, K. S., and Ruttner, L. E., 1978b, "Application of Heat Pipes in Electronic Modules," Proc. 3rd Int. Heat Pipe Conf., Palo Alto, CA, pp. 367–372.

Ogushi, T., Murakami, M., Masumoto, H., and Yoshida, H., 1988, "Study on Newly Developed Heat Pipe Heat Sink," Proc. ASME Natl. Heat Trans. Conf., Houston, TX, pp. 517–522.

Peterson, G. P., 1986, "Analytical Development and Computer Modeling of a Bellows Type Heat Pipe for the Cooling of Electronic Components," Report No. 86-WA / HT-69, American Society of Mechanical Engineers, New York.

Peterson, G. P., 1987, "Analysis of a Heat Pipe Thermal Switch," Proc. 6th Int. Heat Pipe Conf., Vol. 1, Grenoble, France, pp. 177–183.

Peterson, G. P., 1988a, "Heat Pipes in the Thermal Control of Electronic Components," Proc. 3rd Int. Heat Pipe Symp., Tsukuba, Japan, pp. 2–12.

Peterson, G. P., 1988b, "Investigation of Miniature Heat Pipes," Final Report, Wright-Patterson Air Force Base, Contract No. F33615-86-C-2723, Task 9.

Peterson, G. P., 1992, "An Overview of Micro Heat Pipe Research," invited review article, *Applied Mechanics Review*, Vol. 45, No. 5, pp. 175–189.

Ptacnik, J., and Polasek, F., 1988, "Present State of Heat Pipe Technology in the Countries of Mutual Economic Assistance," Proc. 3rd Int. Heat Pipe Symp., Tsukuba, Japan, pp. 13–36.

Ruttner, L. E., 1978, "Application of Heat Pipes in the Navy's Standard Electronic Module," Proc. of the Nat'l Elect. Packg. Conf., Andover, MA, pp. 162–166.

Sekhon, K. S., Nelson, L. A., and Fritz, J. E., 1977, "Transistor Cooling by Heat Pipes Having a Wick of Dielectric Powder," U. S. Patent 4,047,198.

Shah, R. K., and Giovannelli, A. D., 1987, *Heat Pipe Heat Exchanger Design Theory, Heat Transfer Equipment Design*, Hemisphere, Washington, DC.

Token, K., 1986, "Trends in Aircraft Thermal Management," Proc. Printed Wiring Board Heat Pipe Workshop, Hughes Aircraft Company, Electron Dynamics Division, Torrance, CA.

Watari, T., and Murano, H., 1985, "Packaging Technology for the NEC SX Supercomputer," Proc. 1985 IEEE Electronic Components Conference, San Diego, CA, pp. 192–198.

Wayner, P. C., and Parks, C. J., 1985, "Effect of Liquid Composition on Enhanced Flow Due to Surface Shear in the Contact Line Region: Constant Vapor Pressure Boundary Condition," *Heat Transfer in Electronic Equipment*, V. D. Kirk (Ed.), ASME HTD-47, American Society of Mechanical Engineers, New York, pp. 57–63.

Wolf, D. A., 1981, "Flexible Heat Pipe Switch," Final Report, Contract No. NAS5-25568, Mod. 4, National Aeronautics and Space Administration, Washington, DC.

Yoshida, K., Ogiwara, S., Murase, T., and Ishida, S., 1984, "Flat Plate Heat Pipes for Cooling Devices," *Research and Development of Heat Pipe Technology*, Vol. 1, JaTec Publishing, Tokyo, Japan, pp. 174–178.

Zorbil, V., and Stulc, P., 1984, "Heat Pipes in Computer Modules," *Research and Development of Heat Pipe Technology*, Vol. 1, JaTec Publishing, Tokyo, Japan, pp. 336–345.

Zorbil, V., Stulc, P., and Polasek, F., 1988, "Enhancement Cooling of the Boards with Integrated Circuits by Heat Pipes," Proc. 3rd Int. Heat Pipe Symp., Tsukuba, Japan, pp. 273–279.

Appendix

A

DIMENSIONAL EQUIVALENTS AND PHYSICAL CONSTANTS

Parameter	SI units	English Equivalent
Acceleration	$1\ m/s^2$	$4.2520 \times 10^7\ ft/h^2$
Area	$1\ m^2$	$1550.0\ in.^2$
		$10.764\ ft^2$
Energy	1 J	9.4787×10^{-4} Btu
Force	1 N	$0.2481\ lb_f$
Heat transfer rate	1 W	3.4123 Btu / h
Heat flux	$1\ W/m^2$	$0.3171\ Btu/h \cdot ft^2$
Heat generation rate	$1\ W/m^3$	$0.09665\ Btu/h \cdot ft^3$
Heat transfer coefficient	$1\ W/m^2 \cdot K$	$0.17612\ Btu/h \cdot ft^2 \cdot °F$
Kinematic viscosity	$1\ m^2/s$	$3.875 \times 10^4\ ft^2/h$
Latent heat	1 J / kg	$4.2995 \times 10^{-4}\ Btu/lb_m$
Length	1 m	39.370 in.
		3.2808 ft
Mass	1 kg	$2.2046\ lb_m$
Mass density	$1\ kg/m^3$	$0.062428\ lb_m/ft^3$
Mass flow rate	1 kg / s	$7936.6\ lb_m/h$
Mass transfer coefficient	1 m / s	1.1811×10^4 ft / h
Pressure and stress[a]	$1\ N/m^2$	$0.020886\ lb_f/ft^2$
		$1.4504 \times 10^{-4}\ lb_f/in.^2$
		4.015×10^{-3} in. water
		2.953×10^{-4} in. Hg
	$1.0133 \times 10^5\ N/m^2$	1 standard atmosphere
	$1 \times 10^5\ N/m^2$	1 bar
Specific heat	$1\ J/kg \cdot K$	$2.3886 \times 10^{-4}\ Btu/lb_m \cdot °F$
Temperature	K	$(\frac{5}{9})°R$
		$(\frac{5}{9})(°F + 495.67)$
		°C + 273.15
Temperature difference	1 K	1°C
		$(\frac{9}{5})°R = (\frac{9}{5})°F$
Thermal conductivity	$1\ W/m \cdot K$	$0.57782\ Btu/h \cdot ft \cdot °F$
Thermal resistance	1 K / W	$0.52750\ °F/h \cdot Btu$
Viscosity (dynamic)[b]	$1\ N \cdot s/m^2$	$2419.1\ lb_m/ft \cdot h$
		$5.8016 \times 10^{-6}\ lb_f \cdot h/ft^2$
Volume	$1\ m^3$	$6.1023 \times 10^4\ in.^3$
		$35.314\ ft^3$
		264.17 gal
Volume flow rate	$1\ m^3/s$	$1.2713 \times 10^5\ ft^3/h$
		$2.1189 \times 10^3\ ft^3/min$
		1.5850×10^4 gal / min

[a]The SI name for the quantity pressure in pascals (PA) having units N/m^2 or $kg/m \cdot s^2$.
[b]Also expressed in equivalent units of $kg/s \cdot m$.

Appendix

B

PROPERTIES OF SOLID MATERIALS

Composition	Melting Point (K)	Properties at 300 K ρ (kg/m^3)	c_p $(J/kg \cdot K)$	k $(W/m \cdot K)$	$\alpha, \times 10^6$ (m^2/s)	Properties at Various Temperatures $k(W/m \cdot K) c_p (J/kg \cdot K)$ 100 K	200 K	400 K	600 K	800 K	1000 K	1200 K	1500 K	200 K	2500 K
Aluminum															
Pure	933	2702	903	237	97.1	302	237	240	231	218					
Alloy 2024-T6 (4.5% Cu, 1.5% Mg, 0.6% Mn)	755	2770	875	177	73.0	65	163	186	186						
						473	787	925	1042						
Alloy 195, cast (4.5% Cu)		2790	883	168	68.2			174	185						
Beryllium	1550	1850	1825	200	59.2	990	301	161	126	106	90.8	78.7			
						203	1114	2191	2604	2823	3018	3227	3519		
Cadmium	594	8650	231	96.8	48.4	203	99.3	94.7							
						198	222	242							
Chromium	2118	7160	449	93.7	29.1	159	111	90.9	80.7	71.3	65.4	61.9	57.2	49.4	
						192	384	484	542	581	616	682	779	937	
Copper															
Pure	1358	8933	385	401	117	482	413	393	379	366	352	339			
						252	356	397	417	433	451	480			
Commercial bronze (90* Cu, 10% Al)	1293	8800	420	52	14		42	52	59						
							785	460	545						
Phosphor gear bronze (88% Cu, 11% Sn)	1104	8780	355	54	17	41	65	74							
						—	—	—							
Cartridge brass (70% Cu, 30% Zn)	1188	8530	380	110	33.9	75	95	137	149						
							360	395	425						
Constantan (55% Cu, 45 Ni)	1439	8920	384	23	6.71	17	19								
						237	362								
Germanium	1211	5360	322	59.9	34.7	232	96.8	43.2	27.3	19.8	17.4	17.4			
						190	290	337	348	357	375	395			
Gold	1336	19300	129	317	127	327	323	311	298	284	270	255			
						109	124	131	135	140	145	155			
Iron															
Pure	1810	7870	447	80.2	23.1	134	94.0	69.5	54.7	43.3	32.8	28.3	32.1		
						216	384	490	574	680	975	609	654		
Armco (99.75% pure)		7870	447	72.7	20.7	95.6	80.6	65.7	53.1	42.2	32.3	28.7	31.4		
						215	384	490	574	680	975	609	654		

Carbon Steels															
Plain carbon (Mn ≤ 1%, Si ≤ 0.1%)		7832	434	60.5	17.7			56.7	48.0	39.2	30.0				
								487	559	685	1169				
AISI 1010		7832	434	63.9	18.8			58.7	48.8	39.2	31.3				
								487	559	685	1168				
Carbon–silicon (Mn ≤ 1%, 0.1% < Si ≤ 0.6%)		7817	446	51.9	14.9			49.8	44.0	37.4	29.3				
								501	582	699	971				
Carbon–manganese–silicon (1% < Mn ≤ 1.65%,		8131	434	41.0	11.6			42.2	39.7	35.0	27.6				
0.1% < Si ≤ 0.6%)															
								487	559	685	1090				
Chromium (low) steels															
$\frac{1}{2}$Cr–$\frac{1}{4}$Mo–Si (0.18% C, 0.65% Cr, 0.23% Mo, 0.6% Si)		7822	444	37.7	10.9			38.2	36.7	33.3	26.9				
								492	575	688	969				
1 Cr–$\frac{1}{2}$Mo (0.16% C, 1% Cr, 0.54% Mo, 0.6% Si)		7858	442	42.3	12.2			42.0	39.1	34.5	27.4				
								492	575	688	969				
1 Cr–V (0.2% C, 1.02% Cr, 0.15% V)		7836	443	48.9	14.1			46.8	42.1	36.3	28.2				
								492	575	688	969				
Stainless steels		8055	480	15.1	3.91			17.3	20.0	22.8	25.4				
AISI 302								512	559	585	606				
AISI 304	1670	7900	477	14.9	3.95	9.2	12.6	16.6	19.8	22.6	25.4	28.0	31.7		
						272	402	515	557	582	611	640	682		
AISI 316		8238	468	13.4	3.48			15.2	18.3	21.3	24.2				
								504	550	576	602				
AISI 347		7978	480	14.2	3.71			15.8	18.9	21.9	24.7				
								513	559	585	606				
Lead	601	11340	129	35.3	24.1	39.7	36.7	34.0	31.4						
						118	125	132	142						
Molybdenum	2894	10240	251	138	53.7	179	143	134	126	118	112	105	98	90	86
						141	224	261	275	285	295	308	330	380	459
Nickel															
Pure	1728	8900	444	90.7	23.0	164	107	80.2	65.6	67.6	71.8	76.2	82.6		
						232	383	485	592	530	562	594	616		
Nichrome (80% Ni, 20% Cr)	1672	8400	420	12	3.4			14	16	21					
								480	525	545					
Inconel X-750 (73% Ni, 15% Cr, 6.7% Fe)	1665	8510	439	11.7	3.1	8.7	10.3	13.5	17.0	20.5	24.0	27.6	33.0		
						—	372	473	510	546	626	—	—		
Niobium	2741	8570	265	53.7	23.6	55.2	52.6	55.2	58.2	61.3	64.4	67.5	72.1	79.1	
						188	249	274	283	292	301	310	324	347	

Composition	Melting Point (K)	Properties at 300 K ρ (kg/m^3)	c_p (J/kg·K)	k (W/m·K)	α, ×10^6 (m^2/s)	Properties at Various Temperatures k(W/m·K)c_p(J/kg·K) 100 K	200 K	400 K	600 K	800 K	1000 K	1200 K	1500 K	200 K	2500 K
Alloy 60Pt–40 Rh (60% Pt, 40% Rh)	1800	16630	162	47	17.4			52	59	65	69	73	76		
								—	—	—	—	—	—		
Silicon	1685	2330	712	148	89.2	884	264	98.9	61.9	42.2	31.2	25.7	22.7		
						259	556	790	867	913	946	967	992		
Silver	1235	10500	235	429	174	444	430	425	412	396	379	361			
						187	225	239	250	262	277	292			
Tin	505	7310	227	66.6	40.1	85.2	73.3	62.2							
						188	215	243							
Titanium	1953	4500	522	21.9	9.32	30.5	24.5	20.4	19.4	19.7	20.7	22.0	24.5		
						300	465	551	591	633	675	620	686		
Tungsten	3660	19300	132	174	68.3	208	186	159	137	125	118	113	107	100	95
						87	122	137	142	145	148	152	157	167	176

Appendix

C

PROPERTIES OF WORKING FLUIDS*

* Reprinted from P. Dunn and D.A. Reay, Heat Pipes, Copyright 1976, Pgs. 272–281, with kind permission from Elsevier Science Ltd, The Boulevard, Langford Lane, Kidlington OX5 1GB, UK.

Temperature (°C)	Latent Heat (kJ / kg)	Liquid Density (kg / m^3)	Vapor Density (kg / m^3)	Liquid Thermal Conductivity (W / m°C)	Liquid Viscosity (cP)	Vapor Viscosity (cP, ×10^2)	Vapor Pressure (bars)	Vapor Specific Heat (kJ / kg°C)	Liquid Surface Tension (N / m × 10^2)
					Freon-11				
−60	211.9	1672	0.04	0.121	1.19	0.86	0.02	0.476	2.95
−40	204.0	1622	0.04	0.115	0.98	0.88	0.05	0.497	2.70
−20	19608	1578	1.04	0.111	0.70	0.95	0.16	0.516	2.40
0	190.0	1533	2.59	0.108	0.55	1.01	0.42	0.532	2.18
20	183.4	1487	5.38	0.100	0.44	1.08	0.93	0.546	1.92
40	175.6	1439	10.07	0.097	0.37	1.14	1.82	0.561	1.66
60	167.5	1389	16.85	0.094	0.32	1.20	3.14	0.576	1.40
80	159.0	1334	30.56	0.089	0.28	1.25	5.85	0.590	1.14
100	146.9	1265	49.04	0.076	0.25	1.31	9.53	0.607	0.90
120	134.4	1192	67.53	0.064	0.23	1.37	13.21	0.623	0.63
140	117.0	1105	110.66	0.055	0.22	1.49	18.92	0.646	0.37
					Pentane				
−20	390.0	663.0	0.01	0.149	0.344	0.51	0.10	0.825	2.01
0	37803	644.0	0.75	0.143	0.283	0.53	0.24	0.874	1.79
20	366.9	625.5	2.20	0.138	0.242	0.58	0.76	0.922	1.58
40	355.5	607.0	4.35	0.133	0.200	0.63	1.52	0.971	1.37
60	342.3	585.0	6.51	0.128	0.174	0.69	2.28	1.021	1.17
80	329.1	563.0	10.61	0.127	0.147	0.74	3.89	1.050	0.97
100	295.7	537.6	16.54	0.124	0.128	0.81	7.19	1.088	0.83
120	269.7	509.4	25.20	0.122	0.120	0.90	13.81	1.164	0.68

				Acetone					
−40	660.0	860.0	0.03	0.200	0.800	0.68	0.01	2.00	3.10
−20	615.6	845.0	0.10	0.189	0.500	0.73	0.03	2.06	2.76
0	564.0	812.0	0.26	0.183	0.395	0.78	0.10	2.11	2.62
20	552.0	790.0	0.64	0.181	0.323	0.82	0.27	2.16	2.37
40	536.0	768.0	1.05	0.175	0.269	0.86	0.60	2.22	2.12
60	517.0	744.0	2.37	0.168	0.226	0.90	1.15	2.28	1.86
80	495.0	719.0	4.30	0.160	0.192	0.95	2.15	2.34	1.62
100	472.0	689.6	6.94	0.148	0.170	0.98	4.43	2.39	1.34
120	426.1	660.3	11.02	0.135	0.148	0.09	6.70	2.45	1.07
140	394.4	631.8	18.61	0.126	0.132	1.03	10.49	2.50	0.81
				Freon-113					
−50	173.0	1720	0.15	0.120	2.300	0.85	0.01	0.600	2.86
−30	167.8	1683	0.32	0.119	1.604	0.90	0.03	0.613	2.60
−20	165.4	1664	0.46	0.118	1.323	0.92	0.05	0.619	2.47
−10	163.2	1643	0.77	0.118	1.108	0.94	0.09	0.626	2.34
0	160.6	1621	1.26	0.117	0.942	0.97	0.12	0.632	2.21
10	158.0	1599	1.95	0.108	0.812	0.99	0.19	0.644	2.08
20	155.2	1576	3.00	0.098	0.707	1.02	0.37	0.656	1.96
30	152.3	1553	4.34	0.097	0.622	1.04	0.55	0.664	1.84
40	149.2	1529	6.02	0.095	0.553	1.07	0.79	0.669	1.73
50	145.9	1503	8.79	0.094	0.502	1.09	1.11	0.674	1.62
70	139.4	1452	14.34	0.091	0.401	1.13	2.04	0.691	1.40

Temperature (°C)	Latent Heat (kJ / kg)	Liquid Density (kg / m^3)	Vapor Density (kg / m^3)	Liquid Thermal Conductivity (W / m°C)	Liquid Viscosity (cP)	Vapor Viscosity (cP, ×10^2)	Vapor Pressure (bars)	Vapor Specific Heat (kJ / kg°C)	Liquid Surface Tension (N / m × 10^2)
					Heptane				
−20	384.0	715.5	0.01	0.143	0.69	0.57	0.01	0.83	2.42
0	372.6	699.0	0.17	0.141	0.53	0.60	0.02	0.87	2.21
20	362.2	683.0	0.49	0.140	0.43	0.63	0.08	0.92	2.01
40	351.8	667.0	0.97	0.139	0.34	0.66	0.20	0.97	1.81
60	341.5	649.0	1.45	0.137	0.29	0.70	0.32	1.02	0.62
80	331.2	631.0	2.31	0.135	0.24	0.74	0.62	1.05	1.43
100	319.6	612.0	3.71	0.133	0.21	0.77	1.10	1.09	1.28
120	305.0	592.0	6.08	0.132	0.18	0.82	1.85	1.16	1.10
					Ethanol				
−30	939.4	825.0	0.02	0.177	3.40	0.75	0.01	1.25	2.76
−10	928.7	813.0	0.03	0.173	2.20	0.80	0.02	1.31	2.66
10	904.8	798.0	0.05	0.170	1.50	0.85	0.03	1.37	2.57
30	888.6	781.0	0.38	0.168	1.02	0.91	0.10	1.44	2.44
50	872.3	762.2	0.72	0.166	0.72	0.97	0.29	1.51	2.31
70	858.3	743.1	1.32	0.165	0.51	1.02	0.76	1.58	2.17
90	832.1	725.3	2.59	0.163	0.37	1.07	1.43	1.65	2.04
110	786.6	704.1	5.17	0.160	0.28	1.13	2.66	1.72	1.89
130	734.4	678.7	9.25	0.159	0.21	1.18	4.30	1.78	1.75

Methanol

−50	1194	843.5	0.01	0.210	1.700	0.72	0.01	1.20	3.26
−30	1187	833.5	0.01	0.208	1.300	0.78	0.02	1.27	2.95
−10	1182	818.7	0.04	0.206	0.945	0.85	0.04	1.34	2.63
10	1175	800.5	0.12	0.204	0.701	0.91	0.10	1.40	2.36
30	1155	782.0	0.31	0.203	0.521	0.98	0.25	1.47	2.18
50	1125	764.1	0.77	0.202	0.399	1.04	0.55	1.54	2.01
70	1085	746.2	1.47	0.201	0.314	1.11	1.31	1.61	1.85
90	1035	724.4	3.01	0.199	0.259	1.19	2.69	1.79	1.66
110	980	703.6	5.64	0.197	0.211	1.26	4.98	1.92	1.46
130	920	685.2	9.81	0.195	0.166	1.31	7.86	1.92	1.25
150	850	653.2	15.90	0.193	0.138	1.38	8.94	1.92	1.04

Flutec PP2

−30	106.2	1942	0.13	0.637	5.200	0.98	0.01	0.72	1.90
−10	103.1	1886	0.44	0.626	3.500	1.03	0.02	0.81	1.71
10	99.8	1829	1.39	0.613	2.140	1.07	0.09	0.92	1.52
30	96.3	1773	2.96	0.601	1.435	1.12	0.22	1.01	1.32
50	91.8	1716	6.43	0.588	1.005	1.17	0.39	1.07	1.13
70	87.0	1660	11.79	0.575	0.720	1.22	0.62	1.11	0.93
90	82.1	1599	21.99	0.563	0.543	1.26	1.43	1.17	0.73
110	76.5	1558	34.92	0.550	0.429	1.31	2.82	1.25	0.52
130	70.3	1515	57.21	0.537	0.314	1.36	4.83	1.33	0.32
160	59.1	1440	103.63	0.518	0.167	1.43	8.76	1.45	0.01

Temperature (°C)	Latent Heat (kJ / kg)	Liquid Density (kg / m³)	Vapor Density (kg / m³)	Liquid Thermal Conductivity (W / m°C)	Liquid Viscosity (cP)	Vapor Viscosity (cP, ×10²)	Vapor Pressure (bars)	Vapor Specific Heat (kJ / kg°C)	Liquid Surface Tension (N / m × 10²)
				Flutec PP9					
−30	103.0	2098	0.01	0.060	5.77	0.82	0.00	0.80	2.36
0	98.4	2029	0.01	0.059	3.31	0.90	0.00	0.87	2.08
30	94.5	1960	0.12	0.057	1.48	1.06	0.01	0.94	1.80
60	90.2	1891	0.61	0.056	0.94	1.18	0.03	1.02	1.52
90	86.1	1822	1.93	0.054	0.65	1.21	0.12	1.09	1.24
120	83.0	1753	4.52	0.053	0.49	1.23	0.28	1.15	0.95
150	77.4	1685	11.81	0.052	0.38	1.26	0.61	1.23	0.67
180	70.8	1604	25.13	0.051	0.30	1.33	1.58	1.30	0.40
225	59.4	1455	63.27	0.049	0.21	1.44	4.21	1.41	0.01
				Water					
20	2448	998.0	0.02	0.603	1.00	0.96	0.02	1.81	7.28
40	2402	992.1	0.05	0.630	0.65	1.04	0.07	1.89	7.00
60	2359	983.3	0.13	0.649	0.47	1.12	0.20	1.91	6.66
80	2309	972.0	0.29	0.668	0.36	1.19	0.47	1.95	6.26
100	2258	958.0	0.60	0.680	0.28	1.27	1.01	2.01	5.89
120	2200	945.0	1.12	0.682	0.23	1.34	2.02	2.09	5.50
140	2139	928.0	1.99	0.683	0.20	1.41	3.90	2.21	5.06
160	2074	909.0	3.27	0.679	0.17	1.49	6.44	2.38	4.66
180	2003	888.0	5.16	0.669	0.15	1.57	10.04	2.62	4.29
200	1967	865.0	7.87	0.659	0.14	1.65	16.19	2.91	3.89

Thermex (Diphenyl-Diphenyloxide Eutectic)									
100	354.0	992.0	0.03	0.131	0.97	0.67	0.01	1.34	3.50
150	338.0	951.0	0.22	0.125	0.57	0.78	0.05	1.51	3.00
200	321.0	905.0	0.94	0.119	0.39	0.89	0.25	1.67	2.50
250	301.0	858.0	3.60	0.113	0.27	1.00	0.88	1.81	2.00
300	278.0	809.0	8.74	0.106	0.20	1.12	2.43	1.95	1.50
350	251.0	755.0	19.37	0.099	0.15	1.23	5.55	2.03	1.00
400	219.0	691.0	41.89	0.093	0.12	1.34	10.90	2.11	0.50
450	185.0	625.0	81.00	0.086	0.10	1.45	19.00	2.19	0.03
Mercury									
150	308.8	13230	0.01	9.99	1.09	0.39	0.01	1.04	4.45
250	303.8	12995	0.60	11.23	0.96	0.48	0.18	1.04	4.15
300	301.8	12880	1.73	11.73	0.93	0.53	0.44	1.04	4.00
350	298.9	12763	4.45	12.18	0.89	0.61	1.16	1.04	3.82
400	296.3	12656	8.75	12.58	0.86	0.66	2.42	1.04	3.74
450	293.8	12508	16.80	12.96	0.83	0.70	4.92	1.04	3.61
500	291.3	12308	28.60	13.31	0.80	0.75	8.86	1.04	3.41
550	288.8	12154	44.92	13.62	0.79	0.81	15.03	1.04	3.25
600	286.3	12054	65.75	13.87	0.78	0.87	23.77	1.04	3.15
650	283.5	11962	94.39	14.15	0.78	0.95	34.95	1.04	3.03
750	277.0	11800	170.00	14.80	0.77	1.10	63.00	1.04	2.75

Temperature (°C)	Latent Heat (kJ / kg)	Liquid Density (kg / m^3)	Vapor Density (kg / m^3)	Liquid Thermal Conductivity (W / m°C)	Liquid Viscosity (cP)	Vapor Viscosity (cP, $\times 10^2$)	Vapor Pressure (bars)	Vapor Specific Heat (kJ / kg°C)	Liquid Surface Tension (N / m $\times 10^2$)
					Lithium				
1030	20500	450	0.005	67	0.24	1.67	0.07	0.532	2.90
1130	20100	400	0.013	69	0.24	1.74	0.17	0.532	2.85
1230	20000	430	0.028	70	0.23	1.83	0.45	0.532	2.75
1330	19700	420	0.057	69	0.23	1.91	0.96	0.532	2.60
1430	19200	410	0.108	68	0.23	2.00	1.85	0.532	2.40
1530	18900	405	0.193	65	0.23	2.10	3.30	0.532	2.25
1630	18500	400	0.340	62	0.23	2.17	5.30	0.532	2.10
1730	18200	398	0.490	59	0.23	2.26	8.90	0.532	2.05
					Cesium				
375	530.4	1740	0.01	20.76	0.25	2.20	0.02	1.56	5.81
425	520.4	1730	0.01	20.51	0.23	2.30	0.04	1.56	5.61
475	515.2	1720	0.02	20.02	0.22	2.40	0.09	1.56	5.36
525	510.2	1710	0.03	19.52	0.20	2.50	0.16	1.56	5.11
575	502.8	1700	0.07	18.83	0.19	2.55	0.36	1.56	4.81
625	495.3	1690	0.10	18.13	0.18	2.60	0.57	1.56	4.51
675	490.2	1680	0.18	17.48	0.17	2.67	1.04	1.56	4.21
725	485.2	1670	0.26	16.83	0.17	2.75	1.52	1.56	3.91
775	477.8	1655	0.40	16.18	0.16	2.28	2.46	1.56	3.66
825	470.3	1640	0.55	15.53	0.16	2.90	3.41	1.56	3.41

				Sodium					
500	5370	828.1	0.003	70.08	0.24	0.18	0.01	9.04	1.51
600	4243	805.4	0.013	64.62	0.21	0.19	0.04	9.04	1.42
700	4090	763.5	0.050	60.81	0.19	0.20	0.15	9.04	1.33
800	3977	757.3	0.134	57.81	0.18	0.22	0.47	9.04	1.23
900	3913	745.4	0.306	53.35	0.17	0.23	1.25	9.04	1.13
1000	3827	725.4	0.667	49.08	0.16	0.24	2.81	9.04	1.04
1100	3690	690.8	1.306	45.08	0.16	0.25	5.49	9.04	0.95
1200	3577	669.0	2.303	41.08	0.15	0.26	9.59	9.04	0.86
1300	3477	654.0	3.622	37.08	0.15	0.27	15.91	9.04	0.77
				Potassium					
350	2093	763.1	0.002	51.08	0.21	0.15	0.01	5.32	9.50
400	2078	748.1	0.006	49.08	0.19	0.16	0.01	5.32	9.04
450	2060	735.4	0.015	47.08	0.18	0.16	0.02	5.32	8.69
500	2040	725.4	0.031	45.08	0.17	0.17	0.05	5.32	8.44
550	2020	715.4	0.062	43.31	0.15	0.17	0.10	5.32	8.16
600	2000	705.4	0.111	41.81	0.14	0.18	0.19	5.32	7.86
650	1980	695.4	0.193	40.08	0.13	0.19	0.35	5.32	7.51
700	1960	685.4	0.314	38.08	0.12	0.19	0.61	5.32	7.12
750	1938	675.4	0.486	36.31	0.12	0.20	0.99	5.32	6.72
800	1913	665.4	0.716	34.81	0.11	0.20	1.55	5.32	6.32
850	1883	653.1	1.054	33.31	0.10	0.21	2.34	5.32	5.92

T (°C)	p (bars)	L (kJ/kg)	σ (10^{-3} N/m)	ρ_v (kg/m^3)	ρ_l (kg/m^3)	η_v (10^{-5} N·s/m^2)	η_l (10^{-5} N·s/m^2)	λ_l (W/m·K)
Diphenylmixture (Dowtherm) (T_b = 258°C, T_m = 12°C)								
100	0.006	345	31.6	0.035	995	0.68	101.0	0.126
150	0.051	329	26.5	0.24	953	0.77	60.3	0.119
200	0.245	314	21.8	0.99	912	0.87	40.7	0.110
250	0.843	291	17.3	3.20	871	0.97	29.7	0.104
300	2.330	264	12.9	8.70	825	1.07	22.7	0.096
350	5.200	235	8.9	20.0	772	1.17	18.2	0.090
400	10.43	207	5.0	42.0	709	1.26	14.9	0.083

T (K)	p (bars)	L (kJ/kg)	σ (10^{-3} N/m)	ρ_v (kg/m^3)	ρ_l (kg/m^3)	η_v (10^{-5} N·s/m^2)	η_l (10^{-5} N·s/m^2)	λ_l (W/m·K)
Rubidium (T_b = 685°C, T_m = 39.66°C)								
500	0.00015	898	81.6	0.32	1385.7	1.5	3.220	29.69
600	0.0033	877	75.7	5.94	1339.7	1.72	2.580	27.72
700	0.0294	859	69.8	45.5	1293.7	1.932	2.170	25.86
800	0.1513	837	63.9	208.4	1247.8	2.167	1.894	24.01
900	0.5378	816	58.0	671.1	1201.8	2.396	1.687	22.15
1000	1.4700	793	51.3	1684.0	1155.8	1.618	1.528	20.30
1100	3.3200	771	44.5	3500.0	1109.8	2.833	1.401	18.44
1200	6.4300	749	37.7	6289.3	1063.8	3.053	1.298	16.50
1300	10.6300	727	30.9	9861.9	1017.8	3.260	1.213	14.70
1400	17.1600	706	26.0	14927.0	971.9	3.457	1.140	13.00
1500	25.1400	683	19.0	20487.0	925.9	3.662	1.078	11.00

T (°C)	p (bars)	L (kJ/kg)	σ (10^{-3} N/m)	ρ_v (kg/m^3)	ρ_l (kg/m^3)	η_v (10^{-5} N·s/m^2)	η_l (10^{-5} N·s/m^2)	λ_l (W/m·K)
				Freon-21 (T_b = 8.90°C, T_m = −135°C)				
−60	0.0253	269	29.81	0.147	1.554	89	0.849	0.132
−40	0.0954	262	26.99	0.510	1.510	95	0.597	0.123
−20	0.2847	253	24.17	1.410	1.470	100	0.444	0.116
0	0.7085	243	21.35	3.310	1.420	106	0.345	0.109
20	1.5300	233	18.35	6.810	1.380	112	0.272	0.102
40	2.955	220	15.71	12.690	1.330	118	0.229	0.095
60	5.216	206	12.89	21.930	1.280	124	0.200	0.087
80	8.567	191	10.07	35.710	1.220	130	0.195	0.080
100	13.283	174	7.25	55.860	1.160	136	0.080	0.072
120	19.666	155	4.43	85.470	1.080	142	0.170	0.060
				Freon-22				
−100	0.0199	269.29	28.1	0.1196	1.557	80.0	6.00	0.1487
−80	0.1034	257.43	24.8	0.561	1.514	87.5	5.00	0.1385
−60	0.3752	245.42	21.5	1.865	1.465	95.0	4.14	0.1283
−40	1.0540	232.92	18.5	4.885	1.412	101.7	3.49	0.1181
−20	2.4560	219.40	15.0	10.821	1.351	110.4	3.02	0.1079
0	4.9830	204.28	11.7	21.285	1.285	118.7	2.67	0.0977
20	9.0970	186.89	8.7	38.550	1.214	126.8	2.40	0.0875
40	15.3150	166.22	5.8	66.225	1.132	134.5	2.19	0.0772
60	24.2360	149.94	3.3	111.65	1.030	142.1	2.00	0.646

T (°C)	p (bars)	L (kJ / kg)	σ (10^{-3} N / m)	ρ_v (kg / m^3)	ρ_l (kg / m^3)	η_v (10^{-5} N · s / m^2)	η_l (10^{-5} N · s / m^2)	λ_l (W / m · K)
					Ethane			
−120	0.096	530	21.23	0.230	582	49.0	2580	0.149
−100	0.600	506	17.93	0.921	562	55.0	1800	0.137
−80	1.700	480	14.60	2.600	540	61.0	1360	0.125
−60	3.700	450	11.30	6.200	516	67.0	1100	0.113
−40	7.200	414	8.00	12.700	488	73.0	900	0.100
−20	14.000	368	4.60	25.500	454	79.0	760	0.088
0	25.000	304	1.20	46.000	414	85.5	660	0.077
20	38.000	200	0.08	85.000	360	91.0	600	0.066
					Silver			
1500	0.01008	298	827.5	0.0076	8.782	61.69	2.88	
1600	0.02420	298	810.1	0.01698	8.683	64.69	2.47	
1700	0.05300	298	792.1	0.03548	8.585	67.69	2.08	
1800	0.10800	298	775.3	0.06823	8.485	70.69	1.75	
1900	0.20600	298	757.9	0.12300	8.385	73.69	1.44	
2000	0.38300	298	740.5	0.21880	8.289	76.69	1.17	
2100	0.63500	298	723.1	0.35480	8.190	79.69	0.90	
2200	0.86000	298	705.7	0.57540	8.092	82.69	0.67	
2300	1.36000	298	638.0	0.87100	8.000	85.69	0.44	
2400	2.53000	298	680.0	1.23000	7.894	88.69	0.24	
2500	3.84000	298	665.0	1.82000	7.796	91.69	0.05	

				Lead			
1400	0.0986	920	347.28	0.147	9.27	7.46	0.9122
1500	0.2108	920	335.88	0.296	9.14	7.90	0.8847
1600	0.4200	920	324.48	0.559	9.01	8.34	0.8586
1700	0.8010	920	313.08	1.011	8.89	8.78	0.8352
1800	1.3620	920	301.68	1.635	8.76	9.21	0.8143
1900	2.3100	920	290.28	2.648	8.63	9.66	0.7958
2000	3.7410	920	278.88	4.106	8.51	10.10	0.7794
2100	5.5500	920	260.00	5.817	8.37	10.54	0.7590
2200	7.2000	920	248.00	8.256	8.25	10.98	0.7410
2300	11.8500	920	237.00	11.480	8.12	11.42	0.7230
2400	16.7500	920	225.00	15.600	7.99	11.86	0.7050
2500	22.6000	920	214.00	20.280	7.86	12.30	0.6870

T (K)	p (bars)	L (kJ / kg)	σ (10^{-3} N / m)	ρ_v (kg / m^3)	ρ_l (10^3 kg / m^3)	η_v (10^{-5} N · s / m^2)	η_l (10^{-4} N · s / m^2)
				Calcium			
1000	0.00032	3885	372	1.549×10^{-4}	1.392	1.35	16.75
1100	0.00174	3885	362	7.674×10^{-4}	1.370	1.46	13.32
1200	0.00714	3885	352	2.891×10^{-3}	1.348	1.57	11.02
1300	0.02360	3885	342	8.8×10^{-3}	1.321	1.68	9.40
1400	0.06580	3885	332	9.2×10^{-2}	1.293	1.79	8.10
1500	0.15980	3885	322	5.8×10^{-2}	1.265	1.90	7.20
1600	0.34780	3885	312	1.1×10^{-1}	1.248	2.01	6.50
1700	0.69050	3885	302	2.1×10^{-1}	1.220	2.12	6.00
1800	1.27000	3885	292	3.7×10^{-1}	1.196	2.23	5.60
1900	2.19100	3885	282	5.8×10^{-1}	1.179	2.34	5.08
2000	3.58000	3885	272	8.2×10^{-1}	1.165	2.45	4.67

AUTHOR INDEX

Camarda, C. J., 13, 176
Campanile, A., 260
Carey, V. P., 19, 20, 21, 22, 25, 26, 30, 31, 33, 38, 41, 192
Carlen, R. A., 318
Caron, R., 263

SUBJECT INDEX

Printed in the United States
20473LVS00005B/54